Paleoenvironmental Reconstruction in Arid Lands

Editors

A.K. Singhvi
E. Derbyshire

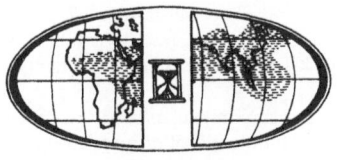

IGCP-349

CRC Press
Taylor & Francis Group
Boca Raton London New York

CRC Press is an imprint of the
Taylor & Francis Group, an **informa** business

A BALKEMA BOOK

CRC Press
Taylor & Francis Group
6000 Broken Sound Parkway NW, Suite 300
Boca Raton, FL 33487-2742

ISBN-13: 978-9-054-10710-1 (hbk)

Visit the Taylor & Francis Web site at
http://www.taylorandfrancis.com

and the CRC Press Web site at
http://www.crcpress.com

Foreword

This volume is one outcome of IGCP Project 349 on "Desert margins and palaeomonsoons of the Old World: 135,000 yr B.P. to the present" which ran, under the leadership of Edward Derbyshire (United Kingdom), Ashok Singhvi (India) and An Zhisheng (China) for the five-year period 1993-1997. In many ways it is a representative product of a successful IGCP project concerned with a topic of global importance. In looking at desert margins and their temporal changes in the last 135,000 years, IGCP 349 brought together Quaternary scientists from over 40 nations. International meetings, workshops and training sessions were held in Egypt, Saudi Arabia, the United Kingdom, north China, Jordan, and Mauratania, and several regional meetings were held in India and Germany.

The International Geological Correlation programme (IGCP) is a joint endeavour by UNESCO's Earth Sciences Division and the International Union of Geological Sciences (IUGS). Its aim is to encourage and facilitate international co-operation in research on geological problems and to promote the wise use of the Earth as both the human habitat and the source of our natural resources. This programme currently involves over 50 individual projects, most of which are global in character. Several thousand geoscientists from more than 150 countries contribute to it, a significant proportion of whom hail from the developing countries. Project proposals are welcome from all nations, with the result that many of them are "grass roots" in type. The fundamental operational characteristic of the IGCP is to bring together like-minded geoscientists to advance knowledge by international conferences and workshops, and by running advanced training courses in the developing world. The IGCP is not a research organisation, its emphasis being rather in the areas of exchange of knowledge and experience, and capacity building in less-favoured regions of the world. It has long been regarded as the most successful international programme of its type, and is widely regarded as a highly effective vehicle for transfer of information and training from the developed to the developing nations. Its programme is carefully designed and focused under the theme "Geoscience in the Service of Society".

The contents of this volume on Paleoenvironmental Reconstruction in Quaternary Arid Lands are representative of selected parts of the work of IGCP Project 349. The introductory chapter by E. Derbyshire reviews the ways in which landform evidence has been used to infer past climatic change over a range of environments and in response to number of different geomorphological processes. It concludes that, although unlikely to match the climatic resolution and temporal continuity found in many long-term archives such as sedimentary sequences, landforms continue to provide invaluable data on Quaternary climatic change in the world's drylands, often with a regional detail and spatial continuity that is not available from other methods. The chapter also provides a valuable bibliographic archive on the topic of climatic geomorphology. However, problems do exist for the interpretation of landforms, and these are spelt out clearly by Derbyshire and developed further in the following chapter by K. White. This outlines the methodology of remote sensing and discusses specifically the ways that remote sensing can add considerable insight, through its characteristic synoptic approach, to the recognition of regional patterns of landforms in a dryland palaeoenvironmental framework.

The three succeeding chapters consist of substantial reviews designed to illustrate the use of pedogenic parameters as a key to improved palaeoclimatic reconstruction. In the first of these, R.A. Kemp provides a systematic guide to the use of microscopic thin section analysis of palaeosols, laying emphasis upon the techniques used in this type of research and the methodology of describing the results. Case studies to illustrate the approaches are provided by examples from Quaternary palaeosol sequences in Europe and the loessic region of China. N. Fedoroff and M.A. Courty follow this with a specific discussion of soil-forming processes and soil distribution under conditions of increasing aridity. In a comprehensive and highly informative review they examine the relationship between pedological, erosional and accretionary processes in deserts, and demonstrate the potential of the palaeopedological approach in the reconstruction of the past environments within drylands by reference to inherited soils and palaeosols currently found in deserts. This theme is taken further by S.K. Tandon and S. Kumar in their chapter on the methods and concepts central to a proper understanding of the genesis of calcretes in semi-arid and arid lands. This chapter deals with palaeoenvironments througn longer perioas ot geological time. It clearly outlines the general importance and essentially interdiscriplinary nature of this research by reference to calcrete profile characteristics, hydrology, micromorphology, cathodoluminescence signatures, mineralogy, chemistry, and the environment and rates of formation. It demonstrates the importance of calcretes as an archive of palaeoclimate, paleo-biomass and vegetation type, and palaeohydrology as well as providing potential material for dating arid and semi-arid soil development.

There follows a chapter on the theme of sand dune systems as indicators of climatic change using the case of the Arabian peninusla. K.W. Glennie attempts to show how exposure and flooding of the shelf in concert with growth and decay of the continental-scale ice sheets controlled the details of the dune fields by controlling the available sand supply. Interpretation of climatic change from the landforms and sediments of geomorphic systems is next illustrated in a comprehensive and highly informative review of the activities, products and climatic responses of arid zone rivers. G.C. Nanson and S. Tooth show that relatively small climatic changes may have marked effects upon the rivers of the world's drylands, in respect of their flow regime, sediment transport, and channel style, although they emphasize that in other cases quite significant changes in global climate may leave no record in certain drainage basins. Particular attention is paid to the interpretation of river channels patterns, floodplains and terrace stratigraphies as evidence of palaeoflow regimes which in turn may be used as an indicator of climatic change. Examples from three continents show the degree of detail on changing climate that can be extracted from dryland rivers for comparison with lake, ocean, sea level, ice core and loess records. Attention is also drawn to the limited amount of work in drylands relative to other climatic regions and the difficulties of understanding the controlling factors, when forcing events (i.e. storms) are so variable in magnitude and frequency, and meaningful climatic records are so difficult to obtain, because of the limited spatial distribution of storm events and the relative absence of climatic records consequent upon the limited interest or occupance in such regions.

The high resolution climatic record contained in the cave deposits (speleothems) of drylands is next examined in the chapter by G.A. Brook. Often speleothems are located in regions too dry for them to form, or in saturated conditions much wetter than during their formation. Following an outline of the methodology for sampling these materials, Brook reviews the ways that evidence from speleothems can be used to derive palaeovegetation from entrapped pollen assembalges, and palaeoclimatic information from oxygen, hydrogen and carbon isotope signals. He also illustrates the further value of speleothem arising from the fact that these deposits have potential for direct dating by radiocarbon, U-series or ESR. With these dating controls and palaeoenvironmental indicators, evidence is provided from cave deposits for paleo-temperatures, paleo-precipitation, paleo-vegetation cover and other palaeoenvironmental characteristics, often with fine detail and a high level of temporal resolution.

In two shorter chapters giving the results of recent research into novel approaches to the interpretation of dryland palaeoenvironments S.A.

Ghazanfar illustrates the use of present vegetation distribution patterns to reconstruct former paleoclimatic regimes for the Arabian Peninsula, and H.N. Barakat and C. Rolando review the basic principles of anthracology (the analysis of archaeological charcoal) in palaeoenvironmental reconstruction. Using similarities of taxa, endemism and distribution of relict taxa, Ghazanfar suggests that species of African origin are present in northern Oman and southwest Arabia, and indicates that one or more periods with a climate wetter than the present existed over an extensive region in the relatively recent past. After outlining the methodology of anthracology and demonstrating that wood can now be identified down to species level, rather than just genus or family as was the case in the past, Barakat and Rolando show how, with reference to Neolithic sites, the palaeovegetation of the Sahara Desert and the Sahel may be reconstructed to provide a basis for inferring fluctuations of the Saharan margins. Overall they are able to show that the Sahara was less hostile to human occupation earlier in the Holocene than it is at present. The volume concludes with a chapter by M.J. Head on radiocarbon dating of deposits in the arid zone. This starts with an outline of the method, then reviews the known problems associated with radiocarbon dating of fossil wood, charcoal, plant macros, pollen, peat, phytoliths, organic sediments, soil carbonates, shells and bone, all of which involve different pretreatment procedures, and variations in analytical methodology. Brief case studies are used to illustrate the significance of the procedure recommended. Overall, this is a very cautious, but exceedingly helpful chapter providing a sound background to collecting and interpreting material used for radiocarbon dating.

This book does not attempt to provide a comprehensive account of desert margin change in the Quaternary but, as its title indicates, it provides a selective focus on methods by which such changes may be detected, analysed and documented. Taken together, the eleven chapters provide a wide ranging perspective on some of the problems under investigation and several of the systematic approaches currently in use by those concerned with refining our knowledge of the changing environments in the world's drylands during the latter part of the Quaternary. Dealing with a part of the globe that is highly sensitive to small changes of climate, or other stresses such as human impact, this text brings together a set of excellent reviews that will be of great value to students beginning to understand the subject and to palaeoenvironmental scientists actively concerned with finding solutions to problems of the region. Additionally, it will be an invaluable archive of important references covering the subject of arid lands palaeoenvironmental research. In all these respects this volume achieves the aspirations of the International Geological Correlation Programme.

March 7, 1999   **James Rose**
Editor-in-Chief, Quaternary Sceice Reviews
and
Professor, Department of Geography
Royal Holloway, University of London
Egham, Surrey, TW20 OEX, UK

Acknowledgements

This volume is based on the training course lectures at different annual meetings of IGCP-349. The Editors thank the UNESCO, the Third World Academy of Sciences and the UNEP for financial support that made possible these lectures and hence this volume. The Editors also thank the following institutions for hosting these meetings: The Geographical Society of Egypt and Ain Shams University, Cairo, Egypt; The Xian laboratory for Loess and quaternary Research, Xian, China; The Department of Geological Sciences, Al Ain University, Al-Ain, United Arab Emirates; The Department of Geology, University of Mauritania, Nouakchott, Mauritania and the Department of Geological Sciences, University of Wollongong, Wollongong, Australia.

The Editors would like to thank all the authors for their contributions and for their patience during the editorial process. Finally, the Editors would like to thank their institutions, The Physical Research Laboratory, Ahmedabad, India and the Centre for Quaternary Research, Department of Geography, Royal Hollway, University of London for their logistical support. Finally AKS would like to thank the University of Sheffield, Sheffield, U.K. and the Lever-Hulme Trust, U.K. for logistics support during the final stages of editing ot this volume.

May 15, 1999   **Ashok Singhvi**
                                                   Edward Derbyshire

Contents

Landforms, Geomorphic Processes and Climatic Change in Drylands

EDWARD DERBYSHIRE[*]

ABSTRACT

It has always been difficult to form general theories on how climatic changes affect landform evolution, because of the complex relationship between the energy inputs from climate and tectonics, surface processes and form-process responses. Landscape sensitivity varies widely in response to climatic and tectonic stimuli of different magnitude and frequency, and discrimination between the geomorphological effects of climate and tectonics remains problematical in many dryland regions. Landform criteria used to infer past climatic changes, such as dune fields, alluvial fans, river terraces and palaeochannels, continue to be questioned. Although the data on climatic change derived from landforms are sparser and generally of lower resolution than information from the sedimentary record, geomorphology continues to provide valuable correlative data on changing climates in the world's drylands.

Introduction

The inference that certain landforms or landform groups are specific to particular climatic environments dates back over a century. By the end of the nineteenth century, this general relationship had become accepted and distinctive landform assemblages were regarded as diagnostic of the arid zone (Davis 1905). The first quarter of the twentieth century saw the emergence of a zonal classification of landform associations broadly coinciding with the Earth's bioclimatic zonation. This became the founding tenet of climatic geomorphology (Budel 1948, 1963, 1982; Stoddart 1969; Derbyshire 1973, 1976), according to which variations in the climatically-driven processes ᵢinvolving physical and biochemical reactions at the atmosphere-lithosphere interface give rise to morphoclimatic regions containing distinctive suites of landforms. However, it must be recognized

*Centre for Quaternary Research, Department of Geography, Royal Holloway, University of London, Egham, Surrey TW20 0EX, U.K.

that use of such schemes in palaeoenvironmental reconstruction is hampered by the fact that all global zonal classifications of landforms utilise mixed criteria. For example, the terminology of the best-known morphoclimatic classification (Tricart and Cailleux, 1972) is pedological and biogeographical rather than strictly climatic or geomorphological. One result is that the landform regions it describes are based as much on inferred bioclimatic process systems as upon associations of landforms. The climatogenetic zonation of landform assemblages by Budel (1963: Fig. 1) is based only partly upon climatic criteria, and Strahkov's (1967) map of landform groups is based on dominant types and inferred rates of geomorphological processes.

Geographical scale is a further constraint on the ready application of the zonal climatic factor to the major landform associations. In common with continental classifications of vegetation associations and soil types, all climatically-grouped landform associations appear to break down below the regional scale (Eybergen and Imeson 1989). Moreover, the evident fact that landforms inherited from past climatic regimes, and most clearly those of Pleistocene age, are abundant in many parts of the world underlines the importance of the temporal framework in climatic geomorphology. All attempts to characterize landforms within global zones have to solve the problem posed by variations in spatial and temporal scales, as well as making assumptions about the complex relationship between surface processes and geomorphological form.

Geomorphological Processes and Climate

Various climatic parameters have been used to infer the relative importance and incidence of earth surface processes. As long ago as 1910, W. Penck used 'physiographic' (landform) criteria to classify climate. He proposed a hydrological framework based on arid, humid, and nival climates, but he did not specify any diagnostic landform types. Only in the past few years has this rationale been used to group landforms (Fig. 2a) and the approach used to construct transects showing morphoclimatic zones as in Africa, for example (Fig. 2b: Hövermann 1985; Hövermann et al. 1992). The most obvious effects of changes in temperature are seen when water freezes, but the effects are not restricted to the middle and higher latitudes. For example, seasonal landscapes, from the Mediterranean to the savannas, are also vulnerable to temperature changes because of the influence of these changes upon potential evapotranspiration and vegetation cover. Changes in precipitation are of considerable geomorphological importance, particularly in the subhumid to arid climatic range. Seasonal effects are important in a number of

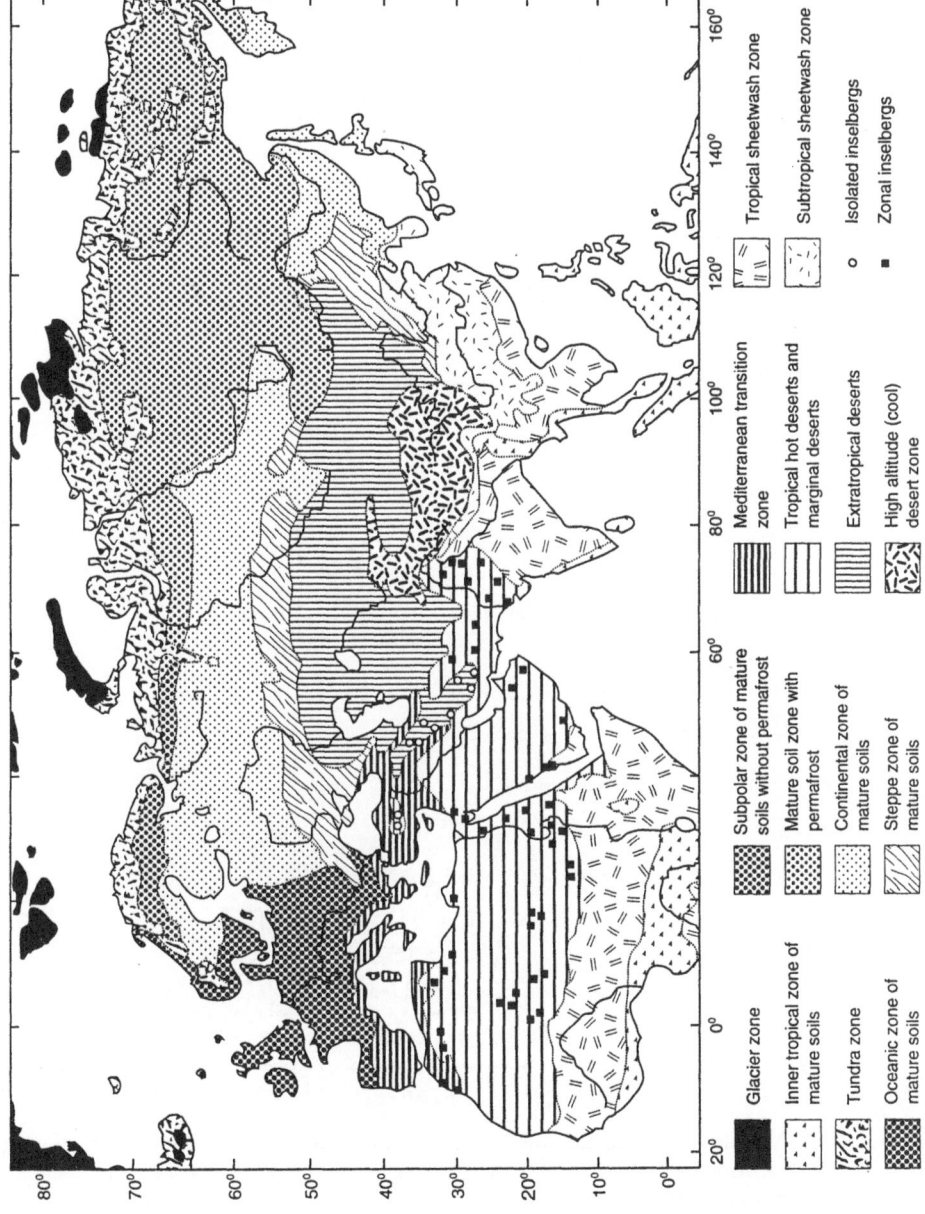

Fig. 1: Climatic-morphological zones of the Old World. Re-drawn after Budel (1948).

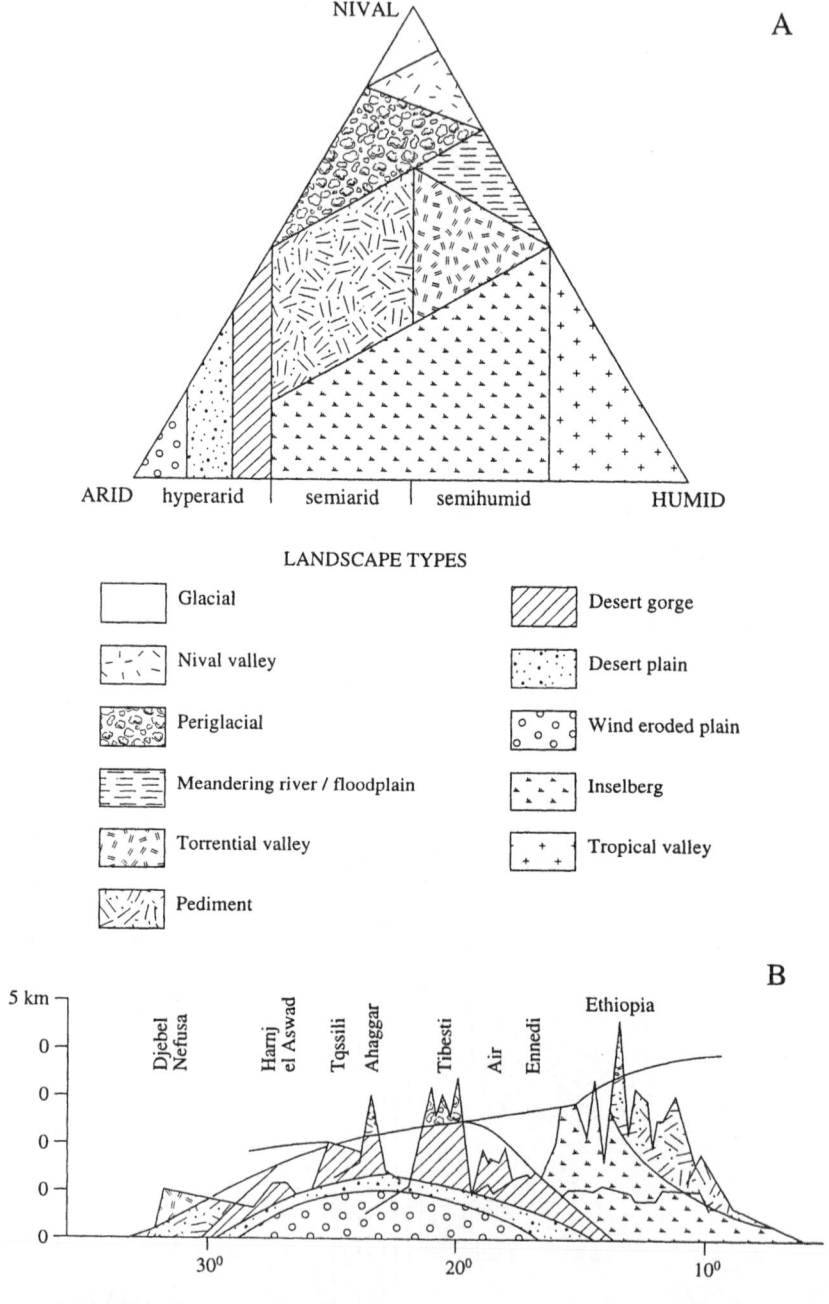

Fig. 2: (A) Morphoclimatic regions derived from generalised hydrological regimes, after Hövermann et al. (1992), and (B) a schematic profile across Africa using the same classification, after Hövermann (1985).

climates, including the drylands. For example, Langbein and Schumm (1958) suggested that fluvial erosion and sediment yield in rivers increases as mean precipitation rises, but only up to a level at which vegetation becomes multi-layered. The inherent complexities in these relationships have since been pointed out (Walling and Webb 1983). Another approach is illustrated by Wilson (1973) who used dynamic classifications of climate to show that fluvial erosion rates are much higher in seasonal, compared to non-seasonal climates. He found that peak erosion rates occur in Mediterranean regimes with mean annual precipitation in the range 1270-1650 mm and in dry continental climates with precipitation values of 250-500 mm. The seasonal rhythm of fluvial erosion and runoff is inferred from graphs of mean monthly sediment yield and mean monthly water yield (Fig. 3).

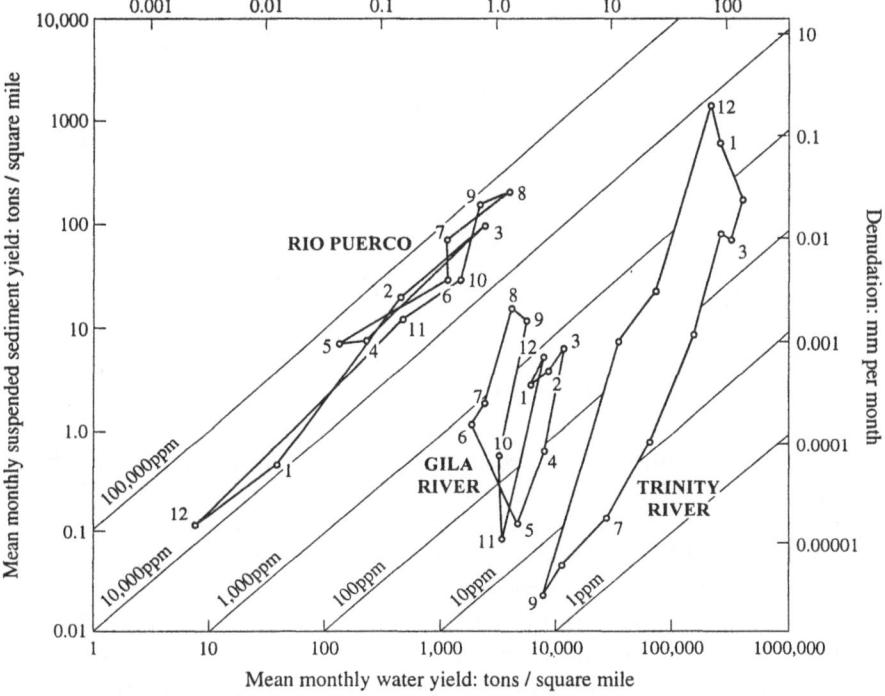

Fig. 3: A sedihydrogram for rivers in three different climatic regimes, after Wilson (1972). The Trinity River, California is Mediterranean, the Gila River, New Mexico is in a semi-arid continental region, and the Rio Puerco, New Mexico is in an arid continental climate.

Generalisations about the role of climatic inputs in landform evolution at regional and local scales have proved difficult to sustain. Assessment of the impact of climate and climatic change on landform processes and threshold linkages requires application of climatic data of demonstrable geomorphological relevance based on the concept of magnitude and frequency of events (Ahnert 1987). Over short time-scales (<100 yr), the relationships between meteorological regime, vegetation cover, rock type and exposure, and sediment yield in some semi-arid to arid terrains may become less clear. For example, Yair and Enzel (1987) have argued that, in some dryland situations, the general relationships inferred by Langbein and Schumm (1958) may even become reversed.

It is reasonable to assume that the relative importance and magnitude of the processes shaping hillslopes are influenced by climate. However, thresholds for different processes vary with factors such as gradient, slope length and vegetation cover. In this case, the results of surface processes may show notable variations in different landform situations within the same climatic regime. In arid landscapes subjected to occasional torrential rainfall, the development of clear process thresholds on steep slopes may be inhibited by slow rates of rock disintegration except on shale or where debris flows occur on talus slopes (Wolman and Gerson 1978). The effectiveness of major episodic climate-induced events is often indicated by the consequent denudation. The ratio between denudation during individual events with different return periods and the mean annual denudation suggests that a rare precipitation event in an arid environment may denude many times the mean annual volume of material. Thus, the ratio of instantaneous to annual denudation may not vary systematically with climate.

Landforms and Climatic Change

The presence of landforms in present-day climatic environments in which they could not have originated, such as the linear dunes in the sub-Saharan savannas of Africa (Grove 1958), carries the inference that many landform assemblages retain the imprint of one or more past climatic environments. Whether a particular climatic change can be regarded as geomorphologically significant depends not only on the magnitude and duration of the change itself but also on the resistance offered by the landforms. The degree to which a landscape will respond to external stimuli, such as climate, by changing its forms is known as landscape sensitivity (Brunsden and Thornes 1979; Thomas and Allison 1993). Shifts from one system-state to another are known as geomorphic thresholds (Schumm 1979, 1980). The time span between a stimulus, such as a change in climate, and the beginning of landform response at a threshold is known as the lag or reaction time, and the time required for the input to result in a characteristic landform is the relaxation time. The relaxation time is the period required for the establishment of a new equilibrium, and the persistence time is the period over which the new equilibrium is sustained (Fig. 4). Highly sensitive landscapes and climatic changes of short duration but high magnitude may yield a complex of transitional forms but few characteristic landforms. Conversely, landforms of low sensitivity such as plateaux will tend to retain their characteristic forms.

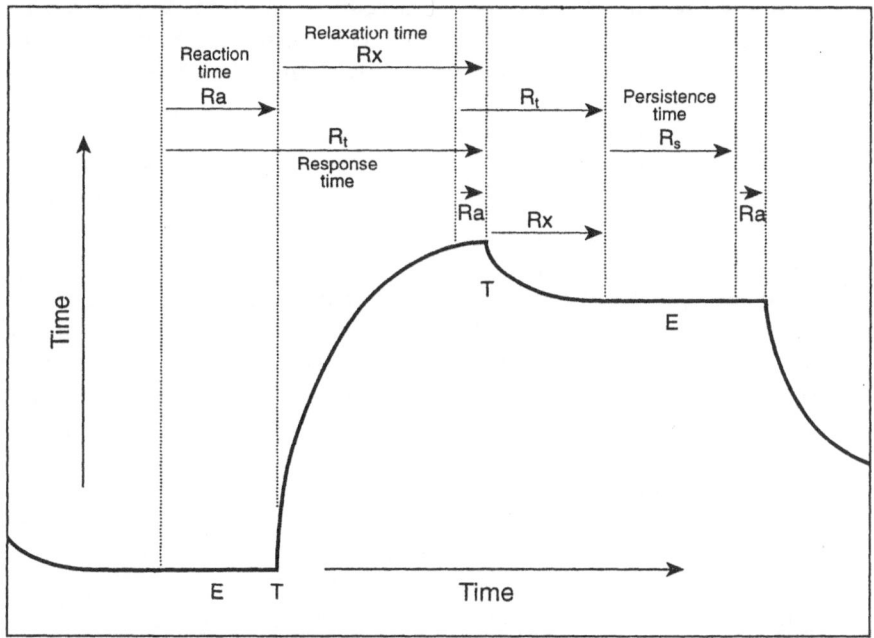

Fig. 4: Changes in the height of a stream bed during aggradation and degradation shown schematically to illustrate response time, reaction time, relaxation time, persistence time and threshold and equilibrium conditions. Re-drawn after Bull (1991).

Quaternary Climatic Change

The complex and subtle relationships between climatic change and morphoclimatic landscape elements are well illustrated by studies of Quaternary environmental change. The alternation of cold (glacial) and warm (interglacial) climates in the recent geological record was first recognized by landform evidence of glaciation and the presence of soils and warm climate fossils between successive (superposed) glacial deposits. This has been confirmed by oxygen-isotope analysis of foraminifera from the oceans, dated using the palaeomagnetic time-scale and radiometric methods (Shackleton et al. 1988), the pattern reflecting variations in global ice volumes, oceanic temperature variations and biological productivity and, indirectly, global temperatures. More recently, cores from the great ice sheets have contributed data on oxygen-isotope ratios, deuterium-hydrogen isotope ratios, the carbon dioxide content of air bubbles, and particulate and dissolved matter (Lorius et al. 1985; Jouzel et al. 1989). This has improved our understanding of the changing atmospheric content of the greenhouse gases, notably carbon dioxide and methane, as an index of the state of the Earth's carbon cycle. Methane shows a range of variation similar to carbon dioxide. The Vostok ice-core record suggests that emissions of methane from low latitude wetlands were greatly influenced by orbitally-controlled changes in monsoonal rainfall (Petit-Maire et al. 1991).

The drylands of the earth provide some of the most detailed terrestrial records of past climatic changes. Lake shorelines and basin fills (Street and Grove 1979; Gasse et al. 1989; Fang 1993) and the thick accumulations of loess, especially those in China and central Asia (e.g. Kukla 1989), have been particularly instructive. Data including magnetic susceptibility and particle size variations derived from closely-spaced sampling of cores and sections show clear similarities to the oceanic oxygen-isotope records (Fig. 5). The record of the last c.850,000 years shows a dominance of the 100,000 year astronomical cycle, the curves suggesting that, in the semi-arid zone, climatic transitions from dry and cold (glacial) to more humid and warmer (interglacial) conditions were relatively rapid events.

These global glacial-interglacial cycles, and the shorter-term climatic changes within such major episodes, had a severe impact upon the landscape, especially in changes in plant and animal associations and soil types. The microfabric (micromorphology) of interglacial and interstadial palaeosols within some Chinese loess sequences clearly show response of pedogenesis to climate and different types and degrees of soil surface degradation as climatic conditions changed (Kemp et al. 1995; Kemp 1999; Derbyshire et al. 1995). However, changes in landscape forms in response to climatic changes have proved more difficult to demonstrate, in part because the components of such complex systems (including slope gradient, slope length, rock type and weathering state, soil type, root density, plant cover, and moisture regime) have different response thresholds. Landforms seldom respond in linear fashion to external stimuli because of their widely divergent sensitivity.

The use of landforms or landform associations as diagnostic indicators ('proxies') of former climatic conditions requires that their 'relict' status be authenticated by comparison with the similar, currently-forming modern landforms within a clearly defined climatic regime. In practice, however, it must be recognized that the specific climatic and process-form characteristics of many landforms have still to be unequivocally

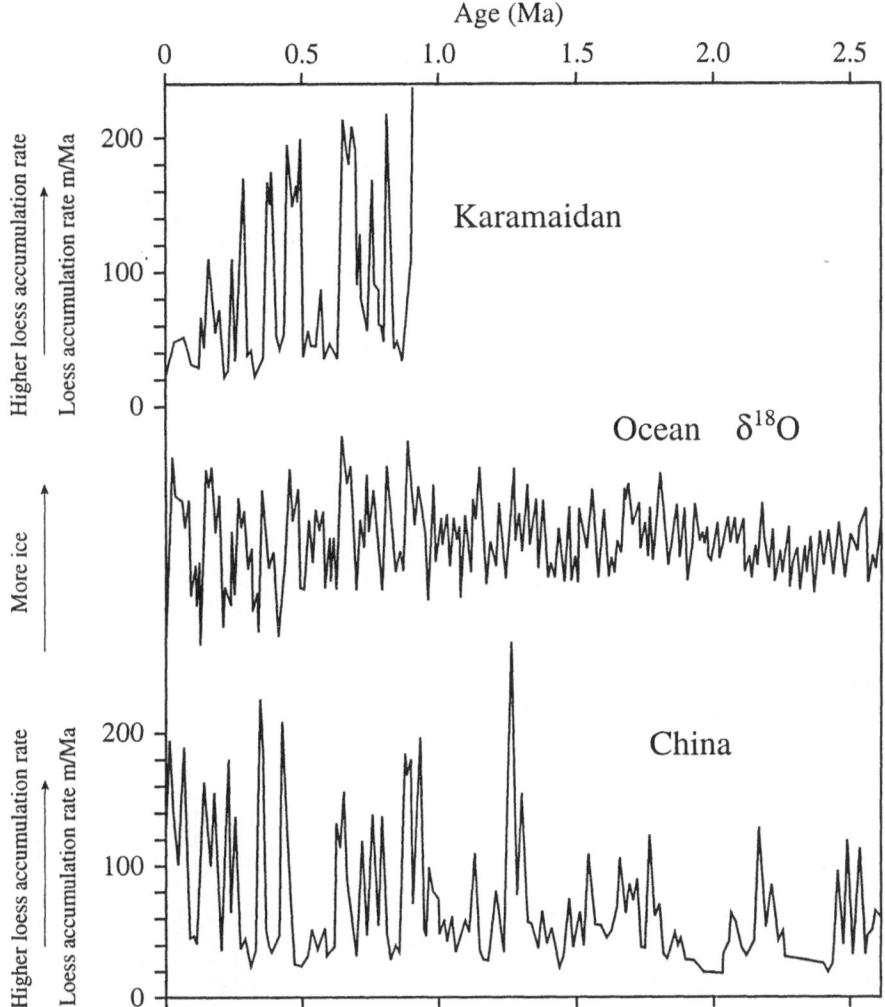

Fig. 5: Loess accumulation rates at Karamaidan in Tadjikistan and on the Loess Plateau of North China compared to the $\delta^{18}O$ oceanic SPECMAP (to 620ka) and ODP 677 (older than 620ka) record, suggesting that high loess accumulation rates are associated with increased ice volume. After Shackleton et al.(1995).

demonstrated. It has long been recognized that, in conformity with the general scientific principle of equifinality, identical landforms may evolve from quite different initial landforms under the action of a number of different surface processes. For example, pediments and alluvial fans occur in a range of climatic environments, including both warm and cool arid to semi-arid, and so are not narrowly diagnostic of a particular climatic regime.

Tectonics, Climate and Landform Change

Landforms change in response to shifts in the energy balance at the lithosphere-atmosphere interface. The most widespread and influential factors underlying such changes are climate and tectonics, two influences that may be intimately related. The similarity in the distribution of drylands and internal drainage regions in Asia illustrates this relationship on a broad scale (Fig. 6). It has recently become evident that the uplift of the Tibetan Plateau, the Himalayas and adjacent mountain ranges has not only diversified the climates of Asia but has greatly influenced climate

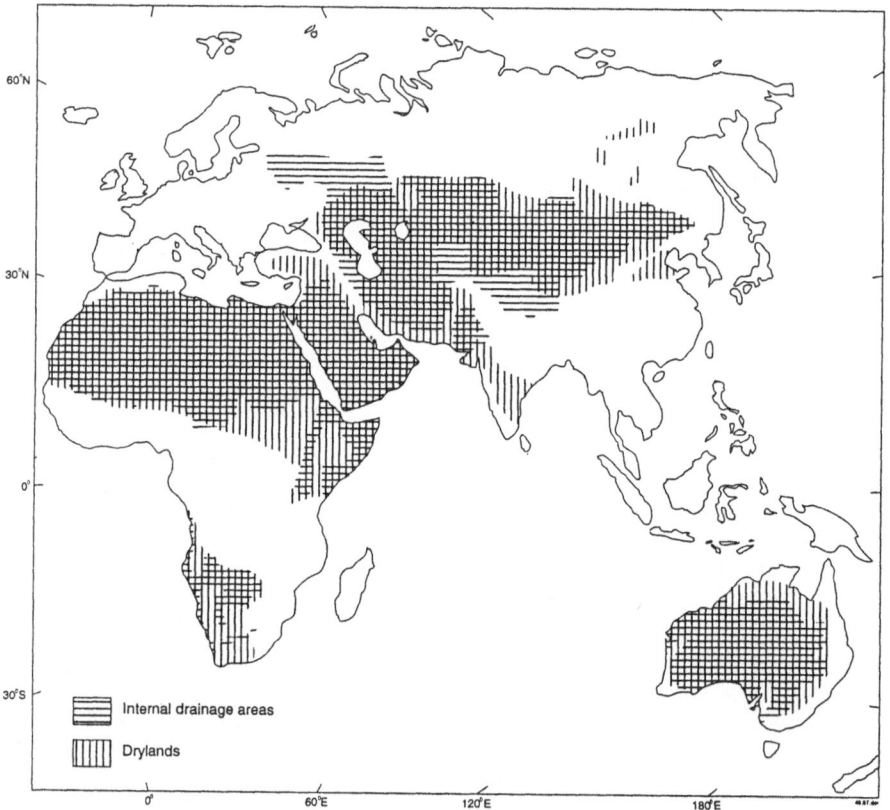

Fig. 6: Internal drainage areas and distribution of drylands in the eastern hemisphere.

on a global scale. It is argued that the rise of the Himalaya has progressively impeded the penetration of oceanic moisture into central Asia, although the time-scale involved remains open to debate (Li et al. 1979; Burbank et al. 1993; Coleman and Hodges 1995; Fort 1996).

Substantial parts of the high and cold Tibetan Plateau are arid or semi-arid regions, much of the north-western part of the Plateau receiving an annual precipitation of less than 200mm. The lake history of the numerous basins between the Himalaya-Karakoram and the Western Kunlun Mountains is one of relatively rapid changes from extremely dry to semi-arid or sub-humid conditions within the last c.13,000 years (Gasse et al. 1991). Many of these lake basins are now arid, extensive salt-rich lacustrine plains and terraces, denuded piedmont deposits and gobi (stony desert). Salt lakes with numerous raised shorelines, and complex barchan dune series with some small dune-dammed lakes, occur in other climatic regions, including the semi-arid parts of north-eastern Tibet (Fig. 7).

Fig. 7: Abandoned shorelines of Da Hai Lian Lake, at an altitude of 2910m, in the Gonghe Basin of north-eastern Tibet.

The considerable morphodynamic energy provided by Asia's tectonic evolution yields high erosional potentials and very high rates of sediment production. Sand and silt-sized rock particles are produced by a wide range of processes including tectonic crushing, glacial grinding, freeze-thaw action, salt weathering and cyclic hydration. The juxtaposition of high mountains and plateaux, in which frost action and glacial erosion are severe, with deep desert basins, mobile sand dunes, thick loess mantles, and large rivers has given parts of Asia a distinctive geomorphology as well as a series of sedimentary accumulations from which high resolution records of climate change can be deduced. Nevertheless, discriminating between the effects of climate and tectonics as expressed in landscape changes is far from straightforward. In many parts of montane Asia, for example, the landform record consists of a complex association of aeolian, fluvial, glacial, periglacial, and mass movement phenomena of different age and sensitivity, and showing varying degrees of modification. This classic geomorphological problem may be illustrated from the mountain desert region of the Karakoram Mountains of northern Pakistan (Derbyshire 1996), an environment characterized by rapid uplift and some of the greatest relative relief on Earth. Despite their currently dry climate (with valley floor mean precipitation totals of only 100 mm/yr in places), these mountains show clear evidence of multiple Pleistocene glaciation. In fact, they contain some of the longest glaciers outside the polar regions. Aeolian landforms and sediments are commonly associated with these glaciated valleys, and include barchan dunes in intermontane basins (Fig. 8) and variously reworked loess drapes that provide some, if rather loose,

Fig. 8: Field of barchan dunes adjacent to the upper Indus River (extreme right) in the Skardu Basin, Karakoram Mountains, northern Pakistan.

dating control on the subjacent moraines (Owen et al. 1992). On a broader scale, however, the geomorphological role of climate remains a contentious issue in this region. For example, Molnar and England (1990) suggested that, for every 1 km thickness of crust removed by glacial, glaciofluvial and fluvial erosion, average elevation would have been reduced by only 0.17 km because of isostatic rebound. Such isostatic rebound would produce peaks with relative elevations greater than those of the preceding landscape, but with a lower average elevation, i.e. relative relief would be greatly increased. In contrast Summerfield and Kirkbride (1992) doubted that the onset of glaciation necessarily triggers significant increases in denudation rates. They favoured association of the highest erosion rates with maximum precipitation rather than with ice cover. Evidently, more data and improved models incorporating the relative importance of climate and climatic change in such high altitude, low latitude terrains, involving the complex interplay of processes from glacial to aeolian, are required if denudational unloading and the relative importance of glaciation versus uplift are to be better understood.

Climate and Fluvial Systems

Despite the complexity of the relationship between climate change, valley-side slopes and river channel morphology, fluvial systems continue to be investigated as a potentially valuable source of information on past and present climatic regimes, particularly in some arid and semi-arid regions (Reid 1994, Nanson and Tooth, 1999). The use of differences in the size and sinuosity of palaeochannels to infer past climatic regimes illustrates some of the complexities involved. The palaeochannels of the Murrumbidgee River in south-eastern Australia provide a classic example. Palaeochannels with lower sinuosities, coarser bedloads, and steeper gradients than the present streams have been interpreted as indicating a former regime with higher peak flows but lower discharges suggesting greater seasonality and a drier climate in the Late Pleistocene compared to the present (Schumm 1968). It has since been shown, however, that the two contrasting sets of palaeostreams (known as ancestral and prior streams) are closely inter-related and both owe their origin to a greatly enhanced flow regime (Nanson and Tooth, 1997). Nevertheless, some support has been expressed (Porter et al. 1992) for a more general model (Schumm 1977) in which river channels are stable during the optimum and later parts of interglacials, when neither erosion nor sedimentation is significant, with the greatest erosion taking place during the latter parts of glacials and the early stages of interglacials.

Alluvial fans have also contributed to the debate on the climatic evidence stored in fluvial landform-sediment associations. Although found across a broad range of climates, alluvial fans are common in arid and semi-arid regions. Most fans described in the literature consist of fluvial and debris flow deposits. In some high mountain regions with multiple climate-process system regimes, including arid and semi-arid, fans may also include sediments of glacial, glaciofluvial, periglacial, aeolian, lacustrine, and mudflow origin, in addition to debris flow and fluvial facies (Derbyshire and Owen 1990). Alluvial fan morphology has been interpreted in terms of aggradation in wetter phases, and fan entrenchment by debris flow erosion in dry phases (Lustig 1965), to construction by debris flow sedimentation (Beaty 1963) and erosive incision in moister conditions, to entrenchment at intrinsic (i.e. non-climatic) thresholds

(Schumm 1980). However, discriminating between the climatic signal and the tectonic imprint upon alluvial fans remains a considerable challenge (Bull 1991; Dorn 1994). River terraces pose similar problems of discrimination. Examples discussed by Bull (1991) point to some major differences in the conditions in which terrace aggradation occurs in arid compared to humid climates.

Some of the world's driest regions were once crossed by major rivers, as the landform-sediment associations of North Africa and elsewhere testify. For example, sinuous, branching gravel ridges up to 10 km in length stand several metres above eroded bedrock surfaces in parts of the eastern Sahara. These are the remains of bedload sediments of seasonal rivers (wadis) that have become inverted by erosion of the surrounding plains. It has been shown (Butzer and Hansen 1968; Said 1975) that the raised wadis of the Nubian Desert form remnants of an integrated drainage system that was active during the Pleistocene. To the south, in the hyper-arid parts of northern Sudan, is the Wadi Howar, a 640 km long watercourse that flowed into the Nile River between 9,500 and 4,500 yr B.P. (Pachur and Kröpelin 1987). The occurrence of even occasional water flow in such a large channel presupposes substantial rainfall recharge of groundwater in a Sahel-type environment that extended at least 500 km further north than it does today. This interpretation is broadly consistent with a body 'of palaeoclimatic evidence suggesting increased moisture influx from a more vigorous African monsoon (e.g. Petit-Maire 1991). In the Arabian Peninsula, extensive superimposed palaeochannel systems extend for over 200 km across two exhumed fans at the foot of the Eastern Mountains of Oman (Maizels 1990). Both inset and raised channels are present, the channel sequence indicating at least 12 successive generations of palaeochannels in at least six major periods of fluvial activity during the Plio-Pleistocene.

One much-used general model links general or seasonal aridity with a reduced vegetation cover and reduced stream power and channel aggradation, moister climates being associated with channel incision and terracing. In specific cases, however, morphoclimatic interpretation of fluvial systems in the drylands has proved to be far from simple. The uncertainties surrounding current understanding of the specific relationship between climatic impulses and fluvial aggradation render the use of river terraces as indicators of former wetter or drier conditions in the drylands an equivocal tool (Bull 1991). The highly episodic flow regimes characteristic of many dryland stream systems (Baker 1983) render bedload flux so irregular that widely different conditions may eventuate, from well graded to blocked channels (e.g. Hövermann 1985). Such variability is an obstacle to reliable inferences about the form-climate relationship in both modern and ancient arid zone stream systems (cf.

Graf 1983). It is commonly observed that the channels of wadi systems are poorly integrated, the channels sometimes being partly overwhelmed by slope-foot colluvium. These colluvial deposits are so important in parts of southern Africa that they have been used to infer semi-arid conditions in the Late Pleistocene (Watson et al. 1985). Classification of fluvial systems in the Kalahari Desert according to their status within an inferred hydrological time series, led Shaw et al. (1992) to propose that the Middle Kalahari valleys were fluvially active between 15,000 and 12,000 yr B.P. (cf. Shaw and Thomas 1993). It was further suggested that high magnitude flood events in the Kuruman River, in the southern Kalahari, correlate with high-phase events of the Southern Oscillation. In contrast, the lacustrine record from most of Africa's inter-tropical zone appears to indicate that the last major shift from dry to wet climate occurred at about 12,500 yr B.P., with rising lake levels and a significant erosional impact (Street-Perrott et al. 1985).

While the ergs of the Sahara consist mainly of ancient reworked fluvial sediments, those seen in the Taklamakan Desert in western China (area 337,600 km^2) are being actively replenished with meltwater-borne sediments from the adjacent mountain ranges such as the Tian Shan and the western Kunlun Shan. These vitally important oasis waters have migrated considerable distances over the past 2,000 years in terrain conditions with an abundant sand supply subject to active dune encroachment (Coque et al. 1991). The Taklamakan appears to have been subject to recurrent flooding until about 7,500 yr ago (Jäkel and Zhu 1991). There is written and archaeological evidence that the Keriya and Tarim rivers crossed the desert about 2000 years ago, and that this situation recurred during the sixteenth and early nineteenth centuries. Abandoned towns on the lower Keriya River, known from chronicles to have sustained large populations in relative prosperity, have been ^{14}C dated at between 2135 and 2684 yr B.P. (the Western Han Dynasty, c.150-250 B.C.). Maps of Xinjiang drawn in the late Qing dynasty (1644-1911 A.D.) show that, although by then only a seasonal stream, the Keriya River periodically crossed the Taklamakan (Chen 1991). The overwhelming of oases by blown sand is well documented in this desert, and consequent loss of some important towns along the Old Silk Road is clearly documented from written accounts of the time. For example, the town of Loulan, situated west of Lop Nor Lake had a population of more than 1600 during the Han dynasty (206 B.C. - 220 A.D.), but was abandoned in 376 A.D. and is now covered by dunes and yardangs. Although evidently not the only cause, as is shown by the accounts of warfare and civil strife, the lowering of meltwater-recharged piedmont water-tables by excessive extraction of water by burgeoning populations (Derbyshire and Goudie 1997) has been a major factor in the drying up of river courses. A longer-term cause is the oscillating trend towards arid conditions in the Holocene (Pachur et al. 1995).

Wind-blown Sediments, Aeolian Landforms and Climate

The down-wind progression from dunes and sand sheets to thick accumulations of wind-blown silts (known as loess) is clearly seen in the drylands of central and eastern Asia. In conditions optimal for accumulation of airborne dusts, loess may accumulate to considerable thicknesses to form a plateau-like landform, the type example being the Loess Plateau (*Huangtu Guoyuan*) of northern China. Research over the past thirty years has shown, with progressively greater detail and precision, that the alternation of loess and palaeosols in the Chinese Loess Plateau provides a very detailed record of climatic change for the past 2.5 Ma. Loess is made up of a grain skeleton consisting mainly of silt- size quartz particles. Its airfall origin accounts for its loose grain fabric and high porosity (voids ratios of Late Pleistocene loess being in the range 0.7 - 1.0). Although loess clearly has a considerable water-holding capacity and is rarely saturated below the upper few decimetres, the presence of joint systems provides a by-pass type of drainage which may result in local saturation at more clay-rich horizons (palaeosols) and, particularly, at the loess-bedrock contact. When such local saturation occurs, loess may fail by instantaneous hydrocollapse (Derbyshire and Mellors 1988; Derbyshire et al. 1993; Derbyshire, Smalley and Dijkstra 1995). Such metastable behaviour in loess renders it subject to periodically rapid and spectacular erosion. The types of failure include various kinds of landslides (Derbyshire et al. 1991, 1995), toppling, (dry) grain flow and creep. Extensive areas of the Plateau have been deeply dissected by rivers, the resultant valley-side slopes, locally more than 200m high, becoming further degraded by various mass wasting processes in sub-humid to semi-arid regimes (Fig. 9). These fluvial valley systems constitute a complex but relatively little-studied geomorphological record of environmental change (Derbyshire 1983). The erodibility of loess slopes is directly responsible for the very high turbidity (>25 kg/m^3) of the master stream of the region, the Yellow River (Derbyshire and Wang 1994). It also explains why the Loess Plateau, as one of the Earth's youngest major landforms, is undergoing such high rates of denudation that reach their peak close to the arid/semi-arid and semi-arid/sub-humid climatic transitions.

Reference has already been made to the classic work that inferred major shifts in climatic regional boundaries on the basis of relict sand dune systems (Grove 1958; 1969). The arid periods in the tropics and sub-tropics that gave rise to phases of dune-building during the late Quaternary have been broadly correlated with the waxing and waning of

Fig. 9: Degradation of valley-side slopes in thick loess by gullying, toppling and sliding. South of Yanan, Loess Plateau of China.

the continental ice sheets of the Northern Hemisphere (Williams 1985). It is important to note, however, that there is an evident lack of synchrony at the inter-regional scale, such as between the north and south sides of the Sahara and between the Middle Kalahari and the Namib deserts in southern Africa (Tchakerian 1994). Linear dunes fixed by vegetation, and in various stages of destruction by erosion, have been mapped using remotely-sensed imagery in a number of places in Africa, Asia, Australia and North and South America (McKee 1979; White 1999). Most present-day active dune regions have a mean annual precipitation of less than 250mm, but relict dunes have been found in some areas with present-day rainfall of more than 1000mm, implying a shift in the margins of such dune fields by as much as 1200km (Goudie 1983). The orientation of linear trends of palaeodune fields has also been used to infer past patterns of atmospheric circulation (Madigan 1936; Mainguet 1978; Lancaster 1981; Wells 1983; Thomas 1984), although some of the intrinsic assumptions have recently been questioned (Thomas 1992; Livingstone and Thomas 1993).

Inferring the role of climatic change from the geomorphology of dunes and dune fields, however, depends fundamentally upon the certainty with which active features can be distinguished from relict forms. Signs of inactive aeolian landforms include a variety of degradational and reworking features ranging from rain-splash impact and surface crusting to rill and stream erosion, weathering and soil profile formation (Pye 1983).

The presence of a vegetation cover has also been used to infer a relict status for dunes and dune fields. However, it has been pointed out that such vegetation is often discontinuous or sporadic, and that sand transport may reach significant levels even on quite well vegetated dunes (Ash and Wasson 1983), suggesting that a complex series of gradational thresholds may be involved. Episodically active dunes, characterized by negligible sand movement but lacking clear evidence of a long-standing relict condition, may be an expression of such a relationship (Livingstone and Thomas 1993). Such dunes occur in transitional zones within the tropics having a mean annual precipitation of between 100 and 300 mm/yr. Here, there may be important lag effects including persistence of vegetation from wet into dry years and slow re-establishment of a plant cover following a series of dry years, such that rapid shifts from active to inactive dune conditions, and vice-versa, do not occur (Livingstone and Warren 1996). Lowering of dune slopes, together with prolonged surface weathering with soil profile development, have been described from many of the world's drylands. Degraded linear sand ridges, produced by a shift from a relatively arid dune formation regime to semi-arid conditions with monsoonal rains, have been described from the Kimberley area of north-western Australia (Goudie et al. 1993). More than 90 percent of these dunes have side slopes of less than 4°, and their mean height is less than 6m, compared to an estimated height during their formation of 15-18m. Severe reddening of the dune sands, but without recognisable soil profile properties, is explained as a result of in situ crystallization of haematite within kaolinite grain coatings. The enhancement of the silt-plus-clay content to more than 20 percent by weight is explained by weathering of sand size feldspars and other aluminosilicates. These dunes, dated by luminescence to the latest Pleistocene-Holocene, thus appear to have been greatly altered over a very brief geological time-span.

A model of the response of aeolian sand systems to external forcing factors, notably climate, has been proposed by Kocurek (1997), using the Sahara as an example (Fig. 10). Most Saharan sand seas lie in topographical basins, the sand deriving from alluvial-fluvial sources in the uplands with lesser amounts derived from dry lake beds. Sediment production occurs mainly during relatively humid climatic phases, and reaches a climax during transition from relatively humid to arid conditions at a time when the streams are prograding farthest into the basins, allowing the bulk of these sediments to remain in storage within the alluvial system. Sediment availability is controlled largely by vegetation, being theoretically nil during phases of peak humidity and unlimited during the most arid conditions. Transport capacity is at its greatest during glacial-arid periods. Significant influxes of sand to the aeolian system are associated with a lowering of the water-table and a decrease in vegetation cover as climatic

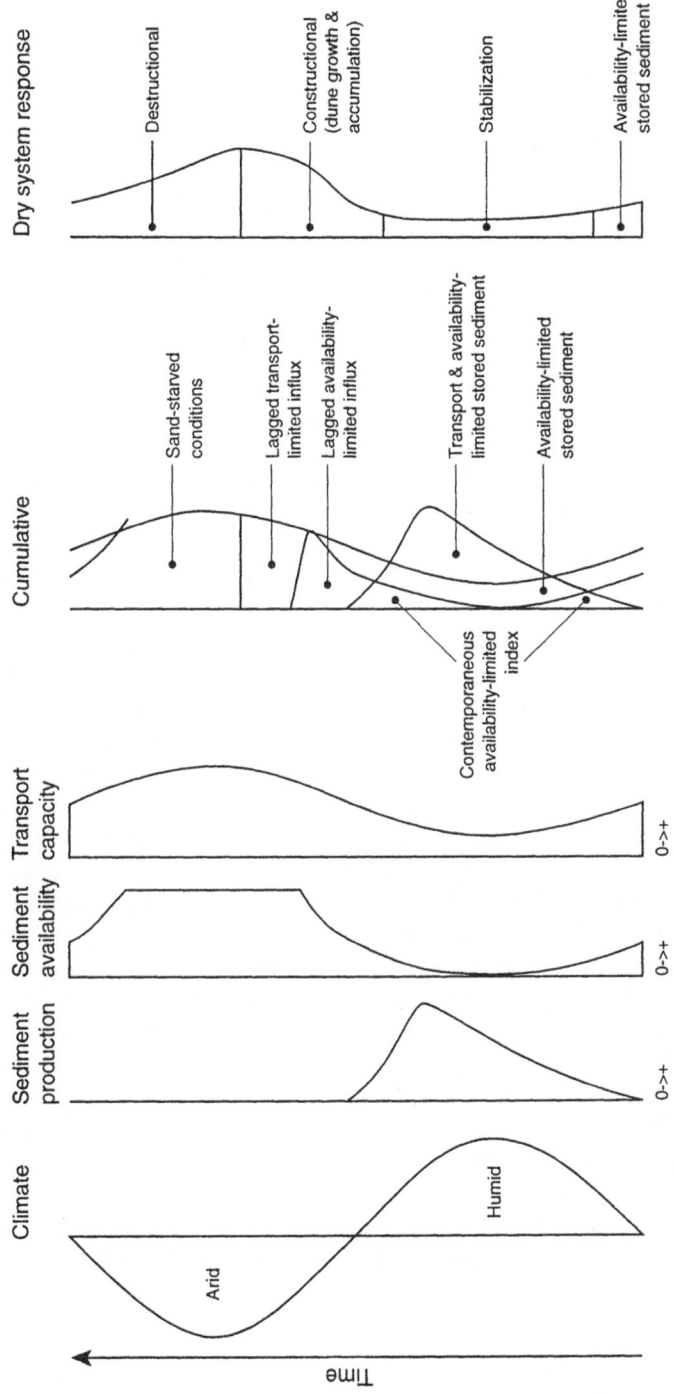

Fig. 10: Schematic diagram showing sediment supply and aeolian dry system response during a climatic cycle. Aspects of sediment supply are shown separately in the left-hand part of the diagram, and then superimposed to show the cumulative pattern and response. After Kocurek (1998).

conditions shift towards greater aridity. After a lag period, this material is then deflated. Dune construction and sand accumulation occurs and, during the aridity peak, the only constraint upon influx to the aeolian system is the transporting capacity of the wind regime. Should arid conditions persist long after the sediment supply has been exhausted, a negative budget is established. Dune building ceases, initially at the windward end of the sandy tract and then progressively to leeward, to form an erosional boundary to the accumulation. Because sand is derived from the dunes, net deflation is slow, and degrading, reworked dunes persist into the succeeding humid cycle when vegetation may stabilise them.

Conclusions

It may be inferred from a wide range of sedimentary proxy indicators of past climatic change that temporal and spatial variations in geo-morphological processes were of global significance throughout the Quaternary. All long-term records of climatic change covering the last few million years indicate a climatic cyclicity broadly similar to the solar orbital and precessional rhythms, with shorter-term, including rapid, climatic changes occurring in both glacial and interglacial periods (cf. Street-Perrott and Perrott 1990). In the tropical zone, including the sensitive subtropical savannas and desert margins, critical changes in erosion-rate were controlled by precipitation changes under the influence of variations in the monsoonal circulation (Kutzbach 1983). The current astronomically forced trend, with a decline in the summer insolation peak since about 9,000 yr B.P. and an expected minimum in about another 5,000 yr time (Berger and Tricot, 1986), is towards deterioration. The effects of the recent anthropogenically-induced increase in the greenhouse gases, on the other hand, constitute a contrary trend. The values for atmospheric carbon dioxide and methane already exceed the values calculated for the Last Interglacial using the Vostok ice-core (Chapellaz et al. 1990). While it is certainly true that all models providing estimates of the amount of global warming to be expected from these opposing trends differ in a number of respects (cf. Street-Perrott and Roberts 1994), there are areas of broad agreement, e.g. that oceanic temperatures will rise and, with them, sea levels.

The effects of such trends on arid and semi-arid landscapes may be estimated by analogy with documented changes during the Holocene, although it has been pointed out that currently available palaeohydrological models are insufficiently constrained to provide a basis for precise analogues (cf. Roberts and Barker 1993). Any rise in the sea-surface temperatures within the tropics would cause conditions in the semi-arid and arid zones to become moister, so changing the ecological and surface process balances. It is likely that parts of the Sahara would witness the re-establishment of many of the lakes, steppic grasses, and animals known to have been present between 9,000 and 7,000 yr ago (Petit-Maire and Riser, 1988). Research findings indicate that Holocene monsoon-driven climatic changes in dry regions as far apart as tropical North Africa and Tibet were closely synchronous (Gasse et al. 1996). This suggests that any future climatic amelioration would have an impact on the surface process balance and, eventually, the landforms of great expanses of the world's deserts. Hydrological changes would result in recharge of groundwaters, raising of groundwater tables and some reactivation of currently-dry palaeochannels and lake basins. Enhanced melting of glaciers would cause increased runoff, with significant replenishment of groundwater in the alluvial fan oases within the central Asian drylands. However, down-wasting of glaciers in the more arid mountain systems, such as parts of the Karakoram, might ultimately render many of the directly-fed meltwater irrigation networks unworkable because of a combination of increasing distances from glacier snouts and an enhanced tendency to stream incision.

It is concluded that information on palaeogeomorphological responses to climatic change is sparse, relative to that derived from the sedimentary record, and that this situation reflects the inherent complexity of the multi-factorial relationship between landforms and climate across a range of temporal and spatial scales. Nevertheless, geomorphological data continue to provide valuable correlative information on climate, especially when used in conjunction with soundly-based stratigraphy and sedimentology.

References

Ahnert, F. (1987). 'An approach to the identification of morphoclimates', In: *International Geomorphology*, V. Gardiner (ed.) 2: 159-188 John Wiley and Sons, Chichester and New York.

Ash, J.E. and Wasson, R.H. (1983). Vegetation and sand mobility in the Australian desert dunefield. *Zeitschrift für Geomorphologie Supplementband* 45: 7-25.

Baker, V.R. (1983). 'Late Pleistocene fluvial systems', In: *Late Quaternary environments of the United States, vol. 1 The Late Pleistocene*, H.E. Wright Tr. (ed.) pp. 115-129, University of Minnesota Press, Minneapolis.

Beaty, C.B. (1963). Debris flow, alluvial fans and a revitalized catastrophism. *Zeitschrift for Geomorphologie Supplementband* 21: 39-51.

Berger, A. and Tricot, C. (1986). 'Global climatic changes and astronomical theory of palaeoclimates', In: *Earth Rotation: Solved and Unsolved Problems*, A. Cazenave (ed.) pp. 111-129 Reidel, Dordrecht.

Brunsden, D. and Thornes, J.B. (1979). 'Landscape sensitivity and change', *Transactions of the Institute of British Geographers*, 4: 463-484.

Büdel, J. (1948). 'Das System der Klimatischen Geomorphologie', *Verhandlungen Deutscher Geographie*, 27: 65-100.

Büdel, J. (1963). 'Klima-Genetische Geomorphologie', *Geographische Rundschau*, 7: 269-285.

Büdel, J. (1982). *Climatic Geomorphology*, (trans. L. Fischer and D.Busche), Princeton University Press, Princeton N.J.

Bull, W.B. (1991). *Geomorphic responses to climatic change*. Oxford University Press, New York and Oxford, 326pp.

Burbank, D.W., Derry, L. and France-Lanord, C. (1993). Lower Himalayan detrital sediment delivery despite an intensified monsoon at 8 Ma. *Nature* 364: 48-50.

Butzer, K.W. and Hansen, C.L. (1968). *Desert and River in Nubia*, University of Wisconsin Press, Madison.

Chapellaz, J., Barnola, J.M., Raynaud, D., Korotkevich, Y.S. and Lorius, C. (1990). 'Ice record of atmospheric methane over the past 160,000 years', *Nature*, 345: 127-131.

Chen, H. (1991). 'The change of eco-environment and the rational utilization of water resources in the Keriya River valley', In: 'Reports on the 1986 Sino-German Kunlun Shan Taklimakan-Expedition', D. Jäkel and Z. Zhu (eds.) *Die Erde*, Erg.H 6: 133-147.

Coleman, M. and Hodges, K. (1995). Evidence for Tibetan plateau uplift before 14 Myr ago from a new minimum age for east-west extension, *Nature* 374: 49-52.

Coque, R., Gentelle, P. and Coque-Delhuille, B. (1991). 'Desertification along the piedmont of the Kunlun chain (Hetian-Yutian sector) and the southern border of the Taklamakan Desert (China): preliminary geomorphological observations (1)', *Revue de Géomorphologie Dynamique*, 15: 1-27.

Davis, W.M. (1905). The Geographical Cycle in an Arid Climate, *Geol*, 13: 381-407.

Derbyshire, E. (ed.) (1973). *Climatic Geomorphology*, Macmillan Press, London, 296pp.

Derbyshire, E. (ed.) (1976). *Geomorphology and Climate*, John Wiley and Sons, Chichester, 512pp.

Derbyshire, E. (1983). On the morphology, sediments and origin of the Loess Plateau of central China, In: *Mega-geomorphology*, R. Gardner and H. Scoging (eds.) Oxford University Press, Oxford, pp. 172-194.

Derbyshire, E. (1996). 'Quaternary glacial sediments, glaciation style, climate and uplift in the Karakoram and northwest Himalaya: review and speculations', In: Environmental Changes in the Tibetan Plateau and Surrounding Areas, *Palaeogeography, Palaeoclimatology, Palaeoecology*, F. Gasse and E. Derbyshire (eds.) 120: 147-157.

Derbyshire, E. and Mellors, T.W. (1988). Geological and geotechnical characteristics of some loess and loessic soils from China and Britain: a comparison. *Engineering Geology* 25: 135-175.

Derbyshire, E. and Owen, L.A. (1990). Quaternary alluvial fans in the Karakoram Mountains, In: *Alluvial Fans: A Field Approach*, A.H. Rachocki and M. Churel (eds.) pp. 27-53 John Wiley and Sons Ltd., Chichester and New York.

Derbyshire, E., Wang, J.T., Jin, Z.X., Billard, A., Egels, Y., Kasser, M., Jones, D.K.C., Muxart, T. and Owen, L. (1991). Landslides in the Gansu loess of China, *Catena*, Supplement 20: 119-145

Derbyshire, E., Dijkstra, T.A., Billard, A., Muxart, T., Smalley, I.J. and Li, Y.J. (1993). Thresholds in a sensitive landscape: the loess region of central China, In: *Landscape Ssensitivity*, D.S.G.Thomas and R.J.Allison (eds.) John Wiley and Sons Ltd., Chichester, pp. 97-127.

Derbyshire, E. and Wang, J.T. (1994). China's Yellow River Basin, *In*: N.Roberts (ed.) *The Changing Global Environment*, Blackwell, Oxford, 417-439.

Derbyshire, E., Kemp, R. and Meng, X. (1995). Variations in Loess and Palaeosol Properties as Indicators of Palaeoclimatic Gradients across the Loess Plateau of North China, In: Aeolian Sediments in the Quaternary Record, E. Derbyshire (ed.) *Quat. Sci. Revi.* 14: 691-699.

Derbyshire. E., Smalley, I.J. and Dijkstra, T.A. (eds.) (1995). *Genesis and Properties of Collapsible Soils*, NATO ASI Series C: Mathematical and Physical Sciences, Vol. 468, NL, Kluwer, Dordrecht, p. 413.

Derbyshire, E. and Goudie, A.S. (1997). 'The drylands of Asia', In: *Arid Zone Geomorphology: process, form and change in drylands*, (1) S.G. Thomas (ed.) 2nd Edn., John Wiley and Sons, Chichester 487-506.

Dorn, R.I. (1994). The role of climatic change in alluvial fan development, In: *Geomorphology of Desert Environments*, (A.1) Abrahams, A.D. and Parsons, A.J. (eds.) pp. 593-615, Chapman and Hall, London.

Eybergen, F.A. and Imeson, A.C. (1989). Geomorphological processes and climatic change, *Catena* 16: 307-319.

Fang, J. (1993). Lake evolution during the last 3,000 years in China and its implications for environmental change, *Quat. Research*, 39: 175-185.

Fort, M. (1996). Late Cenozoic environmental changes and uplift on the northern side of the central Himalaya: a reappraisal from field data, In: Environmental Changes in the Tibetan Plateau and Surrounding Areas, F. Gasse and E. Derbyshire (eds.) *Palaeogeography, Palaeoclimatology, Palaeoecology*, 120: 123-145.

Gasse, F., Ledée, V., Massault, M. and Fontes, J. (1989). Water-level fluctuations of Lake Tanganyika in phase with oceanic changes during the last glaciation and deglaciation, *Nature*, 342: 57-59.

Gasse, F. and 13 co-authors (1991). A 13,000 year climate record from western Tibet, *Nature*, 353: 742-745.

Gasse, F., Fontes, J.C., Van Campo, E. and Wei, K. (1996). Holocene environmental changes in Bangong Co bason (western Tibet). Part 4: Discussion and Conclusions, In: Environmental Changes in the Tibetan Plateau and Surrounding Areas', F. Gasse and E. Derbyshire (eds.) *Palaeogeography, Palaeoclimatology, Palaeoecology*, 120: 79-92.

Goudie, A.S. (1983). The arid Earth, In: *Mega-geomorphology*, R.Gardner and H.Scoging (eds.), pp.152-171, Oxford University Press, Oxford.

Goudie, A.S., Stokes, S., Livingstone, I., Bailiff, I.K. and Allison, R.J. (1993). Post-depositional modification of the linear sand ridges of the West Kimberley area of northwest Australia, *Geog. J.* 159: 306-317.

Graf, W.L. (1983). Flood related channel change in an arid region, *Earth Surface Processes and Landforms* 8: 125-140.

Grove, A.T. (1958). The ancient ergs of Hausaland, and similar formations on the south side of the Sahara, *Geog. J.* 124: 528-533.

Grove, A.T. (1969). Landforms and climatic change in the Kalahari and Ngamiland, *Geog. J.* 135: 191-212.

Hövermann, J. (1985). Das System der klimatischen geomorphologie auf landschaftskundlicher Grundlage, *Zeitschrift fur Geomorphologie Supplementband* 56: 143-153.

Hövermann, J., Lehmkuhl, F. and Sussenberger, H. (1992). Neue Befunde zur Palaoklimatologie Nordafrikas und Zentralasiens, *Abhandlungen der Braunschweigischen Wissenschaftlichen Gesellschaft* 43: 127-150.

Jäkel, D. and Zhu, Z. (1991) Reports on the 1986 Sino-German Kunlun Shan Taklimakan-Expedition, *Die Erde* 6, 200pp. and map volume.

Jouzel. J. and thirteen co-authors (1989). Global change over the last climatic cycle from the Vostok ice core record, *Quaternary International* 2: 15-24.

Kemp, R.A. (1999). Soil micromorphology as a technique for reconstructing paleoenvironmental change. (*This volume*).

Kemp, R.A., Derbyshire, E., Meng, X.M., Chen, F.H. and Pan, B.T. (1995). Pedosedimentary reconstruction of a thick loess-palaeosol sequence near Lanzhou in north-central China, *Quat. Res.* 43: 30-45.

Kemp, R.A., Derbyshire, E., Chen, F. and Ma, H. (1996). Pedosedimentary development and palaeoenvironmental significance of the S1 palaeosol on the northeastern margin of the Qinhgai-Xizang (Tibetan) Plateau, *J. of Quat. Sci.* 11: 95-106.

Kocurek, G. (1998). Aeolian system response to external forcing factors - a sequence stratigraphic view of the Saharan region, In: *Quaternary Deserts and Climatic Change*, A.S. Alsharhan, K.W. Glennie, G.L. Whittle and C.G. St. C. Kendall (eds.) Balkema Dordrecht, *in press*.

Kukla, G. (1989). Long continental records of climate - an introduction, *Palaeogeography, Palaeoclimatology, Palaeoecology* 72: 1-9.

Kutzbach, J. (1983). Monsoon rains of the Late Pleistocene and early Holocene: patterns, intensity and possible causes of changes, In: *Variations in the Global water Budget*, pp. 371-389 F.A. Street-Perrott, M. Beran and R. Ratcliffe (eds.) Reidel, Dordrecht.

Lancaster, N. (1981). Palaeoenvironmental implications of fixed dune systems in southern Africa, *Palaeogeography, Palaeoclimatology, Palaeoecology* 33: 327-346.

Langbein, W.B. and Schumm, S.A. (1958). Yield of sediment in relation to mean annual precipitation, *Trans. Am. Geophys. Union* 39: 1076-1084.

Li, J.J., Wen, S., Zhang, Q., Wang, F., Zheng, B.X., and Li, B. (1979). A discussion on the period, amplitude and type of uplift of the Qinghai-Xizang Plateau, *Scientia Sinica* 22: 1314-1328.

Livingstone, I. and Thomas, D.S.G. (1993). Modes of linear dune activity. and their palaeoenvironmental significance: an evaluation with reference to southern African examples, In: *The Dynamics and Environmental Context of Aeolian Sedimentary Systems*, K. Pye (ed.) Geological Society Special Publication 72: 91-101., London.

Livingstone, I. and Warren, A. (1996). *Aeolian Geomorphology*, Longman, 211p. Harlow, U.K.

Lorius, C., Jouzel, J., Ritz, C., Merlivat, L., Barkov, N.I., Korotkevich, Y.S. and Kotlyakov, V.M. (1985). A 150,000 year climatic record from Antarctic ice, *Nature* 316: 591-596.

Madigan, C.T. (1936). The Australian sand ridge deserts, *Geog. Rev.* 26: 205-227.

Mainguet, M. (1978). The influence of Trade Winds, local air-masses and topographic obstacles on the aeolian movement of sand particles and the origin and distribution of dunes and ergs in the Sahara and Australia, *Geoforum* 9: 17-28.

Maizels, J. (1990). Raised channel systems as indicators of palaeohydrologic change: a case study from Oman, *Palaeogeography, Palaeoclimatology, Palaeoecology* 76: 241-277.

McKee, E.D. (ed.) (1979). A study of global sand seas, *U. S. Geog. Survey Professional Paper* 1052.

Molnar, R. and England, P. (1990). Late Cenozoic uplift of mountain ranges and global climatic change: chicken or egg?, *Nature* 346: 29-34.

Nanson, G.C. and Tooth, S. (1999). Arid-zone rivers as indicators of climate change, *This Volume*.

Owen, L.A., Derbyshire, E., White, B.J. and Rendell, H. (1992). Loessic silt deposits in the western Himalayas: their sedimentology, genesis and age, *Catena* 19: 493-509

Pachur, H.J. and Kröpelin, S. (1987). Wadi Howar: palaeoclimatic evidence from an extinct river system in the southeastern Sahara, *Science* 237: 298-300.

Pachur, H.J., Wunnemann, B. and Zhang, H. (1995). Lake evolution in the Tengger Desert, northwestern China, during the last 40,000 years, *Quat. Res.* 44: 171-180.

Penck, A. (1910). Versuch einer Klimaklassification auf physiographischer Grundlage, *Preussen Akademie der Wissenschaft Sitz. Der physikalisch-mathematischen* 12: 236-246.

Petit-Maire, N. (ed.) (1991). *Paléoenvironnements du Sahara: lacs holocènes à Taoudenni (Mali)*, Paris, Editions du Centre National de la Recherche Scientifique.

Petit-Maire, N. and Riser, J. (1988). *Le Sahara à l'Holocéne: Mali*, Paris, CCGM, Map at scale 1:1 million.

Petit-Maire, N., Fontugne, M. and Rouland, C. (1991). Atmospheric methane ratio and environmental changes in the Sahara and Sahel during the last 130 K yrs, *Palaeogeography, Palaeoclimatology, Palaeoecology* 86: 197-204.

Petrov, M. P. (1976). *Deserts of the World*, John Wiley and Sons Ltd., New York, 447p.

Porter, S.C., An, Z.S. and Zheng, H.B. (1992) Cyclic Quaternary alluviation and terracing in a nonglaciated drainage-basin on the north flank of the Qinling Shan, central China, *Quaty. Res.* 38: 157-169.

Pye, K. (1983). Early post-depositional modification of aeolian dune sands, In: *Eolian Sediments and Processes*, M.E. Brookfield, T.S. Ahlbrandt (eds.) pp. 197-221, Elsevier, Amsterdam.

Reid, I. (1994). River landforms and sediments: evidence of climatic change, In: *Geomorphology of Desert Environments*, A.D. Abrahams, and A.J. Parsons (eds.) pp. 571-615, Chapman and Hall, London.

Roberts, N. and Barker, P. (1993). Landscape stability and biogeomorphic response to past and future climatic shifts in intertropical Africa, In: *Landscape Sensitivity*, D.S.G. Thomas and R.J. Allison (eds.) pp. 65-82, John Wiley and Sons Ltd., Chichester.

Said, R. (1975). Some observations on the geomorphology of the south Western Desert of Egypt and its relation to the origin of ground water, *Annals of the Geol. Sur. Egypt* 5: 61-70.

Schumm, S.A. (1968). River adjustment to altered hydrologic regimen - Murrumbidgee River and palaeochannels, Australia, *U. S. Geol. Survey Professional Paper* 598.

Schumm, S.A. (1977). *The Fluvial System*, John Wiley and Sons, New York and London.

Schumm, S.A. (1979). Geomorphic thresholds: the concept and its applications, *Transactions of the Institute of British Geographers* 4: 485-515.

Schumm, S.A. (1980). Some applications of the concept of geomorphic thresholds, In: *Thresholds in Geomorphology*, D.R. Coates and J.D. Vitek (eds.) pp. 473-485 Proceedings of the 9th Binghampton Geomorphology Symposium 1978, Binghampton, New York.

Shackleton, N.J., An, Z., Dodonov, A.E., Gavin, J., Kukla, G.J., Ranov, V.A. and Zhou, L.P. (1995). Accumulation rate of loess in Tadjikistan and China; relationship with global ice volumes, In: Wind Blown Sediments in the Quaternary Record, E. Derbyshire (ed.) *Quaternary Proceedings* 4: 1-6.

Shackleton, N.J., Imbrie, J, and Pisias, N.G. (1988). The evolution of oceanic oxygen-isotope variability in the North Atlantic over the past three million years, *Philosophical Transactions of the Royal Society, London, B*, 318: 679-688.

Shaw, P.A. and Thomas, D.S.G. (1993). Geomorphological processes, environmental change and landscape sensitivity in the Kalahari region of southern Africa, In: *Landscape Sensitivity*, D.S.G. Thomas and R.J. Allison (eds.) pp. 83-95, John Wiley and Sons Ltd., Chichester.

Shaw, P.A., Thomas, D.S.G. and Nash, D.J. (1992). Late Quaternary fluvial activity in the dry valleys (mekgacha) of the Middle and Southern Kalahari, southern Africa, *J. Quaty. Sci.* 7: 273-281.

Stoddart, D.R. (1969). Climatic Geomorphology: review and re-assessment, *Progress in Geography* 1: 160-222.

Strahkov, N.M. (1967). In: *Principles of Lithogenesis* vol. 1, Tomkeieff, S.I. and J.E. Hemingway (eds.) Oliver and Boyd, Edinburgh.

Street, A.F. and Grove, A.T. (1979). Global maps of lake-level fluctuations since 30,000 year BP, *Quaty. Res.* 10: 83-118.

Street-Perrott, F.A. and Perrott, R.A. (1990). Abrupt climatic fluctuations in the tropics: the influence of Atlantic Ocean circulation, *Nature* 343: 607-612.

Street-Perrott, F.A. and Roberts, N. (1994). Past climates and future greenhouse warming, In: *The Changing Global Environment*, N. Roberts (ed.) pp. 47-68, Blackwell, Oxford.

Street-Perrott, F.A., Roberts, N. and Metcalfe, S. (1985). Geomorphic implications of late Quaternary hydrological and climatic changes in the Northern Hemisphere tropics, In: *Environmental Change and Tropical Geomorphology*, I. Douglas and T. Spencer, (eds.), pp. 165-183, George Allen and Unwin, London.

Summerfield, M.A. and Kirkbride, M.P. (1992). Climate and landscape response, *Nature* 355: 306.

Tchakerian, V.P. (1994). Palaeoclimatic interpretations from desert dunes and sediments, In: *Geomorphology of Desert Environments*, A.D. Abrahams, and A.J. Parsons (eds.) pp. 631-643 Chapman and Hall, London.

Thomas, D.S.G. (1984). Ancient ergs of the former arid zones of Zimbabwe, Zambia and Angola', *Transactions of the Institute of British Geographers* 9: 75-88.

Thomas, D.S.G. (1992). Desert dune activity: concepts and significance, *J. Arid Environments* 22: 31-38.

Thomas, D.S.G. and Allison, R.J. (1993). *Landscape Sensitivity*, John Wiley and Sons Ltd., Chichester and New York, 347p.

Tricart, J. and Cailleux, A. (1972). *Introduction to Climatic Geomorphology*, (trans. C.J.K. de Jonge), Longman London.

Walling, D.E. and Webb, B.W. (1983). Patterns of sediment yield, In: *Background to Palaeohydrology: a perspective*, K.J. Gregory (ed.) pp. 69-100, John Wiley and Sons Ltd., Chichester.

Watson, A., Price Williams, D. and Goudie, A. (1985). The palaeoenvironmental interpretation of colluvial sediments and palaeosols in Southern Africa, *Palaeogeography, Palaeoclimatology, Palaeoecology* 45: 235-249.

Wells, G.L. (1983). Late-glacial circulation over central North America revealed by aeolian features, In: *Variations in the Global Water Budget*, F.A. Street-Perrott, M. Beran and R. Ratcliffe (eds.) pp. 317-330, Reidel, Dordrecht.

White, K. (1999). Remote sensing for palaeoenvironmental studies in drylands. *This Volume.*

Williams, M.A.J. (1985). Pleistocene aridity in tropical Africa, Australia and Asia, In: *Environmental Change and Tropical Geomorphology*, I. Douglas and T. Spencer (eds.) pp. 219-223, George Allen and Unwin, London.

Wilson, L. (1973). Relationships between geomorphic processes and modern climates as a method in paleoclimatology, In: *Climatic Geomorphology*, E. Derbyshire (ed.) pp. 269-284, Macmillan, London.

Wolman, M.G. and Gerson, R. (1978). Relative scales of Time and Effectiveness of Climate in watershed geomorphology, *Earth Surface Processes* 3: 189-208.

Yair, A. and Enzel, Y. (1987). The relationship between annual rainfall and sediment yield in arid and semi-arid areas. The case of the Northern Negev, In: Geomorphological Models, Theoretical and Empirical Aspects, F. Ahnert (ed.) *Catena*, Supplement 10: 121-136.

2 Remote Sensing for Palaeoenvironmental Studies in Drylands

Kevin White[*]

ABSTRACT

Earth observation remote sensing techniques involve using electromagnetic radiation to gather information about the Earth surface, atmosphere and oceans. Both passive and active remote sensing techniques have been used for analysis of dryland palaeoenvironments. By facilitating geomorphological mapping over large areas of inaccessible terrain, this approach allows extrapolation of detailed field observations and measurements to other areas. Computer processing of satellite-derived data provides enhanced ability to map landforms in considerable detail. Large amounts of data on aeolian and fluvial landforms can be derived in this way, but palaeoenvironmental interpretation of landforms remains problematic due to our poor understanding of the relationships between landforms and the processes that form them. However, the ability to analyse landforms within their synoptic setting enables recognition of regional patterns of landforms, providing insight into the geomorphological response to changing dryland palaeoenvironments.

Remote Sensing Techniques

Remote sensing has been variously defined (Curran,1985; Mather, 1987), but common to all definitions is the notion of collecting information about the Earth (land surface, oceans and atmosphere) without physical contact. The degree of remoteness can vary between information collected just above the surface (photographs of landscape changes over time would be an example of this), through the use of aircraft as platforms for photography or radiation measurement, to various orbital platforms, which can range between 100-36,000 km altitude. The degree of remoteness selected for a particular application depends on the surface area for which information is to be derived and the spatial resolution (degree of detail) of the information required. For example, monitoring of small features, such as barchan dunes, would require data from low orbit or aerial

[*]Department of Geography, University of Reading, Whiteknights, Reading RG6 6AB, U.K.

surveys, whereas large features, such as regional vegetation patterns, can be monitored from higher orbits, at a commensurably coarser scale.

Various systems are used to collect remotely sensed information; some make use of naturally occurring electromagnetic radiation from the Sun (passive systems) while others generate their own radiation with which to illuminate the surface (active systems). Passive systems can make use of a broad part of the electromagnetic spectrum (the range of wavelengths over which electromagnetic radiation is propagated), from short wavelength visible light, through reflected and thermal infrared radiation, to long wavelength microwave radiation (Fig. 1). Both passive and active systems depend on the way this radiation interacts with the Earth's atmosphere, land surface and oceans for gathering information.

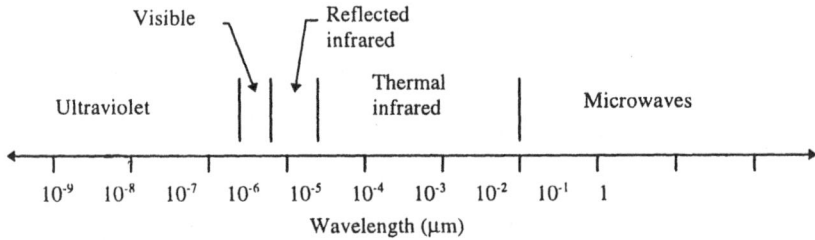

Fig. 1: The part of the electromagnetic spectrum used in remote sensing.

Passive Systems

Passive systems fall into two categories: camera systems which focus incoming radiation, usually in the visible to near infrared part of the spectrum, onto sensitive film to derive an image of the target; and radiometer systems, electrical devices which measure the radiation from a target and return a numerical value (digital number or DN) proportional to the amount of incoming radiation. Radiometers can be used to produce images by using the forward motion of the platform (aircraft or satellite) and a scanning mirror operating at right angles to this (a whiskbroom system, Fig. 2a), or alternatively a linear array of charge-coupled devices aligned at right angles to platform motion (a pushbroom system, Fig. 2b), to measure the variation in radiation over a target area, known as a scene. This produces a two-dimensional matrix of radiation measurements, each individual measurement corresponding to the radiation received from a discrete part of the scene (referred to as a pixel). If an orbital platform is used, the data are downloaded to a surface receiving station using radio communication. To convert these pixel DN values into an image requires the matrix to be displayed on a computer screen, with the

(a)

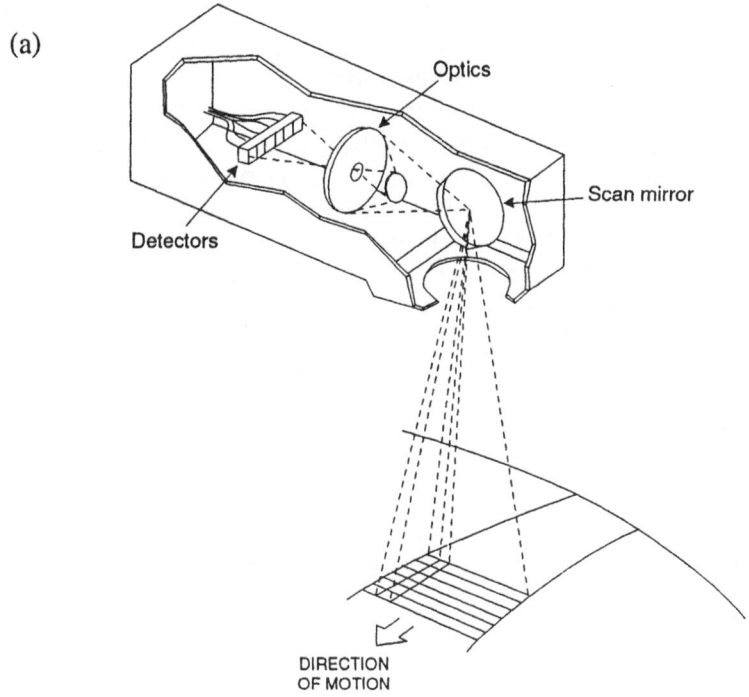

(b)

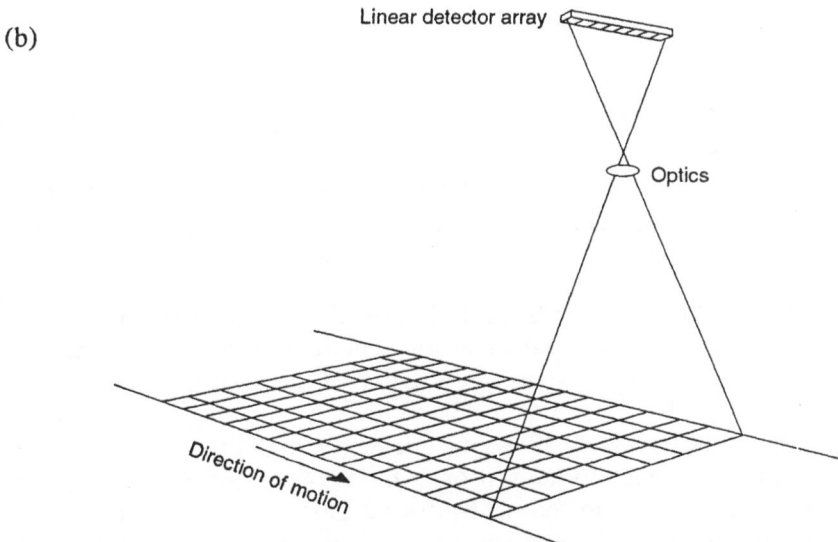

Fig. 2: Passive remote sinsing systems; (a) scanner or whiskbroom system (after Crackwell and Hayes, 1991), (b) multilinear array or pushbroom system (after Lo, 1991).

brightness of each part of the screen being controlled by the pixel DN value. This results in a black and white image showing the distribution of radiance in the imaged area (Mather, 1987).

Whichever method is used to gather the electromagnetic radiation, it is possible to split the incoming radiation into its various wavelength components and image them separately, as is done by 'multispectral' remote-sensing systems. Typically, radiance upwelling from the surface is split into three or more broad spectral bands (for example, visible green, visible red, near-infrared etc.). The DN values for each of the three spectral bands are then used to control the intensity of the three primary colours (red, green and blue) on a colour computer monitor, creating a false colour composite. This is the form in which satellite images are most commonly encountered (Mather, 1987).

Because different surface materials absorb and reflect different proportions of incident solar radiation at different parts of the electromagnetic spectrum, each surface tends to have a distinctive 'spectral signature' when reflectance is measured across the visible to short-wave infrared part of the electromagnetic spectrum (Fig. 3). At regions of the

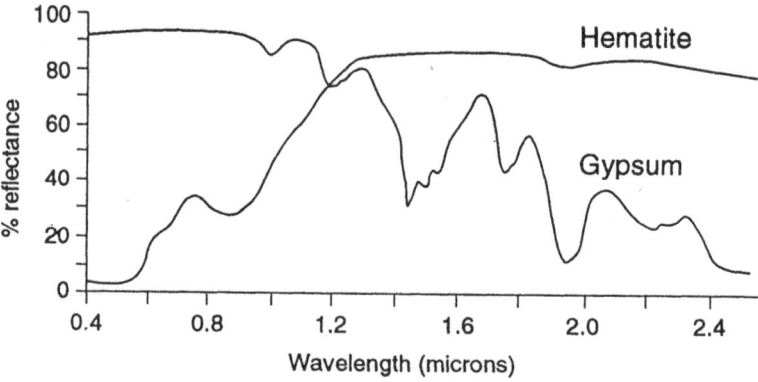

Fig. 3: Spectral reflectance curves of hematite and gypsum, commonly occurring minerals in drylands.

spectrum where energy is strongly absorbed, the reflectance decreases forming an absorption feature in the spectral curve (several features are evident in the short-wave infrared region of the gypsum spectrum in Fig. 3). This would translate to a dark region in an image collected at this wavelength. Conversely, if a surface reflects a high proportion of incoming solar radiation back to the sensor in the wavelength range being sensed (such as vegetation in the near-infrared part of the spectrum), then the area will appear bright in the resultant image. When images of three different spectral bands are combined as a false colour composite, as outlined above, the different spectral behaviour of surfaces is manifested by different colours. By comparing images collected at different wavelengths by multispectral systems, a trained interpreter can identify certain landcover types on the basis of their spectral signatures. More importantly, computer image-processing techniques can be used to classify images into areas of different landcover types (Mather, 1987), or to make estimates of cover types present within individual pixels (Settle and Drake, 1993). The final step in the process of image interpretation for palaeoenvironmental studies normally involves deriving geomorphological maps based on the observed distribution of landcover types (Townshend and Hancock, 1981).

Active Systems

Active remote-sensing techniques usually involve illuminating the surface to one side of nadir (the point on the surface immediately below the platform) with an antenna operating at microwave frequencies, giving the resultant images a strong side-lit appearance. The pulse of microwave energy propagates from the antenna at the speed of light until it is intercepted by the surface. Depending on the orientation, roughness and composition (specifically the dielectric constant) of the surface, a certain proportion of the incident energy gets scattered back towards the antenna (Fig. 4). The antenna then goes into receive mode and measures how

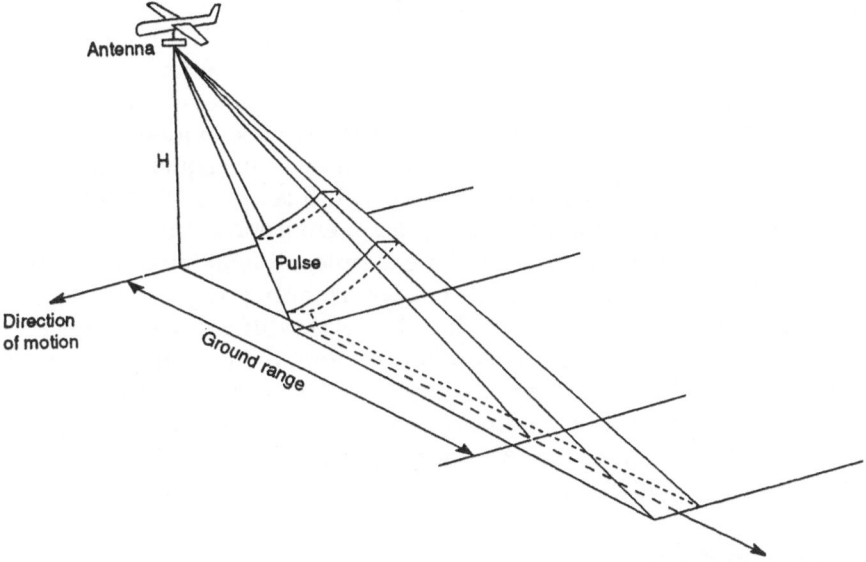

Fig. 4: Active (radar) remote sensing system (after Lillesand and Kiefer, 1994).

much of the pulse is returned. Because of the side-looking viewing geometry, the time taken for the pulse to return to the antenna is a function of distance from the nadir track of the platform. This enables a strip of bright and dark returns along the range of the antenna to be constructed. The forward motion of the platform between pulses builds up a two-dimensional image of backscatter.

Surface orientation control on backscatter makes radar particularly useful at mapping subtle relief (Krul, 1993); the roughness control enables radar to map particle-size distributions of surface materials (Janse, 1993) and structure of vegetation canopies (de Loor, 1993), and the dielectric constant control enables radar to map soil moisture. All these characteristics are of considerable use in geomorphological mapping (Millington et al., 1995).

Dryland Palaeoenvironments: Evidence from Remote Sensing

Both passive and active remote-sensing systems have been used to collect palaeoenvironmental information in drylands. Reconstruction of dryland palaeoenvironments is usually based on geomorphological and sedimentological evidence (Thomas, 1989) but it is through large-scale geomorphological mapping that remote sensing can play a most useful role (Gardner and Scoging, 1983), enabling extraction of information not readily evident from conventional field or laboratory analyses. The relationship between landforms and climate has a long history of study (Budel, 1982) and interpretation of palaeoenvironments from landform interpretation has become an important area of research (Thomas, 1989). Derbyshire has discussed these aspects in detail in Chapter 1.

Palaeoenvironmental interpretation of geomorphological evidence in drylands is based on the identification of landforms, evidence that they are relict (often requiring absolute dating), and the subsequent application of modern analogues. A major problem with this approach results from our imperfect understanding of the precise environmental prerequisites for formation of many landforms; for example, alluvial fans can develop in many disparate environmental settings (from hot, dry deserts to cold pro-glacial terrains). This means that it is often difficult to demonstrate how relict landforms are out of equilibrium with the present geomorphic processes. However, there are many examples of remote-sensing data being used to derive geomorphological evidence of dryland palaeoenvironments (see, e.g., Abrams and Chadwick, 1994; Brookes, 1993; McCauley et al., 1982).

Aeolian Landforms

Sand dunes are characteristic dryland landforms (Lancaster, 1994), and the presence of degraded or inactive dunes is widely used to indicate past arid phases (Tchakerian, 1994). In conjunction with dune dating techniques (Singhvi et al., 1982) this approach can be used to extrapolate dated arid phases over large areas by virtue of the synoptic overview provided by remote sensing (Thomas 1989). Both passive (Paisley et al., 1991) and active (Lancaster et al., 1992) remote-sensing techniques are used to discriminate active and inactive sand dunes. Relict sand seas are easily mapped from remote sensing because of their distinct vegetation zonations. The dune crests may be relatively sparsely vegetated because of their free-draining characteristics, or densely forested because the soft sands favour the development of deep root systems (Thomas, 1984). However, problems of interpretation remain: dunes fixed by dense woodland are inactive but it is less easy to demonstrate this in the case of sparsely or partially vegetated dunes, as appreciable sand movement has been demonstrated with up to 35% vegetation cover (Ash and Wasson, 1983). In Australia, dune stabilisation is now not simply attributed to vegetation cover, but to a decline in wind force since the time of dune formation (Wasson, 1984).

Different dune types have different environmental requirements (specifically sediment supply and wind regimes) for their development, providing further potential for palaeoenvironmental information from remote sensing. Relict dune systems can be used to reconstruct palaeocirculation patterns (Lancaster, 1985 Glennie, this volume). The orientation of relict dunes is compared to the resultant direction of modern sand-moving winds, calculated from wind data. This technique can highlight changes in wind regime but, because of the high seasonal or diurnal directional variability of modern wind regimes, interpretation of palaeowind regimes remains highly problematic (Thomas, 1989). Remote-sensing techniques have also been used to map the mineral composition of the Gran Desierto sand sea, Mexico (Blount et al., 1990) and the Namib sand sea, Namibia (Walden et al., 1996). This has enabled detailed analysis of the Quaternary evolution of the sand sea, particularly with regard to the changing sources of sand (Blount and Lancaster, 1990).

Fluvial Landforms

Despite the ephemeral nature of runoff in drylands, fluvial processes are important in dryland geomorphology. However, fluvial systems are sensitive to climatic changes (Bull, 1991) and interpretation of fluvial landforms can provide evidence of past climates (Reid, 1994).

Active remote-sensing techniques have the ability to identify and map river systems now buried under aeolian sand. A celebrated example is the 'radar rivers' underneath the Selima sand sheet, Egypt (McCauley et al., 1982). Radar is able to detect this feature due to the subtle textural differences of the material infilling the channel, information which is not available from passive systems using visible to infrared wavelengths, in which the channel systems are invisible (Davis et al., 1993). Such features support the existence of past humid phases and analyses of the resultant networks can provide useful palaeohydrological information (Baker, 1983).

Alluvial fans are widely used as palaeoenvironmental indicators (Dorn, 1994) due to their piedmont position between upland areas dominated by erosion and lowland areas dominated by transport and deposition. If a climatic change is of significant magnitude to alter the distribution of these process domains, it will leave a record of this in the landforms of the piedmont zone (White, 1991). For example, a move to a more humid climate would tend to decrease sediment yield from the upland basin due to increased vegetation cover (Bull, 1991). As lower amounts of sediment are delivered to the fan per unit discharge, it tends to undergo erosion (Harvey, 1990), forming a fanhead trench and a new depositional segment at a lower level (Blair and McPherson, 1994). Successive erosional events can form distinctive flights of fan terraces and slope segments (Blissenbach, 1954), analogous to the flights of river terraces that form within drainage basins (Harvey, 1989). Aerial photographs have long been used to map alluvial fan slope segments (Beaumont, 1972; Denny, 1965; Mayer et al., 1984) and the patterns of fan incision (Bowman, 1978; Rust, 1979). Climatic interpretations are still based on detailed aerial photograph interpretations (Abrams and Chadwick, 1994). However, the synoptic capability of satellite imagery enables the regional pattern to be determined, thereby accounting for local tectonic effects along faulted mountain fronts (White, 1987). As different fan segments have experienced different durations of pedogenesis (White et al., 1992) and rock varnish formation (White, 1990), both time-dependent characteristics, they are manifested by different spectral reflectance characteristics (White, 1993). This enables recognition of fan incision events from passive remote-sensing data (Fig. 5), allowing regional patterns of fan development to be mapped from satellite imagery (White, 1991).

Dryland lakes provide further evidence of climatic change as they are sensitive to changes in the regional water balance (Arnow, 1980). Palaeohydrological history is particularly well recorded by closed-basin amplifier lakes (Street-Perrott and Harrison, 1985), as amplifier lakes (i.e., those for which almost all input is from surface runoff and all output is from evaporation) show significant response to short-term precipitation changes. Palaeolake chronologies are reconstructed from shoreline evidence, from which surface area and elevation can be determined (Sack, 1994). A ratio of lake surface area to basin tributary area (called the z ratio), is commonly used, both of which can be determined from aerial photography (Mifflin and Wheat, 1979) or satellite imagery (Lambin et

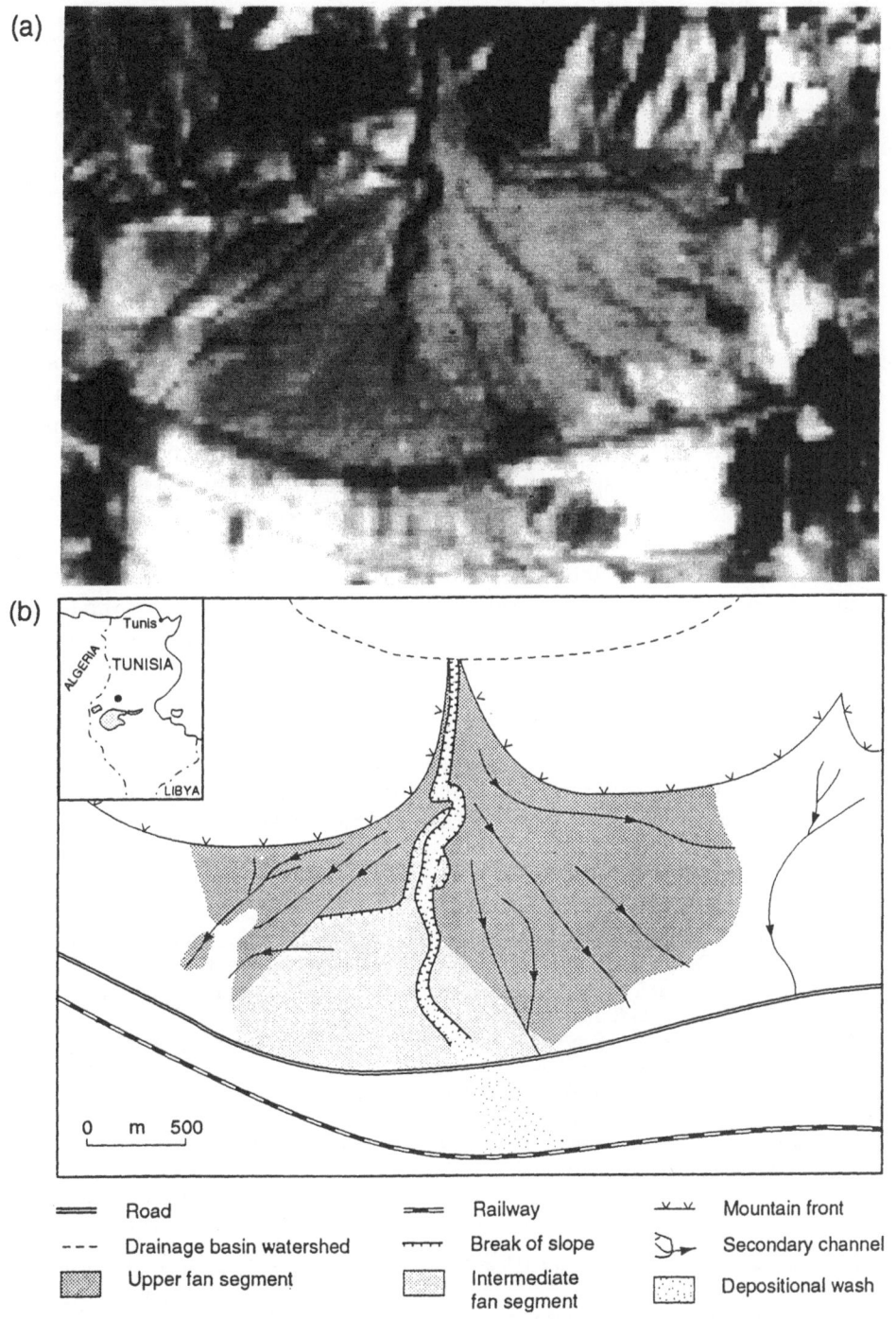

Fig. 5: (a) Landsat Thematic Mapper short-wave infrared image of Oued es Seffaia alluvial fan, Tunisia. (b) Geomorphological map derived from interpretation of Thematic Mapper image, showing pattern of fan dissection.

al., 1995). Satellite remote sensing is particularly useful in this respect as it permits analysis of large and inaccessible areas, such as the surface of playas (Quarmby et al., 1989).

Conclusions

Both active and passive remote-sensing techniques are used for analysis of dryland palaeoenvironments because they facilitate geomorphological mapping over large areas of inaccessible terrain and allow extrapolation of detailed field observations and measurements to other areas. Computer processing of satellite-derived data provides enhanced ability to map surficial materials which make up landforms, while the availability of stereo imagery from space enables three-dimensional geomorphological analysis of landforms. Large amounts of data on aeolian and fluvial landforms can be derived from remotely sensed data, but palaeo-environmental interpretation of landforms remains problematic. However, the ability to analyse landforms within their synoptic setting enables recognition of regional patterns of landforms, providing insight into the geomorphological response to changing dryland palaeoenvironments.

Acknowledgments

Michelle Rogan (University of Canterbury, New Zealand) and Heather Browning (University of Reading, United Kingdom) drew the diagrams.

References

Abrams, M.J. and Chadwick, O.H. (1994). Tectonic and climatic implications of alluvial fan sequences along the Batinah coast, Oman. *J. Geol. Soc., London*, 151: 51-58.

Arnow, T. (1980). Water budget and water-surface fluctuations of Great Salt Lake. *Utah. Geol. Min. Surv. Bull.* 116: 255-263.

Ash, J.E. and Wasson, R.J. (1983). Vegetation and sand mobility in the Australian desert dunefield. *Zeitschrift für Geomorph. Suppl.* 45: 7-25.

Baker, V.R. (1983). Large-scale fluvial palaeohydrology. In: *Background to Palaeohydrology, a Perspective* K.J. Gregory (ed.), pp. 453-478 Wiley and Dons, Chichester.

Beaumont, P. (1972). Alluvial fans along the foothills of the Elburz Mountains, Iran. Palaeogeog., Palaeoclim., Palaeoecol. 12: 251-273.

Blair, T.C. and McPherson, J.G. (1994). Alluvial fan processes and forms. In: *Geomorphology of Desert Environments*, A.D. Abrahams and A.J. Parsons (eds.), pp. 354-402. Chapman and Hall, London.

Blissenbach, E. (1954). Geology of alluvial fans in semiarid regions. *Geol. Soc. Amer. Bull* 65: 175-190.

Blount, G. and Lancaster, N. (1990). Development of the Gran Desierto sand sea, north-western Mexico. *Geology*, 18: 724-728.

Blount, G., Smith, M.O., Adams, J.B., Greeley, R. and Christensen, P.R. (1990). Regional aeolian dynamics and sand mixing in the Gran Desierto: evidence from Landsat Thematic Mapper images. *J. Geophys. Res.* 95: 15463-15482.

Bowman, D. (1978). Determination of intersection points within a telescopic alluvial fan complex. *Earth Surface Proc. Landforms* 3: 265-276.

Brookes, I.A. (1993). Geomorphology and Quaternary Geology of the Dakhla Oasis region, Egypt. *Quat. Sci. Rev.* 12: 529-552.

Budel, J. (1982). *Climatic Geomorphology.* Princeton University Press, Princeton, p. 443.

Bull, W.B. (1991). *Geomorphic Responses to Climatic Change.* Oxford University Press, New York, p. 326.

Cracknell, A.P. and Hayes, L.W.B. (1991). *Introduction to Remote Sensing.* Taylor and Francis, London, p. 293.

Curran, P.J. (1985). *Principles of Remote Sensing,* Longman, London, p. 282.

Davis, P.A., Breed, C.S., McCauley, J.F. and Schaber, G.G. (1993). Surficial geology of the Safsaf region, south-central Egypt, derived from remote sensing and field data. *Remote Sensing Environ.* 46: 183-203.

Denny, C.S. (1965). *Alluvial Fans in the Death Valley Region, California and Nevada.* U.S. Geol. Surv. Profes. 466: 1-62.

Dorn, R.I. (1994). The role of climatic change in alluvial fan development. In: *Geomorphology of Desert Environments* A.D. Abrahams and A.J. Parsons (eds.), pp. 593-615, Chapman and Hall, London.

Gardner, R. and Scoging, H. (1983). *Mega-Geomorphology.* Oxford University Press, Oxford, p. 240.

Harvey, A.M. (1989). The occurrence and role of arid zone alluvial fans. In: *Arid Zone Geomorphology* D.S.G. Thomas (ed.), pp. 136-158, Belhaven, London.

Harvey, A.M. (1990). Factors influencing Quaternary alluvial fan development in southeast Spain. In: *Alluvial Fans—a Field Approach* A.H. Rachocki and M. Church (eds.), pp. 247-269, J.Wiley and Sons, New York.

Janse, A.R.P. (1993). Radar backscatter of soils. In: *Land Observation by Remote Sensing* H.J. Buiten and J.G.P.W. Clevers (eds.), pp. 237-254 Gordon and Breach, Amsterdam.

Krul, L. (1993). Remote sensing in the microwave region. In: *Land Observation by Remote Sensing,* H.J. Buiten and J.G.P.W. Clevers (eds.), pp. 155-174, Gordon and Breach, Amsterdam.

Lambin, E.F., Walkey, J.A. and Petit-Maire, N. (1995). Detection of Holocene lakes in the Sahara using satellite remote sensing. *Photogram. Eng. Remote Sensing* 61: 731-737.

Lancaster, N. (1985). Winds and sand movements in the Namib sand sea. *Earth Surface Proc. Landforms* 10: 607-619.

Lancaster, N. (1994). Dune morphology and dynamics. In: *Geomorphology of Desert Environments* A.D. Abrahams and A.J. Parsons (eds.), pp. 474-505, Chapman and Hall, London.

Lancaster, N., Gaddis, L. and Greeley, R. (1992). New airborne imaging radar observations of sand dunes, Kelso dunes, California. *Remote Sensing Environ.* 39: 233-238.

Lillesand, T.M. and Kiefer, R.W. (1994). *Remote Sensing and Image Interpretation* J. Wiley and Sons, New York, 3rd ed., p. 721.

Lo, C.P. (1991). *Applied Remote Sensing.* Harlow, Longman, 393 p.

Loor, G.P. de (1993). Radar backscatter of crops. In: *Land Observation by Remote Sensing* H.J. Buiten and J.G.P.W. Clevers, (eds.), pp. 203-217 Gordon and Breach, Amsterdam.

Mather, P.M. (1987). *Computer Processing of Remotely Sensed Images.* Wiley and sons, Chichester, p. 346.

Mayer, L., Gerson, R. and Bull, W.B. (1984). Alluvial gravel production and deposition; a useful indicator of Quaternary climatic changes in deserts (a case study in southwestern Arizona). *Catena Suppl.* 5: 137-151.

McCauley, J.F., Schaber, G.G., Breed, C.S., Grolier, M.J., Haynes, C.V., Issawi, B., Elachi, C. and Blom, R. (1982). Subsurface valleys and geoarchaeology of the eastern Sahara revealed by Shuttle Radar. *Science*, 218: 1004-1020.

Mifflin, M.D. and Wheat, M.M. (1979). Pluvial lakes and estimated pluvial climates of Nevada. *Nevada Bur. Mines Geol. Bull.* 94: 1-28.

Millington, A.C., White, K., Drake, N.A., Wadge, G. and Archer, D.J. (1995). Remote sensing of geomorphological processes and surficial material geochemistry in drylands. In: *Advances in Environmental Remote Sensing*, F.M. Danson and S.E. Plummer (eds.), pp. 105-122 Wiley and sons, Chichester.

Paisley, E.C.I., Lancaster, N., Gaddis, L.R. and Greeley, R. (1991). Discrimination of active and inactive sand dunes from remote sensing. Kelso dunes, Mojave Desert, California. *Remote Sensing Environ.* 37: 153-166.

Quarmby, N.A., Townshend, J.R.G., Millington, A.C., White, K. and Reading, A.J. (1989). Monitoring sediment transport systems in a semi-arid area using Thematic Mapper data. *Remote Sensing Environ.* 28: 305-315.

Reid, I. (1994). River landforms and sediments: evidence of climatic change. In: *Geomorphology of Desert Environments*, A.D. Abrahams and A.J. Parsons (eds.), pp. 571-592 Chapman and Hall, London.

Rust, B. R. (1979). Coarse alluvial deposits. In: *Facies Models*, R.G. Walker (ed.), pp. 9-21 Geol. Soc. Canada, Ontario.

Sack, D. (1994). Geomorphic evidence of climate change from desert-basin palaeolakes. In: *Geomorphology of Desert Environments*, A.D. Abrahams and A.J. Parsons (eds.), pp. 616-630 Chapman and Hall, London.

Settle, J.J. and Drake, N.A. (1993). Linear mixing and the estimation of ground proportions. *Internat. J. Remote Sensing* 14: 1159-1171.

Singhvi, A.K., Sharma, Y.P. and Agrawal, D.P. (1982). Thermoluminescence dating of sand dunes in Rajasthan, India. *Nature* 295: 313-315.

Street-Perrott, F.A. and Harrison, S.P. (1985). Lake levels and climate reconstruction. In: *Palaeoclimate Analysis and Modelling*, A.D. Hecht (ed.), pp. 291-340 J. Wiley and sons, New York.

Tchakerian, V.P. (1994). Palaeoclimatic interpretations from desert dunes and sediments. In: *Geomorphology of Desert Environments*, A.D. Abrahams and A.J. Parsons (eds.), pp. 631-643 Chapman and Hall, London.

Thomas, D.S.G. (1984). Ancient ergs of the former arid zones of Zimbabwe, Zambia and Angola. *Trans. Inst. British Geographers*, NS, 9: 75-88.

Thomas, D.S.G. (1989). Reconstructing ancient arid environments. In: *Arid Zone Geomorphology*, D.S.G. Thomas (ed.), pp. 311-334 Belhaven, London.

Townshend, J.R.G. and Hancock, P.J. (1981). The role of remote sensing in mapping surficial deposits. In: *Terrain Analysis and Remote Sensing*, J.R.G. Townshend (ed.), pp. 204-218 George Allen and Unwin, London.

Walden, J., White, K. and Drake, N.A. (1996). Controls on dune colour in the Namib sand sea: preliminary results. *African Earth Sci.* 22, 349-353.

Wasson, R.J. (1984). Late Quaternary palaeoenvironments in the desert dunefields of Australia. In: *Late Cainozoic Palaeoclimates of the Southern Hemisphere*, J.C. Vogel (ed.), pp. 419-432 Balkema, Rotterdam.

White, K. (1987). Piedmont surface mapping in the Tunisian Southern Atlas: use of Landsat Thematic Mapper data. In: *Advances in Digital Image Processing, Proc. 13th Ann. Conf. Remote Sensing Society, Sept. 7-11, 1987, University of Nottingham, Nottingham*, pp. 641-649.

White, K. (1990). *Spectral Reflectance Characteristics of Rock Varnish in Arid Areas.* School of Geography Research Paper no. 46, University of Oxford, 38p.

White, K. (1991). Geomorphological analysis of piedmont landforms in the Tunisian Southern Atlas using ground data and satellite imagery. *Geog. J.* 157: 279-294.

White, K. (1993). Image processing of Thematic Mapper data for discriminating piedmont surficial materials in the Tunisian Southern Atlas. *Internat. J. Remote Sensing* 14: 961-977.

White, K., Walden, J. and Rollin, E.M. (1992). Remote sensing of pedogenic iron oxides using Landsat Thematic Mapper data of southern Tunisia. In: *Remote Sensing; from Research to Operation, Proc. 18th Ann. Conf. Remote Sensing Society Sept. 15-17, University of Dundee, Dundee,* pp. 179-187.

3

Soil Micromorphology as a Technique for Reconstructing Palaeoenvironmental Change

ROB A. KEMP[*]

ABSTRACT

This paper describes how soil micromorphology (microscopic study of soil thin-sections) may be used to help reconstruct palaeoenvironmental change. Emphasis is on the details and approach of the technique, rather than the specific findings of particular studies within confined environments. Thus, most attention is paid to field sampling, thin-section manufacture, the approach to description and interpretation of thin sections, the presentation and interpretation of data from sequences of thin sections. Examples from Quaternary palaeosol sequences in Europe and China are used to demonstrate general principles which could be applied to a wide range of environments including those of the arid and semi-arid zones. The paper concludes with a summary of the types of micromorphological features recorded in semi-arid and arid soils, and illustrates how their presence and distribution have been used to reconstruct palaeoenvironmental change in these drier regions of the world.

Introduction

Soil micromorphology is "concerned with the description, interpretation and to an increasing extent, the measurement of components, features and fabrics in soils" at a scale "beyond that which can readily be seen with the naked eye" (Bullock et al., 1985, p. 9). It includes the examination of aggregates of undistributed soil material with optical microscopes and more high-powered techniques such as scanning electron microscopes, but here it is restricted to the study of thin sections using polarizing microscopes. Soil micromorphology came to prominence with the pioneering work of Kubiena (1938, 1953, 1970), who recognised different types of micromorphological fabrics and related them to particular suites

[*]Centre for Quaternary Research, Department of Geography, Royal Holloway, University of London, Egham, Surrey TW20 OEX, United Kingdom.

of soil processes and soil-forming environments. Although Kubiena's terminology has been superceded by more morphologically-based systems (Brewer, 1976; FitzPatrick, 1984, 1993; Bullock et al., 1985), the essence of his genetically-based approach is still maintained within micromorphological contributions to pedological and palaeopedological studies. Soil micromorphology is a valuable technique for the examination of palaeosols, in particular, as the recognition of individual or suites of micromorphological features provides the basis for relatively detailed reconstructions of formative processes and associated palaeoenvironmental conditions (Mücher and Morozova, 1983; Kemp, 1985, Fedoroff et al., 1990; Bronger et al., 1994; Kemp et al., 1995, 1996).

At its simplest, a buried soil within a sedimentary sequence indicates that there was a period of relative land surface stability favourable to pedogenic processes sandwiched between phases of instability characterised by a dominance of depositional processes. The soil may often provide the only evidence of climatic conditions during this period represented by the depositional hiatus, although the difficulties in isolating the effects of individual factors influencing pedogenic pathways are well known (Kemp, 1985; Catt, 1995a). A buried palaeosol is generally interpreted in terms of the broad pedogenic processes and environments assumed to be currently responsible for that type of soil forming at the present surface (Liu et al., 1987). Although normally based upon field and bulk analytical criteria, there have been some attempts to characterise and utilise micromorphological fabrics for this purpose (Mücher and Morozova, 1983). Bronger and Heinkele (1989), for instance, summarised a number of examples from Central European and Chinese loess-palaeosol sequences where detailed micromorphological comparisons of surface and buried soils have led to the reconstruction of climatic conditions during Pleistocene soil-forming intervals.

The superposition of micromorphological features within polygenetic palaeosols (Catt, 1995b) has provided the basis for reconstructions of ordered sequences of pedogenic (and associated palaeoenvironmental) events (Brewer, 1972; Kemp, 1985). Fedoroff et al. (1990), for instance, detailed a number of calcitic pedofeature associations which reflect cyclical changes in aridity and humidity. Polygenetic palaeosols in temperate latitudes are traditionally interpreted in terms of cycles of stability and instability whereby cryogenic features superimposed upon argillic fabrics are taken to represent changes from 'stable' interglacial to 'unstable' cold stage conditions (Mücher and Morozova, 1983). This is the rationale behind the palaeoclimatic significance attached to complex microstratigraphic associations of fragmented and deformed clay coatings within buried and non-buried 'paleoargillic' soils (Avery, 1985).

Covering and incomplete burial of soils by thin layers of sediments may lead to superimposition or 'welding' of new features on underlying 'fossil' horizons during subsequent 'soil-forming intervals'. Micromorphology has again played a major role in identifying and unravelling the form and sequence of the relevant pedosedimentary processes in pedocomplexes containing welded components (Fedoroff and Goldberg, 1982; Kemp et al., 1994).

The notion of landscape 'stability' with complete cessation of erosion and deposition during 'soil-forming intervals' represented by buried soils, although attractive in principle, is frequently difficult to accept. Accretionary soils form where pedogenesis and sediment deposition effectively take place continuously—and simultaneously—within a geological time framework. Distinct soil horizons amidst these vertical sequences represent situations where the balance between pedogenic and geomorphic inputs swung towards, and remained with, the former for a significant period of time. Some entire loess, colluvial and alluvial sequences therefore may be considered as pedocomplexes (Kemp et al., 1994). Depth functions of micromorphological features have been used to help identify such sequences and characterise the nature and order of pedosedimentary processes and environments (Mücher and Coventry, 1994; Kemp, 1995; Kemp et al., 1995, 1996).

Phases of profile truncation, reworking and redeposition of soil material are frequently included within soil-landscape reconstructions derived from micromorphological records (Fedoroff and Goldberg, 1982; Cremaschi et al., 1990). Many pedocomplexes contain reworked soil materials transported to present locations by fluvial or mass movement activity. In situ pedological fabrics have been superimposed, the transport phase being recognised micromorphologically by the presence, arrangement and depth distribution of 'pedorelicts' (Fedoroff and Goldberg, 1982; Cremaschi et al., 1990).

The main aim of this paper is to describe how soil micromorphology is used to reconstruct palaeoenvironmental change, concentrating on field sampling, thin-section manufacture, the approach to description and interpretation of thin sections, and the presentation and interpretation of data from sequences of thin sections. The links between these steps in the micromorphological technique and associated field and laboratory analyses are illustrated in Fig. 1. Emphasis, therefore, is on the details and approach of the technique rather than the specific findings of particular studies within confined environments. The referenced examples and case studies, which relate to Quaternary palaeosol sequences from Europe and China, are chosen to reflect principles and approaches applicable to a wide range of environments and time-spans. Although there has been limited use of the techniques so far in arid and semi-arid zones, there is clearly considerable potential. The paper will conclude with a brief review of the micromorphology of arid and semi-arid soils,

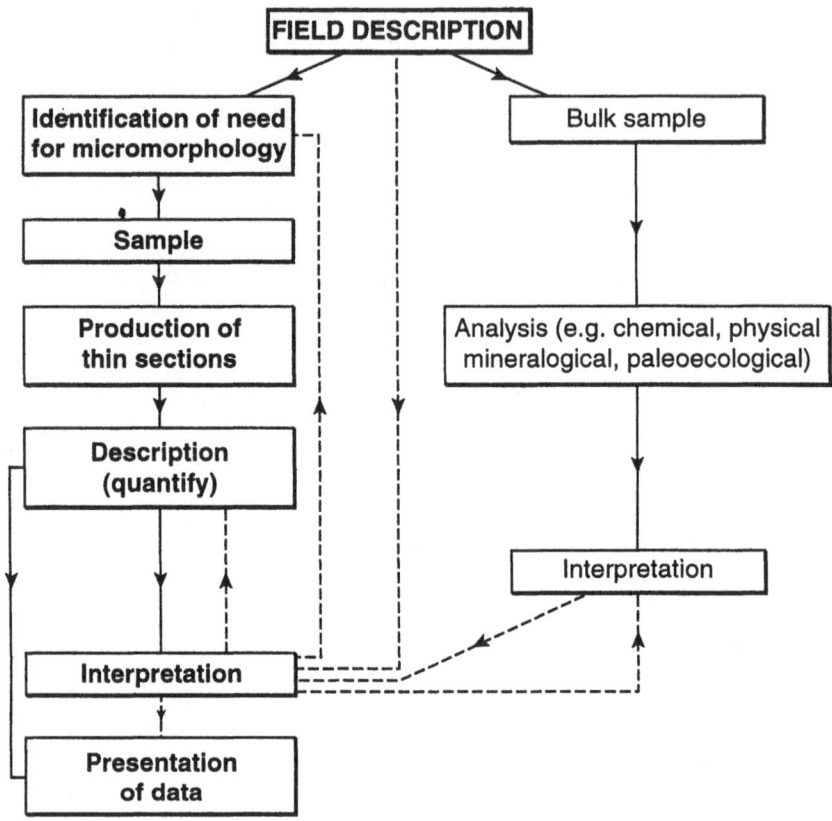

Fig. 1: Flow chart summarizing the various steps in the micromorphological technique and their links to asscciated field descriptions and laboratory analyses.

concentrating on the most common types of features recorded and illustrating how their presence and distribution has helped our understanding of palaeoenvironmental change in these drier regions of the world.

Sampling

The time and expense associated with thin-section production should encourage careful consideration of the initial field sampling strategy. With some exceptions, there is rarely the opportunity to produce thin sections as a matter of routine procedure. Care must be taken to ensure that there is a problem to be solved and, moreover, that micromorphology is the most appropriate technique to tackle the problem (Fig. 1). Samples should not be taken in isolation or collected randomly so as to simply 'characterise' a palaeosol. There is a need to cover the full range of depth-related characteristics from the burying sediments, through the soil and deep into the parent material (if they exist as discrete units) (Mücher and Morozova, 1983). Ideally, thin sections should be 'matched' to bulk samples so as to allow direct comparison of micromorphological and bulk analytical data.

It is traditional to take samples in, or across the boundaries of, each macromorphologically-defined horizon. It is often more useful, however, to sample at fixed vertical intervals so as to record micromorphological features and trends which do not manifest themselves in terms of macromorphological change. At the same time it is important to retain the flexibility to sample across important boundaries or in critical locations when required. Recent research programmes of the Centre for Quaternary Research at Royal Holloway (University of London) have found that a 20-25 cm vertical sampling interval ensures adequate coverage for detailed examination of the thick loess-palaeosol sequences of the Loess Plateau of China (Kemp et al., 1995, 1996), although optimum intervals must reflect the (conflicting) balance between the requirements of the sites and the restrictions imparted by financial budgets or capacities of the thin section laboratories.

Having deduced a complicated pedosedimentary and palaeo-environmental reconstruction from a dense vertical network of thin-sections, it is sometimes revealing to assess the effects of removing one or two thin sections from the data set, i.e. use a less dense sampling strategy. This inevitably results in a much simpler reconstruction! Occasionally, it can make a previously unintelligible pattern appear straightforward! Inadequate sampling may therefore 'produce a story' which does not represent the full record of events. Whilst consideration is normally given to the most appropriate vertical sampling interval, the representativeness of individual thin-sections at a particular level is generally not questioned. It is understandable in terms of budget and capacity restrictions that replicate samples are rarely processed from equivalent depths. On the other hand, it would a matter of concern if overriding emphasis was to be placed upon the characteristics of a single thin section. Palaeoenvironmental reconstructions should be based upon interpretations of sequences of thin sections.

As a thin section should represent a portion of soil material retaining the internal patterns and organisation of its original field position, care must be taken not to disturb the sample during collection. The simplest way of achieving this aim is to use a rectangular metal box, termed a 'Kubiena tin' (approximately $80 \times 60 \times 40$ mm), with a single hinged corner and detachable top and bottom lids. The top lid is removed and the open end gently pressed into a flat (vertical or horizontal) surface whilst simultaneously cutting around the tin with a sharp knife. The tin is carefully cut away from the face when full, excess soil trimmed off and the lid(s) replaced and sealed with masking tape. Location and orientation of the tin with respect to the top of the profile should be recorded.

It is often not feasible to collect stony or very dense material by the technique described above. Lumps or aggregates are obtained by the best means possible and stored between packing materials in bags or boxes. Very loose material, such as dry sand, can be stabilised first by application of a thin slurry of plaster of Paris (FitzPatrick, 1993). Once it has dried and hardened, the undisturbed samples can be removed in tins or as aggregates.

Production of Thin Sections

Production of thin sections from unconsolidated soils and sediments is not a routine technique in most laboratories, even those equipped with standard 'hard-rock' cutting and grinding facilities. Whilst the way in which unconsolidated samples are collected and dried may be easily accommodated within most systems, more major modifications to procedures are required in order for the samples to be impregnated with hard-setting resin and transformed into 'rock-like' blocks. Furthermore, the abrasive slurries commonly used to grind 'hard-rock' sections are often unsuitable for impregnated blocks as fine particles of the abrasive may become embedded within resin infilling the pores, resulting in 'dirty' thin sections. Fixed-abrasive papers or diamond-impregnated plates are therefore recommended for the production of high-quality specimens. Finally, thin sections of soils and sediments are generally required to be larger than standard 'hard-rock' sections (25 × 75 mm) routinely produced in most geological laboratories. Provision therefore must be made for handling 'mammoth-size' sections (e.g. 100 × 75 mm). A brief outline of the technique used at Royal Holloway (University of London) is provided below (Fig. 2). Further details or alternative techniques can be obtained from other more specialised publications (FitzPatrick, 1984, 1993; Murphy, 1986; Courty et al., 1989; Dobrovol'ski, 1991; Lee and Kemp, 1993).

The impregnation of undisturbed soil blocks is normally done using polyester or epoxy resins, substances which are immiscible in water (Tippkötter and Ritz, 1996). Consequently, it is necessary to remove any water present prior to impregnation. The lids are removed from the Kubiena tins and the enclosed blocks or discrete aggregates placed in plastic containers that are about 1 cm larger in all three dimensions than the samples. Most are air-dried for a week, although moisture within clay-rich materials may be displaced by saturation with acetone. This is achieved by filling the containers with acetone up to a level above the top

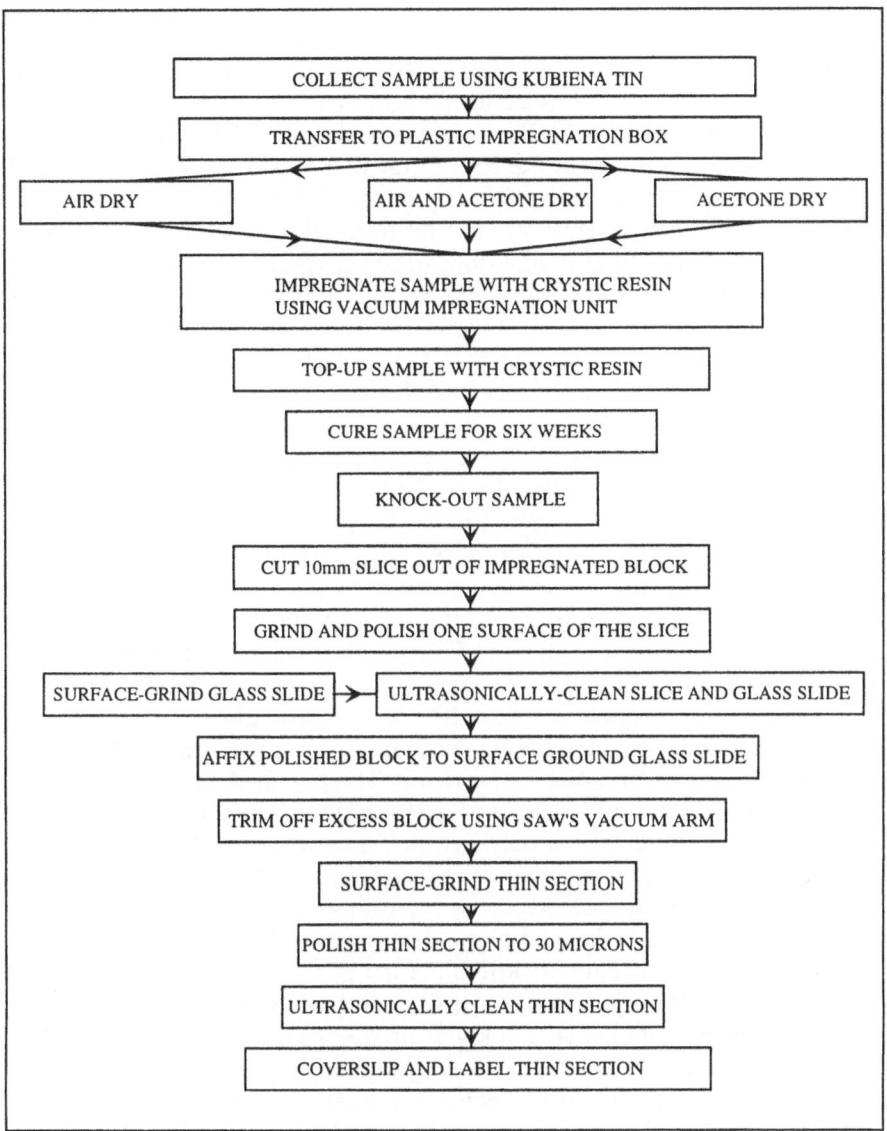

Fig. 2: Flow chart summarizing the stages involved in the preparation of thin sections of unconsolidated sediments and soils (after Lee and Kemp, 1993).

of the blocks. The sealed containers are left for a week and then the displaced water, mixed with excess acetone, discarded. The process is repeated up to six times until no water is present within the acetone waste. More frequent changes of acetone, perhaps combined with the use of a magnetic stirrer in order to encourage rapid displacement of the water, may reduce the length of time involved (Murphy, 1985). The acetone-saturated block is then drained of excess acetone and is ready for impregnation using the same procedures as for air-dried samples.

The Royal Holloway Thin-Section Laboratory uses a crystic (polyester) resin as an impregnating medium (Lee and Kemp, 1993). Once the resin infiltrates into all the pores of the soil, it gradually polymerises and hardens until a solid block is created which is literally 'as hard as rock'. This resin is very viscous and consequently is first thinned with acetone: a catalyst is added to speed the polymerisation reaction. The impregnating mixture is poured into the plastic container around the edge of the soil block. It initially infiltrates into the soil by capillarity. After 5-10 minutes, more impregnating mixture is added so that the soil block is completely covered. A batch of samples is then placed under vacuum overnight. The next morning the vacuum is released, the impregnating mixture topped up and the containers transferred to a fume or gelling cupboard where they remain until complete polymerisation and hardening has taken place. This may take up to six weeks (depending on the amount of catalyst used), though it is acceptable to hasten the final stages of hardening by placing samples in an oven at 60°C overnight.

Once a block is sufficiently hard, it is knocked out of the container and any remanent tin casing pulled off. The block is then cut into several 10-mm thick slices parallel to its long axis using a diamond saw with suitable oil as lubricant and coolant. One slice is selected, ground smooth and polished on a lapping machine using diamond-impregnated plates of progressive fineness, and oil as a lubricant. The polished surface is cleaned in an oil-filled ultrasonic bath and mounted onto a glass microscope-slide, which has been previously ground to a uniform thickness on an automatic surface grinder, with a bonding resin. When the resin has hardened, the slide is placed on a vacuum chuck and the excess soil block cut away using a diamond saw, leaving less than a few millimetres of specimen attached. This thickness is further reduced on the automatic surface grinder, then by hand grinding and polishing on the lapping machine. The final specimen thickness of 25-30 μm is reached when the interference colours of quartz under a microscope are of the first order (i.e. white to grey). The thin section is then cleaned in the oil-filled ultrasonic bath, labelled and a cover slip attached using a bonding resin. Cover slips, however, should not be bonded onto the thin section if SEM and/or microprobe work is intended.

Description and Interpretation of a Thin Section

Initial examination of a thin section should always be at a 1:1 scale and, if the necessary equipment is available, at low magnifications (×2 - ×10) so that mesoscale patterns of organisation can be recognised and related to each other or to the relevant field descriptions (Wilding and Flach, 1985). Some micromorphologists maintain that they prefer to describe and interpret 'blind'—without even reference to field descriptions. This approach is supposed to ensure their objectiveness. Personally, I feel that they are misguided, as one should make use of all ancillary data from the start (Fig. 1).

Most descriptions of thin sections are undertaken using a petrological microscope under a combination of plane, crossed, and circular polarized or incident light conditions at a range of magnifications (×10 - ×400). The beginner, on seeing for the first time a thin section under the microscope, will be confronted with a complex and seemingly incomprehensible array of patterns which appear to offer no immediate insight into the internal composition, origin and development history of the soil. However, any immediate urge to panic should be restrained! Having made a reconnaissance survey of the whole thin section at various scales, noting down characteristics or sketching details of particular fields of view, it is advisable to concentrate on observing and describing the thin section under a number of headings which reflect different aspects of its micromorphology. Concentration on well-defined components tends to clarify particular relationships and allows detailed observations to be made more easily. These systematically written descriptions and sketch diagrams provide the factual base from which interpretations can then be made.

To the non-specialist, the terminology used by micromorphologists is disconcerting to say the least. However, although it could be made more 'user-friendly', it is imperative to have well-defined terms so as to reduce subjectiveness and provide some kind of internal and external consistency. A number of different descriptive systems have been proposed (Brewer, 1976; FitzPatrick, 1984, 1993: Bullock et al., 1985; Dobrovol'ski, 1991). Even when following established systems, however, there have been reports of significant discrepencies between experienced micromorphologists' descriptions of the same thin sections (Murphy et al., 1985).

Although silt and sand-size minerals in thin sections can often be identified by standard petrographic microscope procedures, the composition of finer components (e.g. iron or manganese oxides) generally has to be inferred from the colour displayed under oblique incident light conditions (Bullock et al., 1985). Recent developments and improved accessibility of techniques of microchemical analysis (e.g. EDXRA; Bisdom et al., 1990), however, have led to more confident interpretation of the composition of particular features and grains, and opened up possibilities for monitoring movement of solutes and establishing weathering trends or patterns (Jenkins, 1994; Jongmans et al., 1994).

Micromorphologists frequently maintain the need to describe thin sections first, and only attempt an interpretation once the description is complete (Brewer, 1972; Murphy et al., 1985). In doing so, the description is deemed to represent an unbiased statement of facts, i.e. data. In reality, however, it is very difficult to follow such a procedure rigidly, either because the descriptive system has an inherent genetical base (e.g. 'pedological features' and 'fecal pellets', Brewer, 1976; 'pedofeatures' and 'excrements', Bullock et al., 1985), or the micromorphologist's descriptive judgement is influenced by his/her interpretative experience (Murphy et al., 1985). Brewer (1972) maintained that interpretation is largely deductive and relies heavily on accumulated experience. It is essential therefore that micromorphologists have experience of describing and interpreting thin sections from contemporary soils and sediments before attempting to study thin sections from palaeosols and pedocomplexes (Mücher and Morozova, 1983). Despite this background, there are undoubtedly situations where the pattern of features is so variable and complex that it is impossible to provide a coherent microstratigraphic and associated palaeoenvironmental or pedosedimentary reconstruction. Protagonists must interpret conservatively and ensure the integrity of the approach, by adhering to the standard stratigraphic philosophy. For an ordered sequence of micromorphological features to have any general significance, it has to be applicable to, or at least not be in conflict with, all parts of the thin section (or indeed all other thin sections within the profile).

A major interpretative problem is provided by our apparent acceptance of the uniqueness of some pedofeature - pedogenic process - macroenvironment relationships. The basis for such relationships is often not truly verified in the first place, whilst there is an increasing body of evidence to suggest that different local and regional combinations of pedogenic processes and/or environments may produce similar pedofeatures and horizons (Brewer, 1972; Kemp, 1985). If we know that a particular soil feature at one location has formed in response to a set of processes under specific environmental controls, can we be certain that a similar feature in a buried soil at another location has formed under the same conditions? All earth scientists are aware of the principles of equifinality and the potential uncertainties provided (Valentine and Dalrymple, 1976), yet any reservations are rarely mentioned when presenting a favoured interpretation of particular micromorphological features and fabrics.

Data Presentation and Interpretation of Thin Section Sequences

The presentation of micromorphological data provides a particular challenge to the micromorphologist. Not surprisingly, in view of the terminology and potential level of detail available, there has been a recent tendency to restrict the publication of descriptions and, instead, to present the data in a more 'reader-friendly' fashion in the form of short-hand summaries, indices, quantitative tables, depth functions, schematic diagrams and/or photomicrographs of typical features and associations. These approaches are undoubtedly attractive from a communication viewpoint, yet 'clear cut' or 'black and white' interpretations and reconstructions based on 'massaged' data may create an unjustified impression of order, allowing apparently 'minor' uncertainties and anomolies to be 'lost' from the record.

Specific interpretations of individual thin sections may be aided or confirmed by reference to other thin sections within a vertical sequence. Often a complete palaeoenvironmental reconstruction depends upon considering the complete network of thin sections, although the assimilation and interpretation of descriptions from a sequence of thin sections can sometimes be even more daunting than the interpretation of a single thin section! It is frequently useful to construct depth functions of qualitative or (semi) quantitative expressions of micromorphological features and associations. These depth functions can then be scanned for trends and anomalies so as to build up a picture of how the profile developed from a pedogenic and sedimentary viewpoint. The next step is to assign palaeoenvironmental significance to each development stage. This approach is illustrated by three examples below.

Figure 3 is a schematic summary of the major micromorphological features and associations within a paleoargillic soil developed in, and buried by, sands and gravels in Eastern England (Kemp, 1987). This diagram was constructed after careful assessment of detailed descriptions of all the thin sections within the profile. Clay translocation and freeze-thaw disruption were the major pedogenic processes active as evidenced by the undisturbed and fragmented clay coatings (Figs. 3d-3h) and duplex textural lamellae features (Figs. 3c2 and 3f2) in the Bt and BCt horizons. Close examination revealed a depth-related change in texture of the coatings. Sand grains in the upper part of the Bt horizon were coated by a mixture of fine and coarse clay with intervening voids packed with (translocated) silt grains and fragmented clay coatings (Fig. 3d). The coatings in lower Bt and BCt horizons were composed of fine clay (Figs. 3g and 3h), whereas those in the intervening central parts were compound in nature with the finer textured layer always inside, and therefore ante-dating the coarser layer (Figs. 3e and 3f1). Thin-section c extended across both the Bt horizon and a wedge structure containing a mixture of loose uncoated and coated sand grains (Fig. 3c1). Duplex textural lamellae features within the Bt horizon were clearly disrupted (and therefore postdated) by the wedge infill (Fig. 3c2).

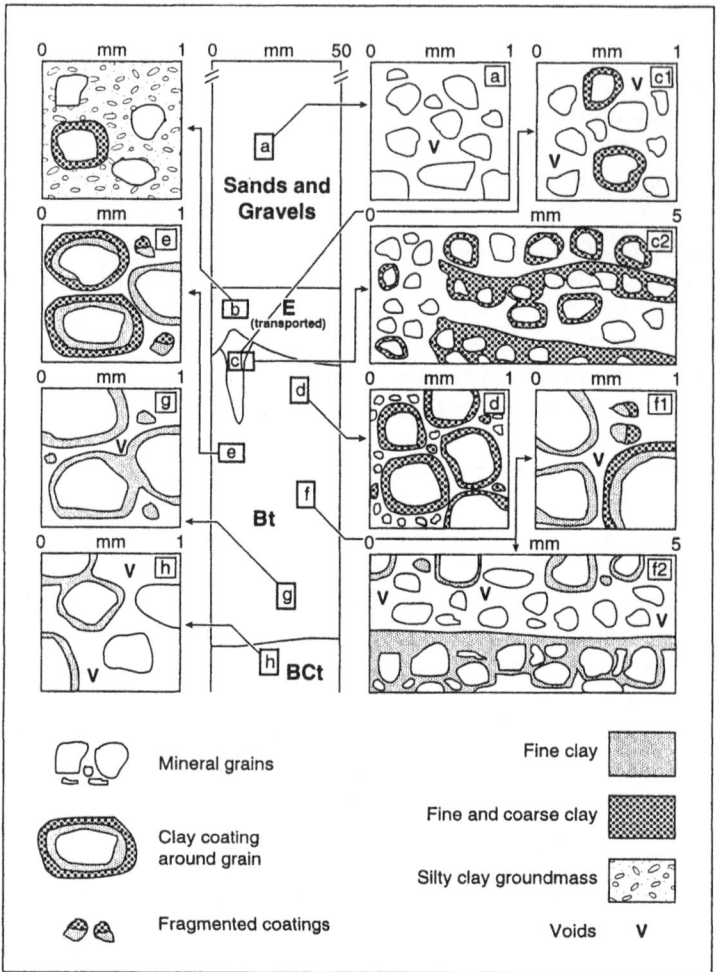

Fig. 3: Schematic summary of the major micromorphological features and associations within a buried palaeosol in eastern England.

On the basis of this diagram and associated field data, it was possible to reconstruct the ordered sequence of pedogenic events leading to the development of the complex depth-related microstratigraphy apparent within the palaeosol. By assigning particular palaeoenvironmental significance to each pedogenic stage, a relatively detailed record of palaeoenvironmental change could be inferred (Kemp, 1987, 1992). Four main pedogenic stages (Fig. 4) were recognised: the first three stages, spanning an interglacial and transition to glacial conditions, were marked by translocation of progressively coarser particles in response to a deteriorating climate. Freeze-thaw processes dominated during stage 3

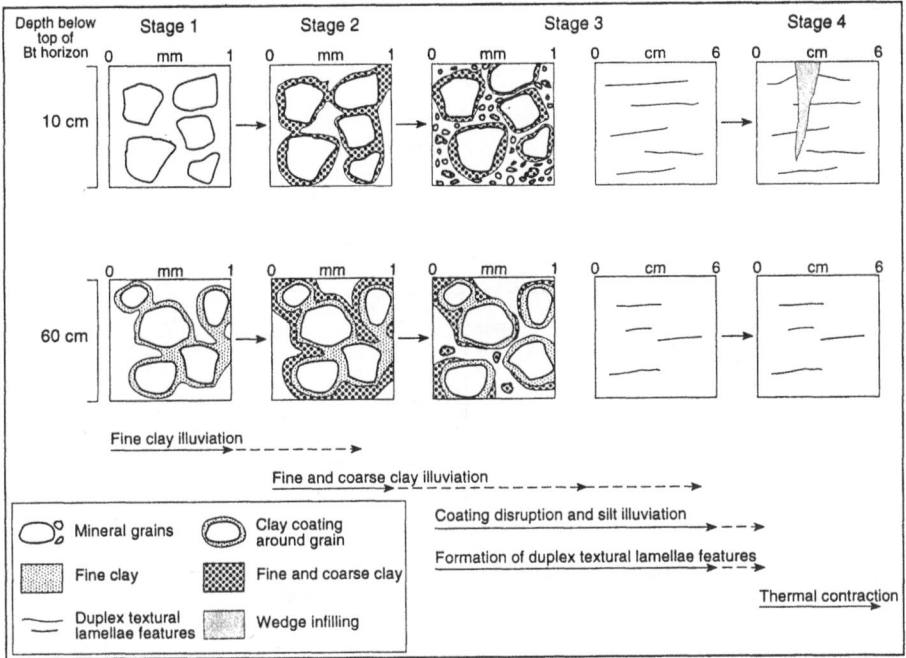

Fig. 4: Schematic summary of the micromorphological development of a buried palaeosol in eastern England (after Kemp, 1992).

with fragmentation of coatings and development of duplex textural lamellae features. Conditions deteriorated further during stage 4 with thermal contraction and wedge formation occurring prior to burial by glacial outwash sediments (Kemp 1987, 1992).

Quantification of particular micromorphological features is normally achieved by point-counting thin sections (FitzPatrick, 1993). This can be a very laborious and time-consuming procedure, particularly if dealing with an extensive network of thin sections down thick sequences. Murphy and Kemp (1984) showed that it can be achieved more rapidly and with sufficient accuracy and precision for most purposes by estimating quantities in comparison to a few point-counted reference thin-sections. Figure 5 represents a schematic reconstruction of the sequence of pedosedimentary stages leading to development of a welded palaeosol or pedocomplex buried beneath loess of Riss age in southern Germany (Kemp et al., 1994). Field examination had failed to recognise the complex nature of the unit. Proportions of translocated clay within a series of thin sections down the pedocomplex were estimated using the approach of Murphy and Kemp (1984). The resulting depth function was unlikely to represent a single buried soil (Fig. 6). Subsequent detailed analysis of the thin sections showed conclusively that the unit had developed over at

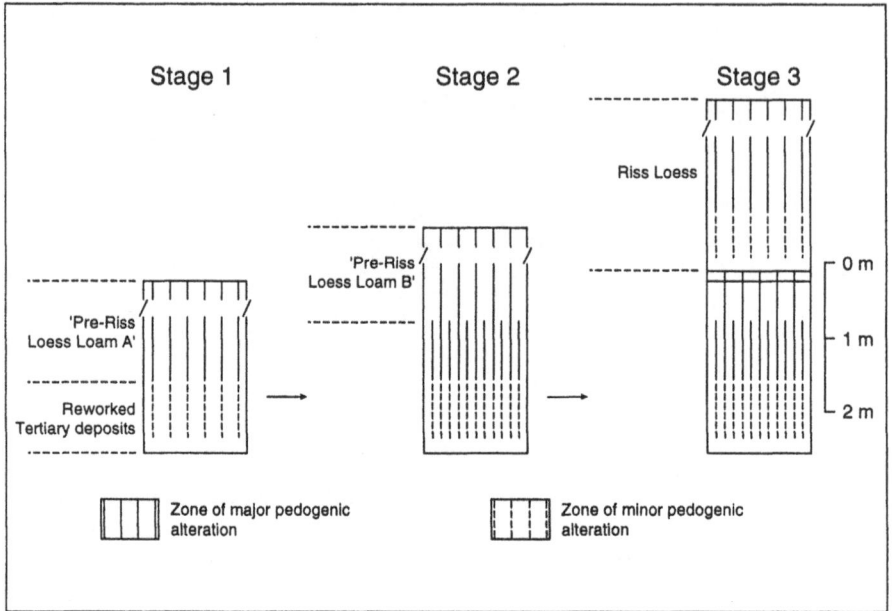

Fig. 5: Schematic reconstruction of the sequence of pedosedimentary stages leading to the development of a buried welded palaeosol or pedocomplex in southern Germany (after Kemp et al., 1994).

least two interglacial-glacial cycles with clay coatings during stage 1 being disrupted by freeze-thaw activity prior to deposition of a thin layer of Loess Loam B (Fig. 5). Renewed clay translocation during stage 2 resulted in clay coatings accumulating within horizons developed in the Loess Loam B and lower down in the Lower Loam A where they were superimposed upon the disrupted argillic fabric of the earlier-formed soil. Once again climate deteriorated during later parts of the stage with many of the illuvial features in the uppermost horizons of the Loess Loam B being disrupted by cryogenic activity. Semi-quantitative changes in levels of disruption within the pedocomplex (Fig. 6) were consistent with this reconstruction. Loess accumulation during stage 3 was sufficient to isolate the pedocomplex from the effects of pedogenesis active at a subsequent major land surface (Fig. 5).

Figure 7 comprises a semi-quantitative depth function of individual micromorphological features down part of the buried S1 pedocomplex from Lanzhou on the semi-arid western edge of the Loess Plateau in China (Kemp et al., 1995). On the basis of vertical trends in these features and associated magnetic susceptibility and calcium carbonate data, a level could be recognised at which there had been a major stable land surface and development of a weak leaching (soil) profile (stage 1; Fig. 8). The

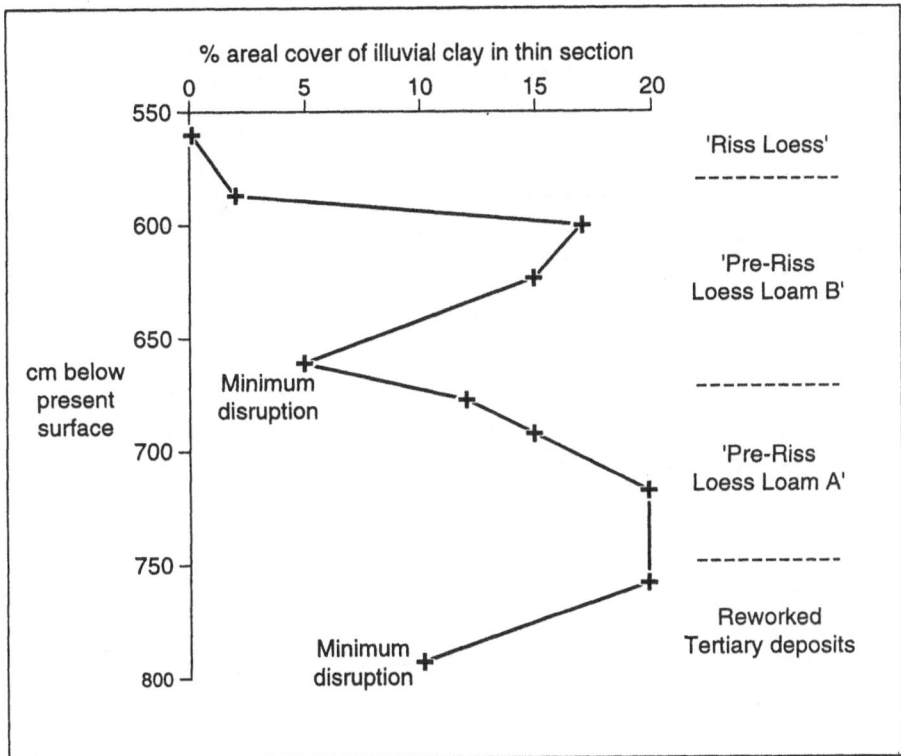

Fig. 6: Depth function of illuvial clay through the buried welded palaeosol or pedo-complex in southern Germany (after Kemp et al., 1994).

continued presence above this level of pedological features, such as faunal excrements and calcitic concentration and depletion features, suggested that pedogenic processes were still significant even during subsequent stages of enhanced loess deposition and surface aggradation. Detailed examination of the depth trends led Kemp et al. (1995) to suggest that this phase of accretionary pedogenesis could be separated into two stages (2 and 3; Fig. 8) marking a climatic transition. Sufficient moisture was available during stage 2 for localised redistribution of calcium carbonate, whilst stage 3 was marked by drier conditions and/or increased rates of dust accumulation. Ephemeral, sparsely-vegetated surfaces during stage 3, however, were subjected to infrequent intense rain leading to formation of crusts. There was some evidence in the form of banded fabrics that freeze-thaw processes may also have become prominent at this time (Figs. 7 and 8).

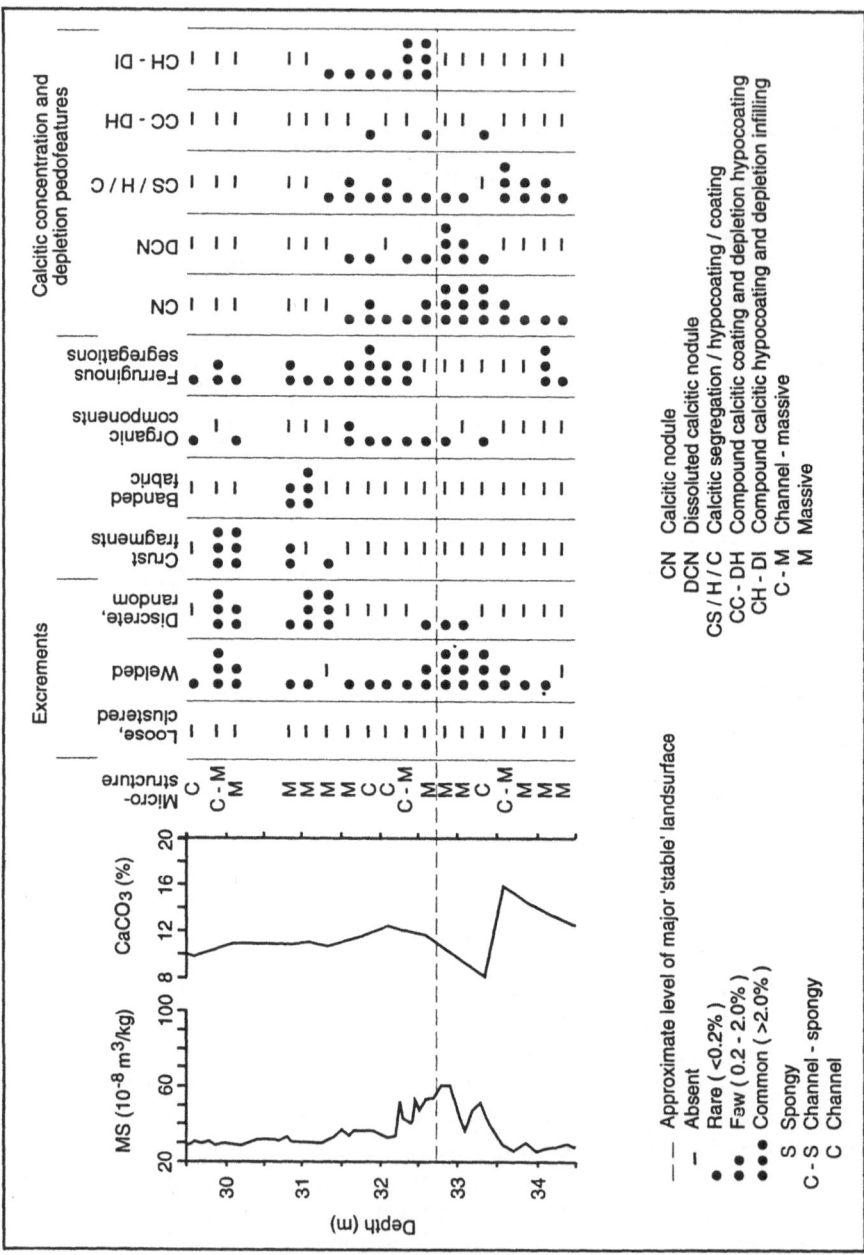

Fig. 7: Semi-quantitative depth function of individual micromorphological features down part of the buried S1 pedocomplex formed in loess on the semi-arid western edge of the Loess Plateau in China (after Kemp et al., 1995).

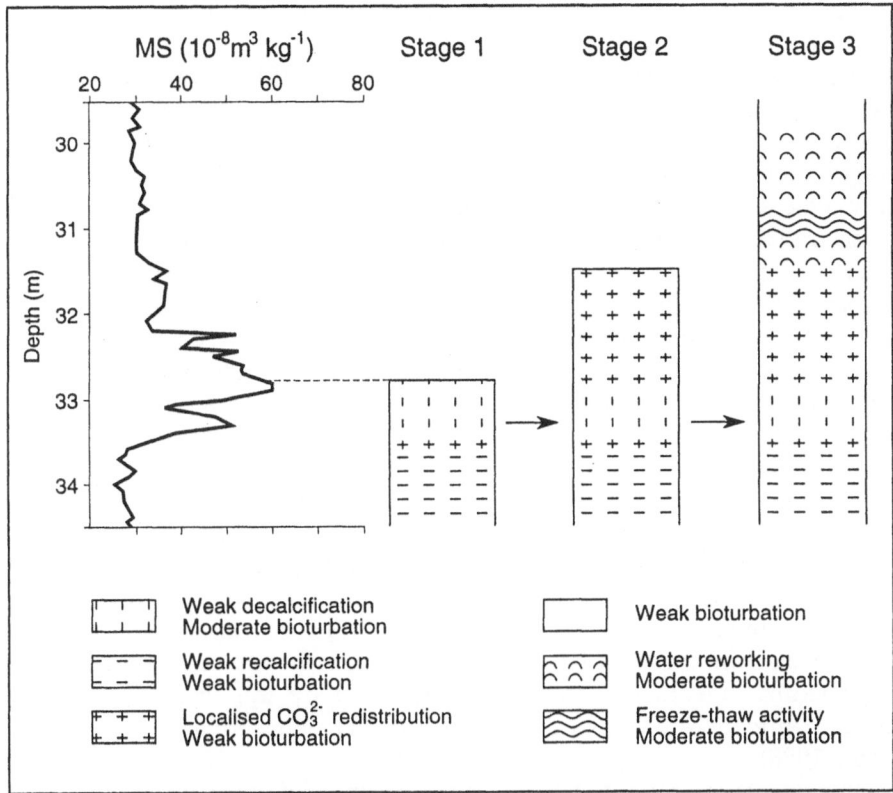

Fig. 8: Schematic reconstruction of the sequence of pedosedimentary stages leading to the development of part of the buried S1 pedocomplex formed in loess on the semi-arid western edge of the Loess Plateau in China (after Kemp et al., 1995).

Micromorphology of Arid and Semi-Arid Soils

Soil Development in Arid and Semi-arid Zones

The major factor influencing soil development in (hot or cold) arid and semi-arid zones of the world is clearly the scarcity of available water (Gerasimova et al., 1996). The moisture conditions, sparse vegetation covers and low organic matter inputs do not generally encourage biological activity, although even limited concentrations of soil fauna may have marked effects on soil morphology through their burrowing, mixing and excremental activities (Courty et al., 1989). Bioturbation may be particularly important in soils where local conditions, e.g. fine textures or topographic location, favour retention of the limited water supply and thus encourage higher faunal populations (Courty and Fedoroff, 1985).

Leaching processes are obviously restricted in dry environments. Indeed, upward migration of salts may even occur under very hot and dry conditions leading to salinisation. Where there is sufficient moisture for limited leaching, however, soluble components tend to be redistributed locally or to shallow depths where they reprecipitate in the form of 'secondary' concentrations of minerals such as gypsum or, more commonly, calcite. Much of our understanding of the mechanisms and forms of secondary carbonate accumulations in arid and semi-arid soils is based upon the detailed work undertaken over several decades in the Desert Project in southern New Mexico (Gile et al., 1966, 1995). A major conclusion from this study was that a large proportion of the secondary carbonate in such soils does not originate from the initial parent material, but is derived from aeolian inputs of calcareous dust as the soils develop. An external source for the calcite is supported by the presence of secondary carbonate accumulations in semi-arid soils developed in non-calcareous parent materials. Theoretical calculations and simulations have shown that rates of chemical weathering in such environments are generally insufficient to release significant quantities of the necessary constituent ions (McFadden. 1988). Furthermore, the relatively high quantities of secondary clay minerals in soils developed in coarse-textured parent materials are also unlikely to have been produced in situ under the restricted chemical weathering conditions afforded by the scarcity of moisture, and are therefore also considered to have an aeolian origin (McFadden, 1988).

The sparse vegetative cover typical of arid and semi-arid zones provides the soil with limited protection from the effects of infrequent, yet often intense, rainfall. Slaking and development of surface crusts are common phenomena (Fedoroff et al., 1990). The reduction in infiltration rates afforded by these crusts may induce overland flow and erosion leading to hiatuses in the pedosedimentary record. Many soils at the present land surface in semi-arid and arid zones may not be very old, in that they have only been forming since the most recent stabilisation of the surface. Chronosequence studies (McFadden, 1988) have shown that the characteristics of soils of these regions are determined not only by climatic restrictions and parent material factors, but also by the length of time available for pedogenic development. Indeed, the classic stages of carbonate accumulation established for desert soils by Gile et al. (1966) reflect these controls.

Vertical movement of particles in suspension has been widely reported from arid and semi-arid zones (Fedoroff and Courty, 1987). Indeed, argillic horizons are commonly recorded in such soils (Allen, 1985; Gile et al., 1995), although the conditions for translocation are generally regarded as being unlike those responsible for the classic illuvial clay features in mid-latitude temperate argillic soils (Fedoroff and Courty, 1987). Certainly, many coarse and silty clay coatings probably form very rapidly following migration of poorly sorted suspensions during phases of surface instability and slaking (Courty and Fedoroff, 1985).

Dust inputs have clearly played an important role in the development of many semi-arid and arid soils. Particularly high rates of dust accumulation may result in burial or in the soils developing at actively aggrading surfaces such that they have an accretionary form (Nettleton et al., 1989; Kemp, 1995). Further complications to their pedosedimentary records may be provided by variable dust inputs over time and periodic erosional events leading to hiatuses (Kemp et al., 1995, 1996). Many non-cumulic, accretionary and buried soils or pedocomplexes, particularly those located within marginal climatic zones, may have experienced a record of environmental change during the Holocene or over longer periods. For instance, some aridosols containing secondary carbonate concentrations superimposed upon argillic horizons have been interpreted as reflecting a reduction in moisture availability and leaching, with an associated increase in calcareous dust inputs, in response to a drying of the climate (Gile and Grossman, 1968). A good overview on soils under increasing aridity is provided by Fedoroff and Courty (this volume, p. 73). Although processes associated with increasing humidity are more likely to remove evidence of existing features, there are examples reported where evidence may be retained within the soil (Fedoroff et al., 1990). In all cases, particularly when interpreting such evidence from recent soils and sediments, it is important to distinguish the effects of anthropogenic, as opposed to climatic, changes (Courty et al., 1989).

Despite the obvious potential, there have been relatively few attempts to reconstruct sequences of palaeoenvironmental changes from micromorphological records derived from arid and semi-arid soils. Before discussing some of the limited work that has been undertaken on this theme, the main types of micromorphological features recorded in such soils will be briefly summarized. Clearly, the list of soil micromorphological features discussed below is neither exhaustive nor necessarily unique to arid and semi-arid environments. For instance, a variety of redoximorphic features (Vepraskas et al., 1994), indicative of temporary waterlogging, has been reported from some soils in these regions (Courty and Fedoroff, 1985), whilst equivalents of most of the calcitic and textural concentration features in arid and semi-arid soils may be found in soils developing under different environmental controls and via different pedogenic pathways.

Main Micromorphological Features of Arid and Semi-arid Soils

Clearly it would be impossible in a review of this length to provide a comprehensive summary of all the micromorphological features described from soils of arid and semi-arid regions. The photomicrographs in Figures 9-12, however, are representative of some key features which have been used as evidence for pedogenic development and palaeoenvironmental change.

Crust features vary considerably in their micromorphological form depending upon the exact processes and environmental controls responsible (Bresson and Valentine, 1994). So-called 'depositional' crusts (Fig. 9a) are commonly identified in surface and buried soils of arid and semi-arid regions. Typically they are moderately sorted, layered units (2 mm thick) of clay grading into fine and coarse silt and then into groundmass at their base. Micas and linear organic components, often aligned parallel to the layers and vesicles, are common. Such units have been reproduced experimentally in the laboratory by Mücher and De Ploey (1977) and in the field by Bresson and Boiffin (1990). They form by breakdown of surface aggregates at a sparsely vegetated land surface subjected to intense rainfall, transport of particles by overland flow and differential settling during water ponding. These surface crusts are often subsequently disrupted and mixed by pedoturbation leaving random clusterings of crust fragments embedded within the groundmass (Kemp et al., 1995, 1996) (Fig. 9a).

Linked to the formation of these surface crusts are a variety of coarse-laminated or poorly-sorted textural concentrations coating or infilling large voids in lower horizons (Fig. 9b). These features, which are interpreted as resulting from rapid translocation of particles in suspension from slaked surfaces (Courty and Fedoroff, 1985), differ conspicuously from the classic well-sorted, fine-textured, microlaminated clay coatings typically associated with argillic horizons (Fig. 9c) developed under sub-humid to humid conditions and often forest cover (Courty et al., 1989). Where argillic horizons are identified in arid and semi-arid soils, easily recognisable clay coatings are absent (Nettleton et al., 1969), confined to grain surfaces (Allen, 1985; Gile et al., 1995), variably sorted (Fedoroff and Courty, 1987) and/or interpreted as representing an earlier moister phase of soil development (Gile and Grossman, 1968; Fedoroff and Courty, 1987).

Semi-arid and arid soils clearly receive some form of organic matter inputs where vegetative covers are present, although identifiable organic components such as (degraded) roots are rarely preserved over long timescales. Indirect evidence of floral activity may be provided by channel microstructures associated with root growth; such fabrics contrast markedly with the massive microstructures typical of very arid regimes characterized by rapid rates of sediment accretion (Figs. 9d and 9e). Perhaps the best evidence of biotic activity, however, is provided by faunal excrements (Figs. 9f, 10a and 10b). Whilst attempts to identify the

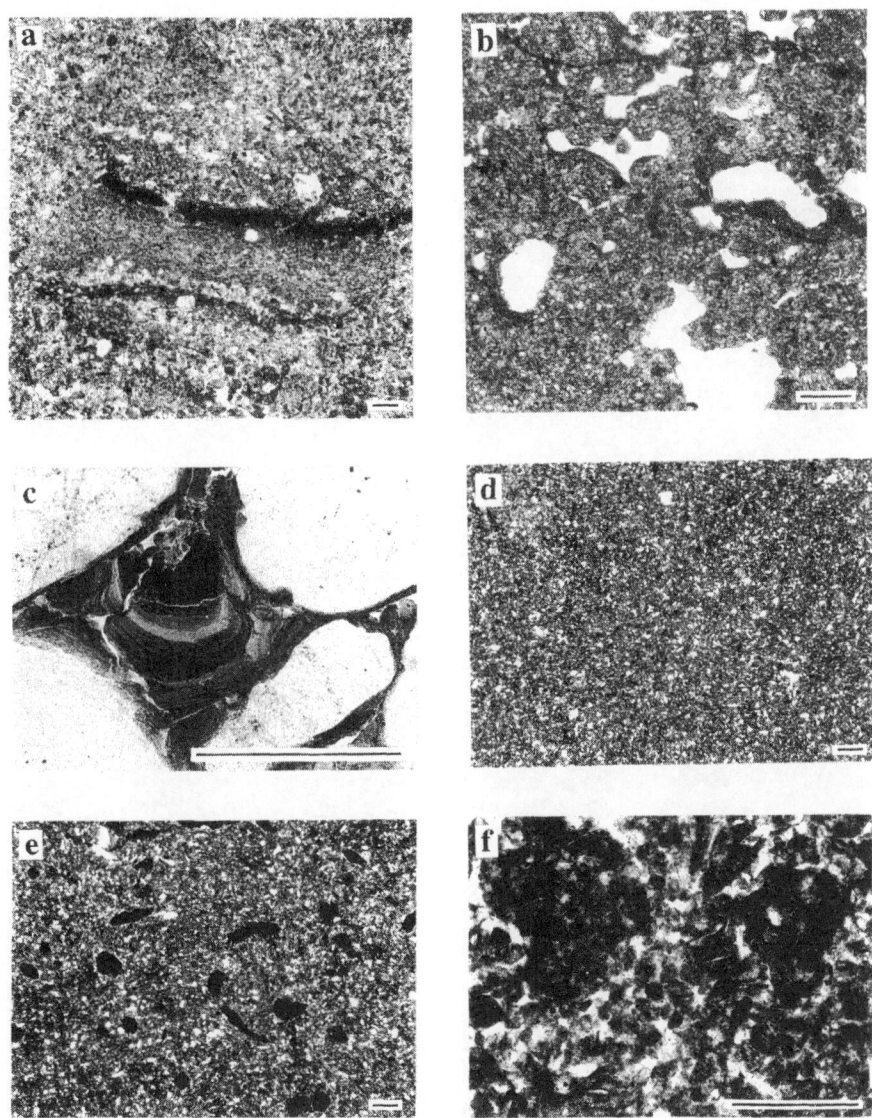

Fig. 9: Photomicrographs of micromorphological features. PPL = plane polarized light;
XPL = crossed polarized light: scale bar = 250 μm. (a) Remnant of a disrupted
crust (PPL). (b) Poorly sorted impure and silty clay coatings around voids (PPL).
(c) Microlaminated clay infilling of a packing void between sand grains (PPL).
(d) Massive structure (XPL). (e) Channel structure (XPL). (f) Welded excrements
(PPL).

presence of particular fauna on the basis of the diagnostic form of their
excrements has met with only limited success, the extent of bioturbation
as inferred from types and degree of development of excremental fabrics

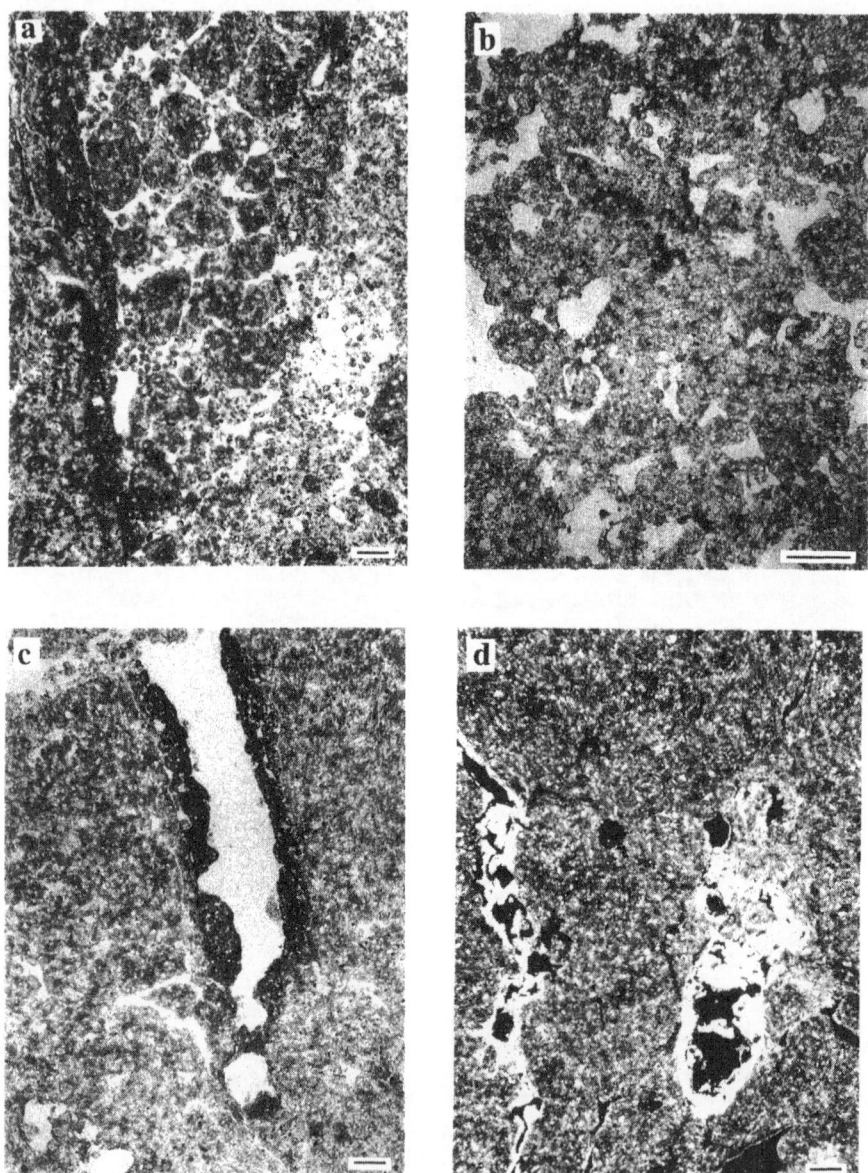

Fig. 10: Photomicrographs of micromorphological features. PPL = plane polarized light; XPL = crossed polarized light; scale bar = 250 μm. (a) Loose excrements (PPL). (b) Partially welded or coalesced excrements (PPL). (c) Calcitic hypocoating along channel (PPL). (d) Calcitic hypocoatings and coatings (XPL).

can give important insights into levels of biotic activity as controlled by changes in available moisture levels (Courty and Fedoroff, 1985; Courty et al., 1989; Fang et al., 1994; Kemp et al., 1995, 1996). Excrements degrade, age or weld quite rapidly (Fig. 10a). Particularly in buried soils, where they may have been also subjected to compaction and consolidation under overlying sediments, they are often difficult to recognize. Here they tend to be poorly defined, often occurring as a dense fabric of welded pellets (Fig. 9f), identifiable either by differences in arrangement of particle sizes or the presence of intervening microvughs or complex packing microvoids.

Probably the most widely reported set of micromorphological features from semi-arid and arid soils are composed of calcite, a reflection of the importance of pedogenic processes of dissolution, migration and reprecipitation in such environments (Courty et al., 1987, 1989; Drees and Wilding, 1987; Wright, 1990; Monger et al., 1991; Kemp, 1995; Gile et al., 1995; Gerasimova et al., 1996). The features vary considerably in both appearance and origin. Our understanding of their genesis and exact environmental controls is still rather limited (Fedoroff et al., 1990), although it is recognized that both lithogenic (inherited) and pedogenic (secondary) forms may occur (Drees and Wilding, 1987). Wright (1990) has proposed that features produced in situ may be either biogenic in origin or formed by physiochemical precipitation. The source of the calcite and direction of mobilization may vary considerably depending upon climate, parent material and groundwater conditions. Vertical leaching from surface to subsurface horizons and upward or lateral migration from a high groundwater table enriched in calcium bicarbonate provide two contrasting mechanisms (Courty et al., 1987), although inputs of calcareous dust or localized movements in response to changes in pH or partial CO_2 pressures may be important controlling influences (Herrero and Porta, 1987).

Courty and Fedoroff (1985) identified four main types of calcitic features in soil of arid and semi-arid regions: coatings, hypocoatings, infillings and nodules. Crystal sizes vary between micrite (< 10 μm), microsparite (10-50 μm) and sparite (> 50 μm) (Bullock et al., 1985). Coatings around grains and along voids comprise pure calcite and often grade into adjacent zones of calcitic impregnation of the groundmass (hypocoatings) (Figs. 10c and 10d). Some coatings, particularly associated with excremental surfaces, are composed of needle-fibre calcite (Figs. 11a and 11b): these are thought to form in association with microbial activity (Phillips and Self, 1987). Some infillings simply may be regarded as more developed forms of coatings or even displacive laminae such as might be found in calcretes (Fig. 11c) (Monger et al., 1991), although Courty and Fedoroff (1985) also include under this category a range of 'calcitic root pseudomorphs' which appear to have replaced and taken the form of root remains (Fig. 11d). Calcitic nodules, generally impregnative concentrations of calcite, have a variety of forms ranging from irregular,

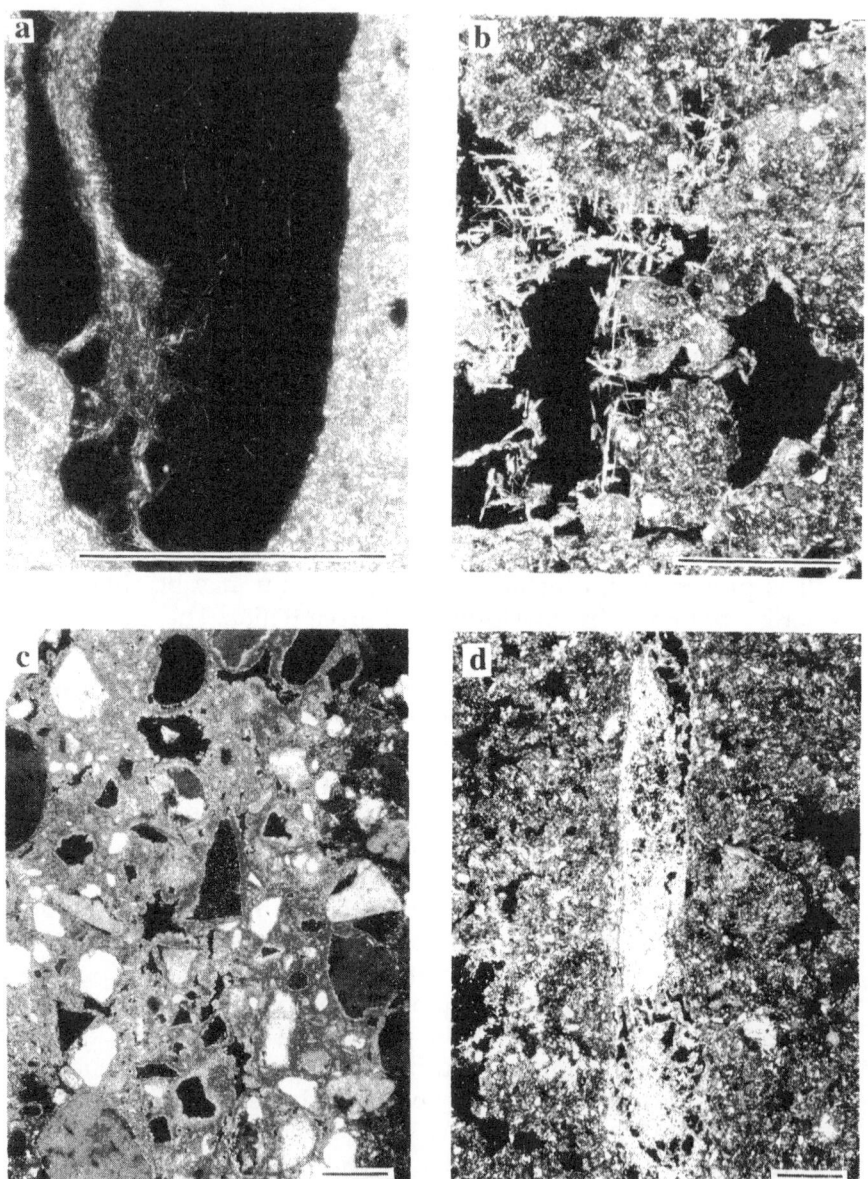

Fig. 11: Photomicrographs of micromorphological features. Crossed polarized light: scale bar = 250 µm. (a and b) Needle-fibre calcite within pore. (c) Framework grains embedded in calcitic cement. (d) Partial infilling of channel by micrite in the form of a root pseudomorph.

diffuse segregations to dense, discrete, sharply bounded features (Fig. 12a). They frequently have complex origins reflecting cyclical phases of

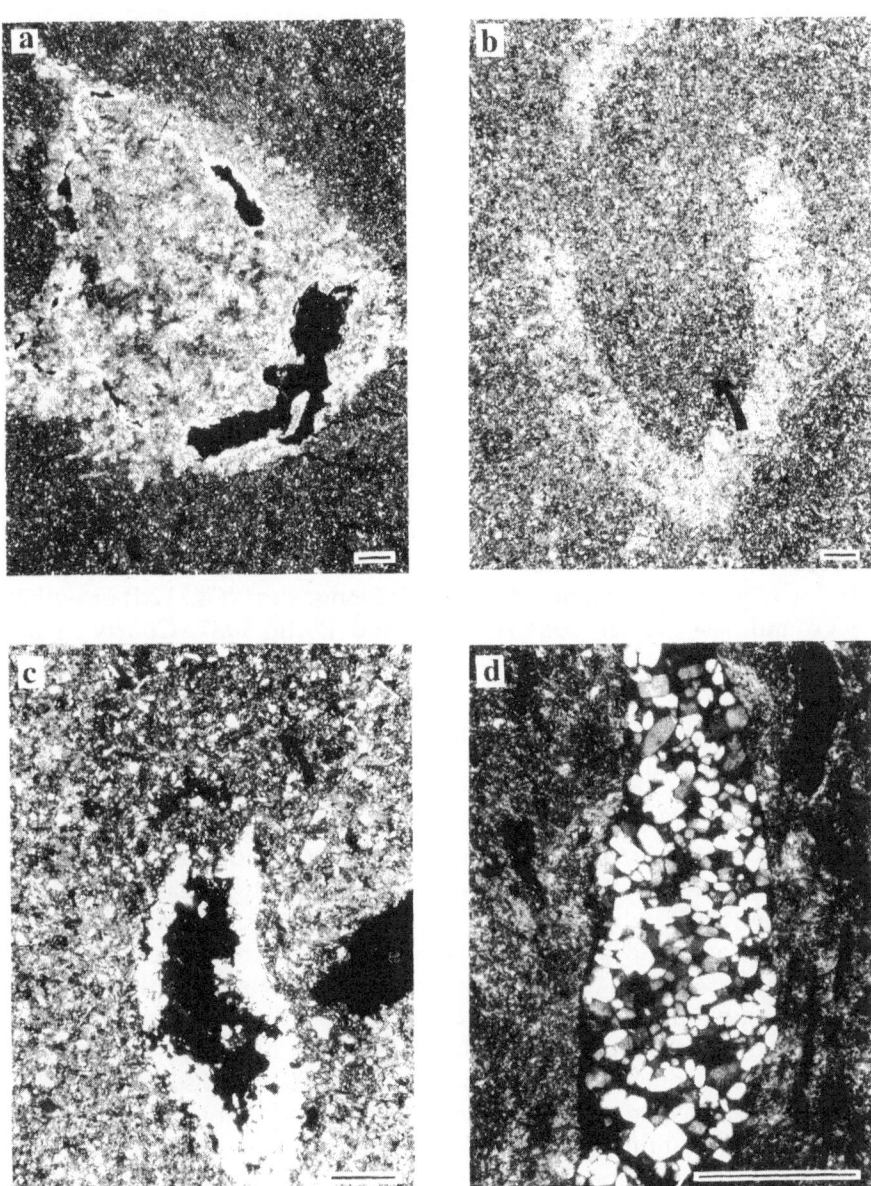

Fig. 12: Photomicrographs of micromorphological features. Crossed polarized light; scale bar = 250 µm. (a) Calcitic nodule. (b) Compound feature comprising zones of calcitic depletion (arrowed) and concentration. (c) Compound feature comprising a coating of sparite grains adjacent to a zone of calcitic depletion (arrowed). (d) Infilling of gypsum crystals.

dissolution and precipitation (Courty et al., 1987; Drees and Wilding, 1987; Fedoroff et al., 1990).

To this list of features reflecting the mobilization of calcite must be added partially dissolved grains of calcite (Courty et al., 1989), non-calcite grains partially or completely replaced by calcite (Drees and Wilding, 1987; Monger et al., 1991), micrite-impregnated and cemented groundmass recognizable in the form of crystallitic b-fabrics (Bullock et al., 1985), individual sparite grains significantly larger than any other component of the groundmass (Bal, 1977; Monger et al., 1991; Kemp, 1995) and a range of compound depletion-concentration features (Figs. 12b and 12c) which form in response to localized dissolution-reprecipitation processes (Bal, 1975; Jaillard and Callot, 1987; Herrero and Porta, 1987; Kemp, 1995).

Concentrations of more soluble salts such as gypsum and halite are commonly observed in aridic soils, yet there have been relatively few attempts to characterize the features from a micromorphological viewpoint (Shahid and Jenkins, 1994). Although secondary gypsum may occur as a microcrystalline mass (Allen, 1985), it normally takes the form of individual or clusters of lenticular crystals along voids (Fig. 12d) or within the groundmass (Allen, 1985; Herrero and Porta, 1987; Courty et al., 1989; Gerasimova et al., 1996).

Soil Micromorphological Evidence of Palaeoenvironmental Change in Arid and Semi-arid Zones

Relatively few studies from semi-arid and arid regions have relied to any extent on soil micromorphology as a means of reconstructing palaeoenvironmental change. One notable exception is that of Fedoroff and Courty (1987), who examined a sequence of thin sections from soils in semi-arid and arid areas of western Africa and northwestern India. They concluded that many fine-textured argillic horizons within soils on desert margins are relict, in that they developed during more humid period of the Early Holocene or the Pleistocene. Nettleton et al. (1989) came to similar conclusions for some argillic horizons in southwest USA largely on the basis that the present maximum depth of wetting is well above zones containing clear micromorphological evidence of clay translocation. In addition they noted at one of their sites in Texas that the clay coatings had been engulfed (or postdated) by calcitic features, presumably during more recent drier periods.

It is fair to say that micromorphology has made only a limited contribution to the reconstruction of soil-landscape changes within the well-documented Desert Project in New Mexico (Gile et al., 1995). One study of a 4.5 m profile in the southern part of the state by Monger et al.

(1991), however, did attempt to provide a complete micromorphological characterization of the various classic zones of pedogenic calcite accumulation. Although the current annual precipitation here is barely 200 mm, it was concluded that the soil probably experienced several periods of moister conditions over its 500,000 year history. The broad details of the genesis of the soil, as summarized by Monger et al. (1991), conformed to earlier macromorphologically-based reconstructions from similar soils.

The recognition of a wind-blown dust source for much of the illuvial clay and secondary calcite in many of the soils examined in the Desert Project has important implications for the reconstruction and interpretation of pedogenic phases based upon the form and depth of macro- and micromorphological feature associations, namely that material is migrating from an aggrading land surface. This factor was emphasized by Allen and Goss (1974), who undertook a detailed micromorphological study of a 6.5 m profile in Texas containing a series of welded and buried soils formed in aeolian sediments. Largely on the basis of the depth distribution and superimposition of argillic and calcitic features through the pedocomplex, they proposed a model of cyclical soil development at a slowly aggrading landscape. Although they recognized that sedimentation and pedogenesis has occurred simultaneously, they envisaged that dust inputs were at a minimum and pedogenic activity at its most intense during relatively humid phases. Intervening arid conditions favoured more rapid accumulation of aeolian sediments and covering of soil horizons. Subsequent reversion to a moister climate resulted in renewed leaching and translocation from a new land surface, often leading to welding of pedogenic profiles.

The essence of this model is also applicable to the Loess Plateau of China where loess-palaeosol sequences of several hundred metres in thickness abound (Derbyshire et al., 1995). Here, there have been alternating periods of geological time when the regional climate was dominated by winds blowing either from the northwest or the southeast. Phases of regional aridity, controlled by the relative dominance of the dry, dust-laden northwestern monsoons, were characterized by high rates of dust deposition, aggrading land surfaces and minimal pedogenic alteration. In contrast, intervening periods dominated by the humid, southeastern monsoons favoured establishment of more 'stable' land surfaces and significant soil development (Derbyshire et al., 1995). Kemp et al. (1995) provided a very detailed documentation of the micro-morphological fabrics throughout a 38 m sequence formed over the last 130,000 years in the semi-arid western part of the plateau. Part of the record is shown in Fig. 7. The equivalent pedosedimentary reconstruction derived from the interpretation of the depth distribution of the features is summarized in Fig. 8. The presence of key micromorphological features such as excrements, crusts and calcitic concentration/depletion features within the 'loess' units illustrates the complexity of the pedosedimentary model, particularly during periods of climatic transition, and emphasizes the difficulties in establishing the upper and lower boundaries of palaeosol units. These points were further highlighted in a similar micromorphological study of the S1 pedocomplex (correlated to Oxygen Stage 5 of the ocean cores) further west on the northeastern margin of Tibetan Plateau by Kemp et al. (1996).

An emphasis on pedosedimentary processes also characterized the micromorphological study of a 19 m thick dune section in the Thar Desert of northern India by Raghavan and Courty (1987). Although discrete palaeosol units could not be recognized in the field, a range of calcitic, textural and biological features were described from thin sections throughout the sequence. On the basis of the depth distribution of these features, these authors concluded that the dune had formed over a significant period of time characterized by alternating arid to sub-humid conditions. These changing climatic environments determined the balance between aeolian deposition of sand, reworking of sediments by floods, redistribution of calcite by fluctuating groundwater levels, biological activity and vertical leaching or particle translocation.

Acknowledgements

I would like to thank Justin Jacyno for drafting the diagrams so expertly and Ed Derbyshire for helpful comments on an earlier draft of this chapter.

References

Allen, B.L. (1985). Micromorphology of aridosols. In: *Soil Micromorphology and Soil Classification*, L.A. Douglas, and M.L. Thompson (eds.), pp. 197-216. Soil Science Society of America, Madison.

Allen, B.L. and Goss, D.W. (1974). Micromorphology of palaeosols from the semiarid Southern High Plains of Texas. In: *Soil Microscopy*, G.K. Rutherford (ed.), pp. 511-525. Limestone Press, Kingston.

Avery, B.W. (1985). Argillic horizons and their significance in England and Wales. In: *Soils and Quaternary Landscape Evolution*, J. Boardman (ed.), pp. 69-86. J. Wiley and Sons, Chichester.

Bal, L. (1975). Carbonate in soil: a theoretical consideration on, and proposal for its fabric analysis. 2. Crystal tubes, intercalary crystals, K fabric. *Netherlands J Agric. Science*, 23: 163-176.

Bal, L. (1977). The formation of carbonate nodules and intercalary crystals in the soil by the earthworm *Lumbricus rubellus*. *Pedobiologia*, 17: 102-106.

Bisdom, E.B.A., Tessier, D. and Schout, I.F.Th. (1990). Micromorphological techniques in research and training. In: *Soil Micromorphology: a Basic and Applied Science*, L.A. Douglas (ed.), pp. 581-603. Elsevier, Amsterdam.

Bresson, L. M. and Boiffin. J. (1990). Morphological characterization of soil crust development stages on an experimental field. *Geoderma*, 47: 301-325.

Bresson, L.M. and Valentine, C. (1994). Soil surface crust formation: contribution of micromorphology. In: *Soil Micromorphology: Studies in Management and Genesis*, A.J. Ringrose Voase and G.S. Humphreys (eds.), pp. 737-762. Elsevier, Amsterdam.

Brewer, R. (1972). The basis of interpretation of soil micromorphological data. *Geoderma*, 8: 81-94.

Brewer, R. (1976). *Fabric and Mineral Analysis of Soils*. Krieger, New York, 482 pp.

Bronger, A. and Heinkele, Th. (1989). Palaeosol sequences as witnesses of Pleistocene climatic history. *Catena Supplement*, 16: 163-186.

Bronger, A., Bruhn-Lobin, N. and Heinkele, Th. (1994). Micromorphology of palaeosols - genetic and palaeoenvironmental deductions: case studies from central China, South India, NW Morocco and the Great Plains of the USA. In: *Soil, Micromorphology: Studies in Management and Genesis*, A.J. Ringrose-Voase and G.S. Humphreys (eds.), pp. 187-206. Elsevier, Amsterdam.

Bullock, P., Fedoroff, N., Jongerius, A., Stoops, G. and Tursina, T. (1985). *Handbook for Soil Thin Section Description*. Waine Research Publications, Wolverhampton, 152 pp.

Catt. J.A. (1995a). Soils in aeolian sequences as evidence of Quaternary climatic change: problems and possible solutions. *Quat. Proc.*, 4: 59-68.

Catt. J.A. (1995b). Report from working group on definitions used in palaeopedology *INQUA/ISSS Palaeopedology Commission Newsletter*, 11: 35-37.

Courty, M.A. and Fedoroff, N. (1985). Micromorphology of recent and buried soils in a semi-arid region of northwestern India. *Geoderma*, 35: 287-332.

Courty, M.A., Goldberg, P. and MacPhail, R. (1989). *Soils and Micromorphology in Archaeology*. Cambridge University Press, 344 pp.

Cremaschi, M., Fedoroff, N., Guerreschi, A., Huxtable, J., Colombi, N., Castelletti, L. and Maspero, A. (1990). Sedimentary and pedological processes in the Upper Pleistocene loess of northern Italy. The Bagaggera sequence. *Quat. Inter.*, 5: 23-38.

Derbyshire, E., Kemp, R.A. and Meng, X.M. (1995). Variations in loess and palaeosol properties as indicators of palaeoclimatic gradients across the Loess Plateau of North China. *Quat. Science Rev.*, 14: 681-697.

Dobrovol'ski, G.V. (1991). *A Methodological Manual of Soil Micromorphology*. International Training Centre for Postgraduate Soil Scientists, Ghent (Belgium), 63 pp.

Drees, L.R. and Wilding, L.P. (1987). Micromorphic record and interpretation of carbonate forms in the Rolling Plains of Texas. *Geoderma*, 40: 157-175.

Fang. X.M., Li, J.J., Derbyshire, E., FitzPatrick, E.A. and Kemp, R.A. (1994). Micromorphology of the Beiyuan loess-palaeosol sequence in Gansu Province, China: geomorphological and palaeoenvironmental significance. *Palaeogeography, Palaeoclimatology, Palaeoecology*, 111: 289-303.

Fedoroff, N. and Courty, M.A. (1987). Morphology and distribution of textural features in arid and semiarid regions. In: *Soil Micromorphology*, N. Fedoroff, L.M. Bresson, and M.A. Courty (eds.), pp. 213-219. Association Francaise pour l'Etude du Sol, Plaisir.

Fedoroff, N. and Goldberg, P. (1982). Comparative micromorphology of two Late Pleistocene palaeosols (in the Paris Basin). *Catena* 9: 227-251.

Fedoroff, N., Courty, M.A. and Thompson. M.L. (1990). Micromorphological evidence of palaeoenvironmental change in Pleistocene and Holocene palaeosols. In: *Soil Micromorphology: a Basic and Applied Science*, L.A. Douglas (ed.), pp. 653-665. Elsevier, Amsterdam.

FitzPatrick, E.A. (1984), *Micromorphology of Soils*. Chapman and Hall, London, 433 pp.

FitzPatrick, E.A. (1993). *Soil Microscopy and Micromorphology*, J. Wiley and Sons, Chichester, 304 pp.

Gerasimova. M.I., Gubin, S.V. and Shoba, S.A. (1996). *Soils of Russia and Adjacent Countries: Geography and Micromorphology*. Wageningen-Moscow, 204 pp.

Gile, L.H. and Grossman, R.B. (1968). Morphology of the argillic horizon in desert soils of southern New Mexico. *Soil Science*, 106: 6-15.

Gile, L.H., Peterson, F.F. and Grossman, R.B. (1966). Morphological and genetic sequences of carbonate accumulation in desert soils. *Soil Science*, 101: 347-360.

Gile, L.H., Hawley, J.W., Grossman, R.G., Monger. H.C., Montoya, C.E. and Mack, G.H. (1995). Supplement to the Desert Project guidebook with emphasis on soil micromorphology. *New Mexico Bureau of Mines and Mineral Resources Bulletin*, 142.

Herrero, J. and Porta, J. (1987). Gypsiferous soils in north-east of Spain In: *Soil Micromorphology*, N. Fedoroff, L.M. Bresson, and M.A. Courty (eds.), pp. 187-192. Association Francaise pour l'Etude du Sol, Plaisir.

Jaillard, B. and Callot, G. (1987). Action des racines sur la ségrégation mineralogique des constituants minéraux du sol. In: *Soil Micromorphology*, N. Fedoroff, L.M. Bresson, and M.A. Courty (eds.) pp. 371-375. Association Francaise pour l'Etude du Sol, Plaisir.

Jenkins, D.A. (1994). Interpretation of interglacial cave sediments from a hominid site in North Wales: translocation of Ca-Fe phosphates. In: *Soil Micromorphology: Studies in Management and Genesis*, A.J. Ringrose-Voase and G.S. Humphreys (eds.) pp. 293-305. Elsevier, Amsterdam.

Jongmans, A.G., van Oort, F., Buurman, P. and Jaunet, A.M. (1994). In: *Soil Micromorphology: Studies in Management and Genesis*, A.J. Ringrose-Voase, and G.S. Humphreys (eds.) pp. 285-291. Elsevier, Amsterdam.

Kemp, R.A. (1985). Soil micromorphology and the Quaternary. *Quat. Res. Assoc. Technical Guide*, 2.

Kemp, R.A. (1987). The interpretation and environmental significance of a buried Middle Pleistocene soil near Ipswich Airport Suffolk, England. *Phil. Transac. of the Royal Soc. of London*, B317: 365-391.

Kemp, R.A. (1992). Soil evidence of Pleistocene environmental change in south-east England. *Seesoil*, 8: 13-28.

Kemp, R.A. (1995). Distribution and genesis of calcitic pedofeatures within a rapidly aggrading loess-palaeosol sequence in China. *Geoderma*, 65: 303-316.

Kemp, R.A., Jerz, H., Grottenthaler, W. and Preece, R.C. (1994). Pedosedimentary fabrics of soils within loess and colluvium in southern England and Germany. In: *Soil Micromorphology: Studies in Management and Genesis*, A.J. Ringrose-Voase and G.S. Humphreys (eds.), pp. 207-219. Elsevier, Amsterdam.

Kemp, R.A., Derbyshire, E., Chen, F.H. and Ma H.Z. (1996). Pedosedimentary development and palaeoenvironmental significance of the S1 palaeosol on the northeastern margin of the Qinghai-Xizang (Tibetan) Plateau. *J. Quat. Science*, 11: 95-106.

Kemp, R.A., Derbyshire, E., Meng, X.M., Chen, F.H. and Pan, B.T. (1995). Pedosedimentary reconstruction of a thick loess-palaeosol sequence near Lanzhou in north-central China. *Quat. Research*, 43: 30-45.

Kubiena, W.L. (1938). *Micropedology*, Collegiate Press. Amsterdam, 243 pp.

Kubiena, W.L. (1953). *The Soils of Europe*. Thomas Murby, London, 317 pp.

Kubiena, W.L. (1970). *Micromorphological Features of Soil Geography*. Rutgers University Press, New Brunswick, 254 pp.

Lee, J.A. and Kemp, R.A. (1993). Thin sections of unconsolidated sediments and soils: a recipe. *Centre for Environmental Analysis and Management (Royal Holloway, University of London) Technical Report*, 2.

Liu, T.S., Zhang, S.X. and Han, J.M. (1987). Stratigraphy and palaeoenvironmental changes in the loess of central China. *Quat. Science Rev*, 6: 489-501.

McFadden, L.D. (1988). Climatic influences on rates and processes of soil development in Quaternary deposits of southern California. In: *Palaeosols and Weathering Through Geologic Time: Principles and Applications*. J. Reinhardt and W.R. Sigleo (eds.) Geological Society of American Special Paper, 126: 153-177.

Monger, H.C., Daugherty, L.A. and Gile, L.H. (1991). A microscopic examination of pedogenic calcite in an aridosol of southern New Mexico. In: *Occurrence, Characteristics and Genesis of Carbonate, Gypsum and Silica Accumulations in Soils*. W.D. Nettleton (ed.). Soil Science Society of America Special Publication, 26: 37-60.

Mücher, H.J. and Morozova, T.D. (1983). The application of soil micromorphology in Quaternary Geology and Geomorphology. In: *Soil Micromorphology*, P. Bullock and C.P. Murphy (eds.), pp. 151-194. AB Academic Publishers, Berkhamstead.

Mücher, H.J. and Coventry, R.J. (1994). Soil and landscape processes evident in a hydromorphic grey earth (Plinthusalf) in semiarid tropical Australia. In: *Soil Micromorphology: Studies in Management and Genesis*, A.J. Ringrose-Voase and G.S. Humphreys (eds.), pp. 221-231. Elsevier, Amsterdam.

Mücher, H.J. and de Ploey, J. (1977). Experimental and micromorphological investigation of erosion and redeposition of loess by water. *Earth Surface Processes and Landforms*, 2: 117-124.

Murphy, C.P. (1985). Faster methods of liquid-phase acetone replacement of water from soils and sediments prior to resin impregnation. *Geoderma*, 35: 39-45.

Murphy, C.P. (1986). *Thin Section Preparation of Soils and Sediments*. AB Academic Publishers, Berkhamstead, 149 pp.

Murphy, C.P. and Kemp, R.A. (1984). The over-estimation of clay and the under-estimation of pores in soil thin sections. *J. Soil Science*, 35: 481-496.

Murphy, C.P., McKeague, J.A., Bresson, L.M., Bullock, P., Kooistra, M.J., Miedema, R. and Stoops, G. (1985). Description of soil thin sections: an international comparison. *Geoderma*, 35: 15-37.

Nettleton, W.D., Flach, K.W. and Brasher, B.R. (1969). Argillic horizons without clay skins. *Soil Science Soc. America*, 33: 121-125.

Nettleton, W.D., Gamble, E.E., Allen, B.L., Borst, G. and Peterson, F.F. (1989). Relict soils of subtropical regions of the United States. *Catena Supplement*, 16: 59-93.

Phillips, S.E. and Self, P.G. (1987). Morphology, crystallography and origin of needle-fibre calcite in Quaternary pedogenic calcretes of south Australia. *Aust. J. Soil Res*, 25: 429-444.

Raghavan, H. and Courty, M.A. (1987). Holocene and Pleistocene pedosedimentary environments in the Thar desert (Didwana, India). In: *Soil Micromorphology*, N. Fedoroff, L.M. Bresson and M.A. Courty (eds.), pp. 639-649. Association Francaise pour l'Etude du Sol, Plaisir.

Shahid, S.A. and Jenkins, D.A. (1994). Mineralogy and micromorphology of salt crusts from the Punjab, Pakistan. In: *Soil Micromorphology: Studies in Management and Genesis*, A.J. Ringrose-Voase and G.S. Humphreys (eds.), pp. 799-810. Elsevier, Amsterdam.

Tippkötter, R. and Ritz, K. (1996). Evaluation of polyester, epoxy and acrylic resins for suitability in preparation of soil thin sections for in situ biological studies. *Geoderma*, 69: 31-57.

Valentine, K.W.G. and Dalrymple, J.B. (1976). Quaternary buried palaeosols: a critical review. *Quat. Res.*, 6: 209-222.

Vepreskas. M.J., Wilding, L.P. and Drees, L.R. (1994). Aquic conditions for Soil Taxonomy: concepts, soil morphology and micromorphology. In: *Soil Micromorphology: Studies in Management and Genesis*, A.J. Ringrose-Voase, and G.S. Humphreys (eds.), pp. 117-131. Elsevier, Amsterdam.

Wilding, L.P. and Flach, K.W. (1985). Micropedology and Soil Taxonomy. In: *Soil Micromorphology and Soil Classification*, L.A. Douglas, and M.L. Thompson, pp. 1-16. Soil Science Soc. Amer., Madison.

Wright, V.P. (1990). A micromorphological classification of fossil and recent calcic and petrocalcic microstructures. In: *Soil Micromorphology: a Basic and Applied Science*, L.A. Douglas (eds.), pp. 401-407. Elsevier, Amsterdam.

4

Soil and Soil Forming Processes under Increasing Aridity

N. Fedoroff[1] and M.A. Courty[2]

ABSTRACT

This chapter aims, (1) to present and discuss soil forming processes and soil distribution under conditions of increasing aridity and, (2) to examine the relationships between pedological, erosional and accretionary processes in deserts. A third section deals with inherited soils and palaeosols presently located in deserts, which developed under varied environmental conditions and which demonstrate the potential of palaeopedology for reconstructing the past environments of deserts.

Introduction

Soil scientists have tended to neglect desert and peridesertic soils, probably because these soils have little or no agricultural value. Earlier investigations on these soils were performed by geomorphologists because of the high intermixing of pedological, erosional and accretionary processes in deserts (Cooke et al.,1993). In the past this led to an underestimation of pedological processes in deserts. However, arid soils are now better documented, and well established classifications of arid soils now exist. In Soil Taxonomy (Soil Survey Staff, 1992), a whole order, the order of Aridisols, is devoted to soils affected by drought. The impact of increasing aridity on soils has been deciphered, including, for example, the process of surface crusting (Casenave and Valentin, 1989). The high sensitivity of arid soils to environmental fluctuations has also been recognized (Courty, 1994; Goodfriend et al., 1996). Recently, signatures of abrupt events like natural fires (Fedoroff and Courty, 1989) or a set of unusual, natural, violent phenomena (Weiss et al., 1993; Fedoroff and Courty, 1995) have been identified in the soils and palaeosols of arid areas.

[1]Département d'Agronomie et Environnement, INA P-G, 78850 Thiverval-Grignon, France
[2]CNRS, USR 708, Dynamique des Milieux et Organisations spatiales, Département d'Agronomie et Environnement, INA P-G, 78850 Thiverval-Grignon, France

Soil Forming Processes and Soil Distribution under Increasing Aridity

An increase of aridity reduces the depth of soil hydration which, in turn, leads to a shift of the active soil horizons towards the soil surface (Dan et al., 1982), except in sand dunes in which almost all the water from occasional rainstorms penetrates deeply into the sand. Penetration to depths of more than 1 m preserves it against evaporation (Blume et al., 1995). In hyper-arid deserts like the Sahara, bare patinated rocks prevail. In these areas when loose material is present, the thickness of the active horizon does not exceed a few centimetres (Dutil, 1971) and surface soil crusting is the main physical soil forming process. In less arid deserts, soils commonly show a mosaic organization with abrupt boundaries resulting from sharp thresholds in the soil water regime, e.g. dense perennial vegetation bands where water and nutrients accumulate/ alternate with crusted barren bands which shed water (Valentin, 1994). Temporarily and permanently waterlogged soils, most of which are salt affected, are common in semi-arid areas, like the Chihuahua desert (northern Mexico). A large extension of waterlogged soils are the result of desert drainage patterns characterized by endoreic depressions (Goudie and Wells, 1995). Soils in these depressions can be caused by, (1) sedimentary and pedo-sedimentary processes induced by floods, the intensity of which can be reinforced by allogenic streams, (2) soluble salt concentrations provided by artesian ground water and, (3) alkaline concentrations. In cold deserts, the impact of water freezing and spring snow melting on the soil cannot be ignored.

Surface Crusting

Surface crusting which is only a cultivation-induced process in humid climates (Bresson and Boiffin, 1990), is the most common soil forming process under arid environments. It controls the soil water regime and has a strong influence on runoff and erosion (Casenave and Valentin, 1989). This process is initiated by the impact of rainfall on bare soil—the splash effect. A rain. drop which hits a mineral soil disperses the soil material at the impact point. However, a fraction of the rain drop infiltrates through the soil. Droplets with suspended particles are splashed in all directions. Accumulation of such rain drop impacts induces surface erosion. After rain, the soil surface is flattened, with a residual roughness. On clay rich material (>15-20% clay) the infiltrating water saturates the uppermost millimetres and disperses clayey materials into slush, which infills the meso-porosity of this micro-layer. This results in an incipient crust, named structural surface crust or slaking crust (Fig. 1a) with common vesicles. On sandy materials, sieving crusts (Fig. 1b) develop (Valentin, 1986) which consist of a layer of loose sand lying upon a thin, dense, hard layer of fine material. Rain drop impact forms micro-craters, the walls of which present a clear sorting of particles (Valentin, 1986) and infiltrating water enhances the downward movement of fine material. The water, rich in the suspended material of the droplets, flows towards micro-depressions between clods, which are then infilled with a cross-bedded alternation of sandy and silty clay layers (Fig. 1c). These cross-bedded crusts are named depositional surface crusts (Valentin, 1994).

Surface crusts can be broken easily, for example, by animal trampling and thus are renewed each year during the rainy season. When preserved over some years, all surface crusts tend to evolve into sedimentary crusts which become stabilized by invading algae and lichens (Plate 1) (Fedoroff and Courty, 1995). However, long-term stabilization of surface crusts is ensured by stone pavements. These pavements commonly develop on alluvium and colluvium rich in gravels and stones. Following its deposition, the fine material is progressively eroded, leaving a concentration of coarse fragments on the soil surface. At a given moment, an equilibrium is reached between abundance of coarse fragments and runoff water energy which leads to the stabilization of the surface crust. These crusts, named pavement crusts (Fig. 1d), consist of three micro-horizons, a loose sand at the surface lying upon a densely packed horizon of grains with abundant vesicles, and a lower one showing a higher content of fine particles with a very low porosity.

Soil erosion by water and soil sealing are counterpoised phenomena. When the soil surface becomes permanently crusted, at a first approximation, most of precipitation is lost as surface runoff (Hairsine and Hook, 1994). Only a small part of rair water infiltrates the soil, either through the textural porosity, or through large voids (vertical cracks and biological channels) which are not sealed by the surface crust. On a pavement crust, Casenave and Valentin (1989) have measured an infiltrability of 0-2 mm/h.

Impact of floods on soils in arid environments

Floods are common in arid environments because of, (1) occasional, sudden and heavy rains, even in the most arid deserts, (2) large extension of surface crusts and outcrops of hard rock, and (3) endoreic character of deserts. The impact of these floods on soils varies according to size, from micro-depressions a few decimetres large (which are flooded for a few minutes to a few hours), to large depressions, some square kilometres in width, in which water is stagnant from a few days to a few months. Micro-depressions flooded by local runoff are characterized by thick

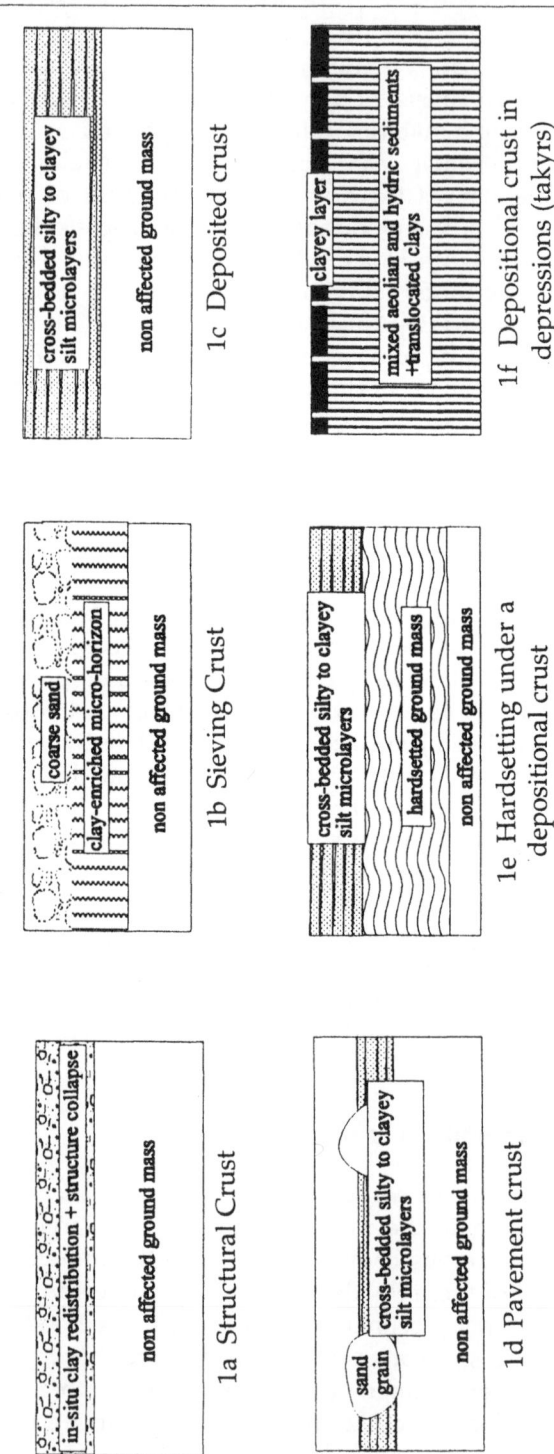

Fig. 1: Main types of surface crusts and related hardsetted horizons.

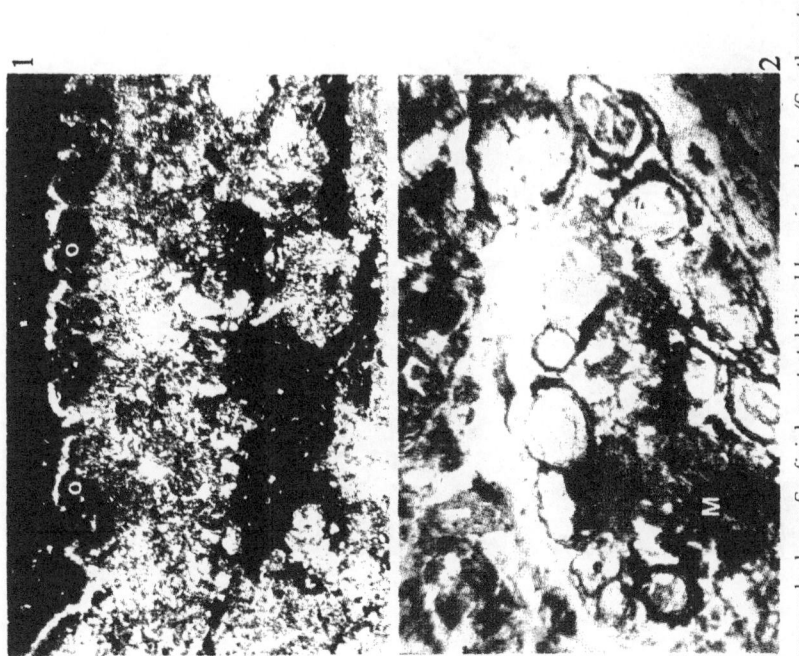

Plate 1: Micrograph shows Surficial crust stabilized by microphytes (Southeastern Spain, Almeria). Exuded calcium oxalate-carbonate on top of living organisms (O). Below a thin compacted, partially impoverished in calcium carbonate sub-horizon lying on a loosely packed ground mass (polarized light, × 65).

Plate 2: Micrograph shows a root mat pseudomorphosed by $CaCO_3$ (Cuotlichan, Altiplano, Mexico). Root cells pseudomorphosed by calcium carbonate (P) and secondary micritization of pseudomirphosed root cells (M) (polarized light, × 190).

Plate 3: Micrograph shows Takyr subsurface fabric (Sinai, Egypt). Silty clay ground mass (s), clay infillings (c) and sand grain accumulation (g) (plain light, × 65).

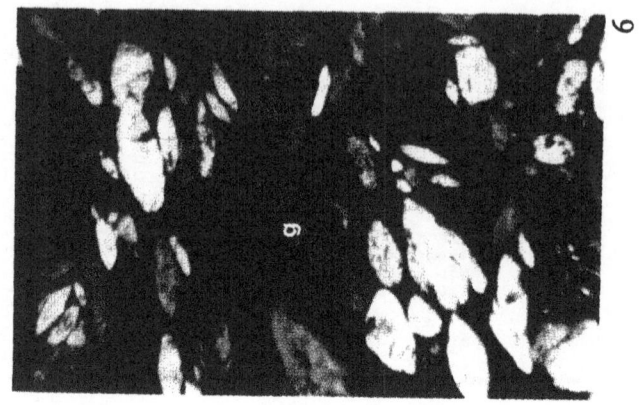

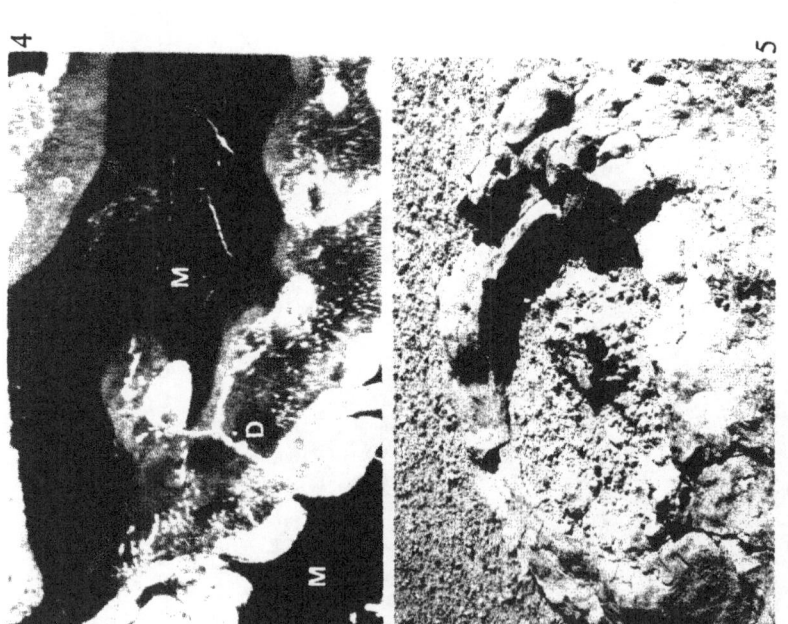

Plate 4: Micrograph shows Carbonate depleted ground mass in an horizon enriched in gypsum (North-eastern Spain). Micritic ground mass (M), calcium carbonate depleted ground mass (D) and assemblages of secondary equigranular calcium carbonate crystals (I) (plain light, × 22).

Plate 5: Photograph shows A fossil karst hollow from a Plio-Villafranchian calcrete (Northern Sahara, Western Grand Erg) of which only outer karstic walls are preserved.

Plate 6: Micrograph shows Subrounded gypsum crystals from a gypsicrete (Syrian desert, El Kowm). Subrounded lenticular gypusm grain (g) (plain light, × 190).

sedimentary surface crusts (Fig. 1e). When the catchment is larger and retains stagnant water much longer, infiltrating water saturates the textural porosity to a greater depth of the subsurface horizon. This leads to a structural collapse and the development of hardsetting just below the sedimentary crust (Fig.1f). In large, takyr-like depressions, in which cracks and curling up plates usually develop during drying, the following fabric has been observed (Plate 3) by Fedoroff and Courty (1987), (1) a silty clay ground mass resulting from the deposition of the flood suspension at the surface, (2) clay coatings and infillings locally merging into intercalations, the result of pedo-sedimentation in the cracks of the takyrs and, (3) irregular accumulations of aeolian sand grains transported by saltation on the depression surface during dry periods, commonly superimposed on the clayey features and less commonly trapped in the cracks.

Particle translocations in soils in arid environments

From the formula of particle translocation (Tl) in soil (Fedoroff, 1991),

$$Tl = f \, (SSm, Swr, Pp, PCs)$$

in which SSm is the soil surface morphology, Swr the soil water regime, Pp the poral pattern and PCs the physico-chemical status. It is possible to infer that an increasing aridity considerably modifies the process of particle translocation because of the leading role played by the soil surface morphology to which the soil water regime is closely linked. Rate and depth of water infiltration in the soil on which particle translocation is dependant does not decrease with increasing aridity, but abruptly as soon as surface crusts appear (Fedoroff and Courty, 1987).

Under subhumid climates (approximating to udic regime of Soil Taxonomy) (Fig. 2.1) the soil surface is covered by plant residues and faecal pellets, and rain drops do not impact the mineral soil. Rain water penetrates the soil through channels and packing voids, causes micro-erosion of walls of pathways and carries away only clay particles. Suspended clays settle during evaporation of capillary water in the B horizon during the dry season forming microlaminated clay coatings. In this case only clays are translocated. The soil profile (Fig. 2.1) consists of an eluvial horizon (e.g. umbric, ochric) overlying an argillic horizon characterized by microlaminated clay coatings (Fig. 2.1) (Fedoroff, 1991). When rainfall decreases or the dry season gets longer (as under ustic regime of Soil Taxonomy), fine silt particles are suspended together with the clay because plant residues do not protect the entire soil surface from the splash effect (Fig. 2.2). However, continuous surface crusts do not form at this stage which indicates that the whole soil remains hydrated. The abundance of textural features increases while microlaminated clay

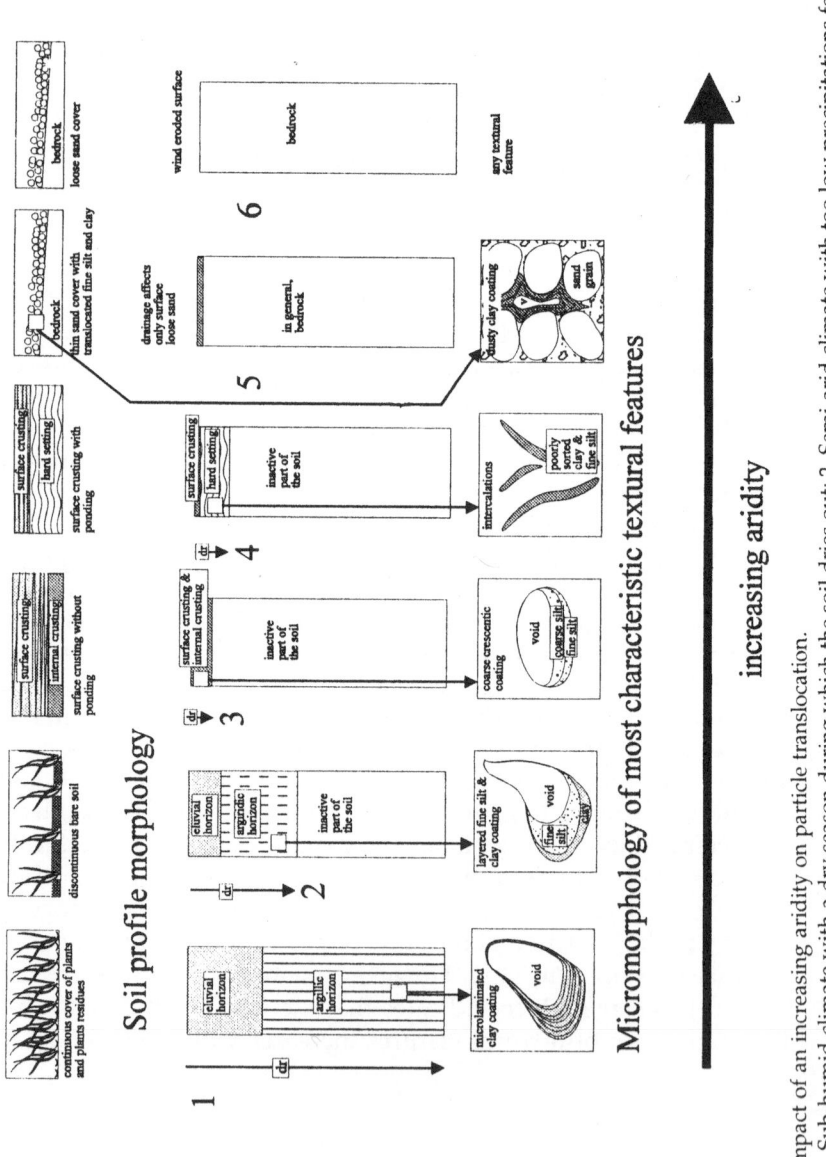

Fig. 2: Impact of an increasing aridity on particle translocation.
1. Sub-humid climate with a dry season during which the soil dries out; 2. Semi-arid climate with too low precipitations for the soil to be fully water percolated every year; 3 and 4. Arid climate with precipitations only affecting the soil surface and subsurface without a sand cover; 3. Rain water runs off; 4. Rain water is concentrated in a micro-depression; 5. Very arid climate (e.g. precipitations less than 50 mm) with a sand cover and dust storms; 6. Hyper arid climate (almost no precipitations, e.g. southern egyptian desert).

by fine silt clay coatings and infillings, with no layering or poorly expressed layering and a medium orientation (Fig. 2.2). In the next stage of aridity, (as in the case of aridic or torric regimes of Soil Taxonomy) surface crusts appear (Fig. 2.3 and 2.4). Infiltration occurs through, either textural porosity of the crust, or large biological pores or cracks. If infiltration occurs through textural porosity, only a thin layer is hydrated (Fig. 2.3). If large pores are present, and all subsurface horizons are hydrated then these pores are infilled with coarse layered crescentic coatings (Fig. 2.3) comparable to sedimentary surface crusts (Courty and Fedoroff, 1985; Fedoroff and Courty, 1987). However, B-horizons remain permanently dry. In micro-depressions where runoff water concentrates, surface horizons are hardsetted as a result of infiltration through the textural porosity (Fig.2.4). Silty intercalations can be observed in such hardsetted horizons (Fig. 2.4). Under the very arid climate (~50 mm annual rainfall) of the Western Great Erg (Algeria), at the base of a thin (~10 cm) aeolian stabilized sand cover lying in the micro-depressions of a calcrete, Fedoroff and Courty (1987) observed abundant, moderately oriented, dusty clay coatings (Fig. 2.5).

Shrink-swell

In contrasting climates (ustic regime) clayey soils, rich in swelling clays, are affected by severe shrinkage during the dry season and by swelling during the rainy season. This annual alternation of shrink and swell leads to differentiation within specific types of soils, the Vertisols. These soils are characterized by cracks during the dry season which can be observed down to a depth of 1.50 m, while shrinkage can reach 50% of the total soil volume, leading to densities of the ground mass close to 2. During the dry season, various materials fall from the soil surface in the cracks, but most of their infilling, which consists of varied materials (from small aggregates to moderately sorted silty clays), occurs at the beginning of the rainy season before the cracks close up. When swelling reaches its maximum, the soil mass increases and the added material is expelled, through the process of churning, specific to Vertisols (Fig. 3.1). Two by-products of vertisol formation must be mentioned, (1) the disruption of the underlying bedrock by penetration of swelling material from the Vertisol in its fissures during the hydration phase and, (2) the grinding of coarse grains and fragments in the ground mass because of high pressure, leading to flowage movements.

With increasing aridity, precipitation cannot completely hydrate the Vertisol, leading to a 'standstill' in the process of vertisol formation. Consequently the cracks remain open permanently which leads to their infilling by allochtonous materials. In the driest areas where loose sands are moved by winds, these infillings are sandy (Fig. 3.3) (Alaily, 1987), elsewhere calcitic infillings have been described (Fig. 3.2), e.g. in Tenerife, Canarias Islands (Rodriguez Hernandez et al., 1979) and in palaeosols in Madeira (Goodfriend et al., 1996).

Biological Activity

In a first approximation, the biological activity in freely drained soils in arid regions depends on energy and nutrients, on the soil water regime, i.e. the depth of water infiltration and water storage capacity. Both of these are contingent on, (1) soil texture and structure, (2) the surface crusting and subsurface hardsetting. Many taxa do not respond to an added moisture in the absence of nutrients (Steinberger, 1995). However the high adaptability of the soil biota to aridity and local soil conditions must be emphasized. Under extreme aridity, as on the bare limestone of the central Negev (Israel), the ecosystem can be only dew contingent (Shachak et al., 1995). These limestone are colonized by endolithic lichens which are foraged by snails when dew is present. The snails ingest rocks in order to eat the lichens, excreting carbonates in faeces; they convert rock into a soil material at a rate of 69.5 -110.4 $g/m^2/year$ (Shachak et al., 1987). The faecal production is closely related to the number of dew days (Jones and Shachak, 1990). In waterlogged soils of deserts, the soil biota is contingent on the amount of soluble salts and their distribution.

Increasing aridity has, undoubted, impact on soil biological activity. However, the precision of the estimation of an aridity increase from soil biological activity is rather low due to the high adaptibility of living organisms and the ecosystems, to unfavourable environmental conditions. The following characteristics should be considered in order to estimate the climatic impact on the soil biota: (1) organic matter content, (2) fabrics of horizons affected by faunal activity, and (3) inventories of the soil fauna. Examination of soil fabrics in thin-section under a polarizing microscope enables, (1) estimation of the degree of soil turbation by fauna (Courty and Fedoroff, 1985), (2) identification of the animals by their feacal pellets and passage features (Bullock et al., 1985), and (3) identification of other partial fabrics superimposed on the excretal features.

Two distinct sequences of the impact of increasing aridity on the soil biota have to be considered, first in cold deserts, located essentially in central Asia, characterized by cold winters with deep soil freezing, and second, in the deserts of the tropics and subtropics, where a long, hot dry season alternates with a short rainy period.

The sequence in cold deserts begins under subhumid climates with a thick, black, stable organic matter rich horizon, a mollic epipedon, characterized in thin-sections by an excretal fabric on which a frost induced

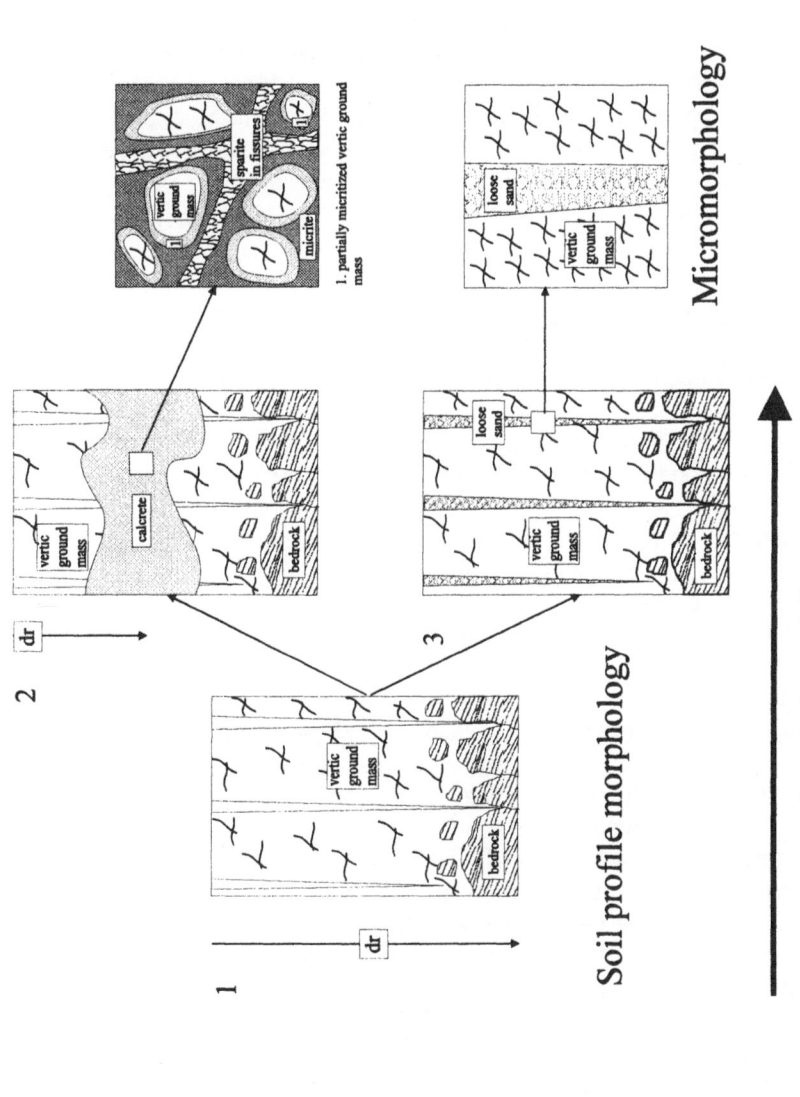

Fig. 3: Impact of an increasing aridity on Vertisols.
1. Sub-humid climate with well contrasted seasons; 2. Semi-arid climate characterized by a calcite accretion; 3. Arid climate.

fabric can be superimposed. These mollic epipedons develop under a continuous grass cover. With increasing aridity, the grass cover becomes discontinuous and the organic matter content decreases. The initial black colour turns to grey, and the thickness of mollic epipedon decreases. However throughout the whole sequence, these epipedons remain characterized by well developed excremental fabrics.

The sequence in hot deserts and their margins, for example, the Sahara and the Sahel, which develops under a short, monsoonal, rainy season and an absence of frost, starts under subhumid climates with a soil marked by an eluvial subsurface horizon displaying a total excremental fabric (Ousseini, 1987), a sombric or umbric epipedon (according to Soil Taxonomy). With an increasing aridity, the amount of organic matter decreases while the excretal fabric persists, even under high aridity (Courty and Fedoroff, 1985).

Chemical Enrichments

In desert margin soils, and to a lesser extent in the heart of deserts, the large excess of evapotranspiration over precipitation is the principal factor in the accretion of gypsum and soluble salts as well as of less soluble minerals, such as carbonates, silica and palygorskite. These accumulations require either a minimum of rainfall which penetrates the soil, but does not drain it, or a water-table of which the capillary fringe reaches the soil surface at some time of the year (Fig. 4). Components of these accumulations may originate from : (1) the parent material, (2) rainwater, (3) dust deposited at the soil surface, and (4) groundwater. Processes, that are responsible for the distribution of these components in the soil can be, (1) dissolution of soluble salts during the rainy season and precipitation during the dry season, (2) dissolution in an upper horizon and precipitation in a concentrated form in a lower horizon, (3) progressive concentration from ions present in the rainwater, and (4) precipitation from ions dissolved in ground water. Distribution of these chemical enrichments in the soil profile depends on the infiltration rate which is controlled by, (1) climatic factors, i.e. mean annual rainfall, distribution of rain through the year and the rainfall intensity, (2) the soil surface and subsurface morphology, and (3) the soil porosity. On stabilized sand-dunes, enriched horizons are deeper and thicker with more dispersed features compared to fine textured soils.

The influence of time, on the development of siliceous, calcic and to a lesser extent of gypsic accumulations in soils has been recognized (Gile et al., 1966; Dan, 1983; Harden et al., 1991). However, the impact of increasing aridity and more generally of climatic fluctuations are still rather poorly documented. The influence of time and climate fluctuations will be considered in the third section of this chapter.

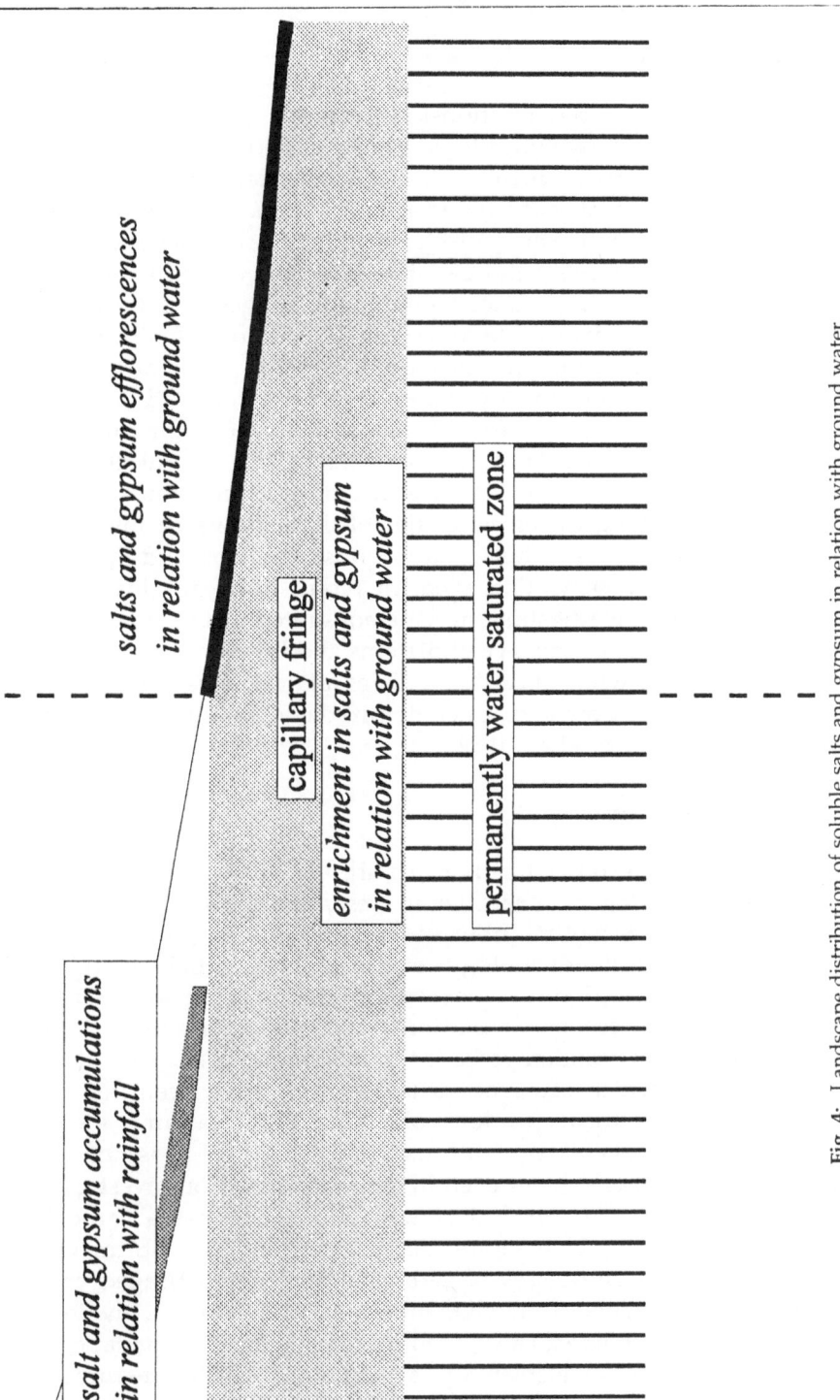

salts and gypsum efflorescences in relation with ground water

capillary fringe

enrichment in salts and gypsum in relation with ground water

permanently water saturated zone

salt and gypsum accumulations in relation with rainfall

Fig. 4: Landscape distribution of soluble salts and gypsum in relation with ground water.

Silica Enrichment

The exact nature of the relationship between silica accretion and aridification is not clearly understood. It is suggested that ancient silcretes formed under wet environments, while it is considered that accretion of silica in recent duricrusts occurred in semi-arid areas (Milnes et al., 1991; Blank and Fosberg, 1991; Thiry, 1992). Soil alkalinity characterized by pH >9 increases silica solubility and therefore favours weathering of silicates and quartz (Pedro, 1994). It is thus tempting to suggest that alkaline soil solution and groundwater favours the lixiviation of silica, while evaporative precipitation of silica occurs in the form of opal and chalcedony. Silica cemented duripans have been reported in soils of subhumid to semi-arid climates of almost the whole Circum Pacific volcanic belt that has inputs of volcanic glasses. Total chemical analysis on these duripans shows a high content of silicium which suggests cementation due to a pedogenic accretion of silica. However, microscopic investigations have not revealed features consisting of opal and chalcedony which could be the cementing agent in recent duripans (Flach et al., 1992). In fact, these investigations have demonstrated that the high silicium content results in the abundance of diatoms and/or phytoliths dispersed in the ground mass while in calcitic features and calcretes, it results from a progressive replacement of calcium carbonate by silica (Blank and Fosberg, 1991; Fedoroff et al., 1994).

Carbonate Accretion

Accumulations of terrestrial carbonates are the most common and the best expressed pedogenic character in desert margins and to a lesser extent in deserts. Scientists from several disciplines (sedimentology, pedology, soil micromorphology, stable isotope geochemistry and numerical modelling) have focused on the process and rates of development of these accumulations. Various forms of carbonate accumulations, from pseudomycelia (a few millimetres in size), to calcretes, the thickness of which can reach a few metres, have been described. Analysis of these forms under polarizing and scanning electron microscopes reveals a tremendous complexity of most of these accumulations, that is evident even in the field. Based on investigations at microscopic scale, three levels of accumulations of terrestrial carbonates can be discerned, (1) elementary calcitic features and fabrics which correspond to basic processes of carbonate accretion, (2) aged calcitic features which derive from elementary calcitic features by in-situ alteration, and (3) complex calcitic fabrics which consist of juxtaposition and superimposition of elementary, synchronous, calcitic features and fabrics. Most of these accumulations are inherited and some can be very old, for example, Callot (1987) described a calcrete in the Great Western Erg of the Sahara which dates back to Plio-Villafranchian. Here, we will consider the basic processes of carbonate accretion, while changes in these with time will be discussed in the third section of this chapter.

In well drained soils, in most cases biological mechanisms are probably, the only mode of carbonate accretion (eg. Wright, 1990; Fedoroff et al., 1994; Amit and Harrison, 1995; Goodfriend et al., 1996). Various organisms can contribute to this fixation, e.g. microbes (Monger et al., 1991), acicular calcite and calcified filaments which are considered to be fungal excreta (Phillips et al., 1987), shells, faunal excreta as well as root pseudomorphs (Fedoroff and Courty, 1994). These biological forms already exist under subhumid climates of mid-latitudes in horizons rich in primary carbonates, for example, in loess. Acicular forms have been mentioned in a wide range of climates, from subhumid to arid. The main factor which controls biological accretion of $CaCO_3$ is abundance of easily soluble calcite in the adjacent material, which can be a primary carbonate or carbonates brought with aeolian dusts. Soil solution becoming oversaturated with Ca^{++} during a drier season is also required.

In temporarily waterlogged soils, carbonates are in the form of nodules whose internal morphology, size of calcitic crystals and abundance of impregnated soil ground mass, depends on soil water regime and porosity. Pure nodules develop in clayey soils in which heavy rainfalls are immediately followed by a high evapotranspiration (Fig. 5.1) while impregnative nodules develop under a more gradual evapotranspiration (Fig. 5.2) (Courty, 1990; Courty, 1994).

In soils affected by a water table, in the permanently saturated zone, dense sparitic, jig-saw, nodules can be observed (Fig. 5.4). Continuous, porous, sparitic fabrics (rarely micritic) exist in the fluctuating zone of the water table (Fig. 5.3). Cooke et al. (1993) also mention floodwater calcretes; however, in most cases these calcretes are difficult to distinguish from groundwater calcretes.

Specific processes of carbonate accretion occur in subsurface horizons and at the surface of the soil. When the upper B-horizon becomes clogged by carbonates which prevents root penetration, root mats develop at the top of the continuous calcic horizon. If the carbonate supply continues, root mats are pseudomorphosed by $CaCO_3$ (Plate 2). Juxtaposition of successive pseudomorphosed root mats may appear in the field as a lamellar fabric.

Undisturbed surface crusts are colonized by lichens, liverworts, algae, fungi and bacteria forming a "microphytic crust" (West, 1990). These organisms produce calcium oxalate by "exsudation" (Plate 1) which ages into calcium carbonate (Fedoroff et al., 1994; Fedoroff and Courty, 1995).

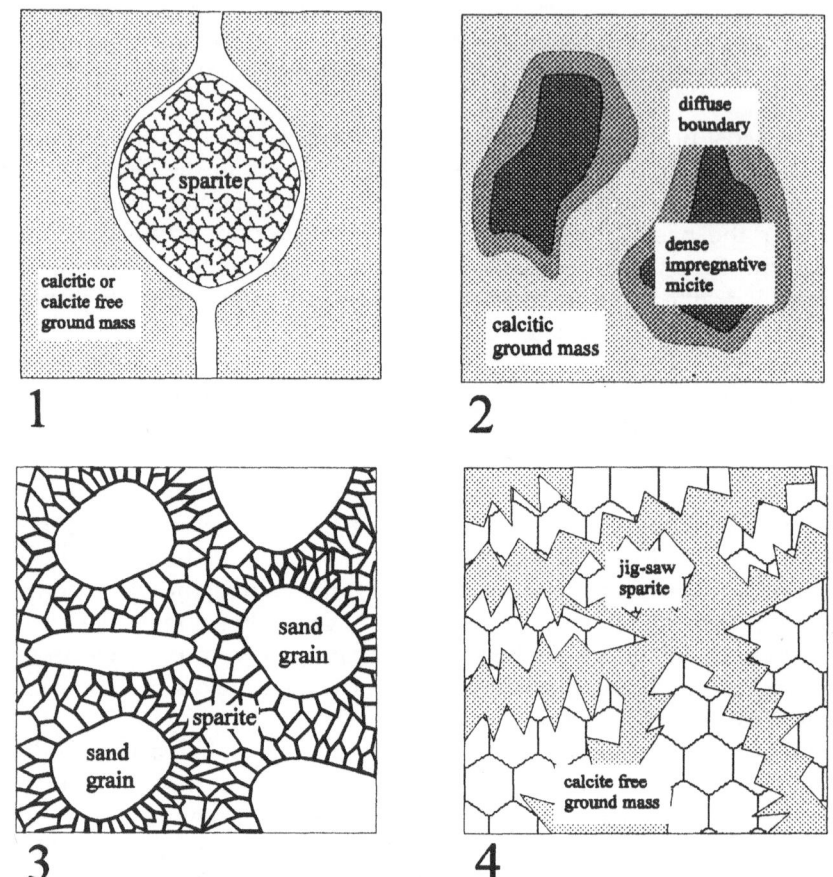

Fig. 5: Carbonate accretion
1 and 2 Temporary waterlogged soils; 1. High evaporative climate; 2. Medium to low evaporative climate; 3 and 4. Permanent waterlogged.

Dust is also frequently trapped into these microphytic crusts. Thus, juxtaposition of aged, successive microphytic dust enriched crusts produce a lamellar crust.

Pellety fabrics which lie just beneath a calcitic lamellar crust have been described in various locations (eg. Elloy and Thomas, 1981 in Algeria; Blank and Fosberg, 1991 in Idaho, USA; Fedoroff et al., 1994 in Altiplano, Mexico). These fabrics consist of a packing of rounded pellets commonly embedded in sparite. Pellets are composed of a nucleus, e.g. a quartz grain or a bioclast, while the cortex is formed by concentric micritic laminae.

Ageing of biogenic forms of carbonates and of porous sparite into a dense micrite is the process which leads to the lithification of carbonate accumulations, known as micritization (Fedoroff et al., 1994; Goodfriend et al., 1996). This micrite replaces the primary features in-situ (Plate 2), for example, acicular calcite is replaced by a thin micrite laminae. Various stages of dissolution of primary calcitic features can also be observed.

Accretion of Gypsum

Secondary gypsum in freely drained soils of arid areas máy be derived from gypsum-rich parent material (Herrero et al., 1992), or originate in coastal areas from rains contaminated by spray. However by far its main source is wind blown gypsum from sebkhas (Coque, 1962). Secondary gypsum which derives from gypsum-rich outcrops appears as soon as evapotranspiration exceeds precipitation. The depth of occurrence in the soil profile of this secondary gypsum is, in a first approximation, related to precipitation versus evapotranspiration (Eisenberg et al., 1982). The fabric of gypcrete resulting from an accretion of wind blown gypsum crystals consists usually of in-situ crystals. However wind eroded subrounded gypsum grains can be observed around sebkhas located in hyper-arid areas such as in the Syrian desert (Plate 6).

In waterlogged soils of arid regions, gypsum crystallizes from groundwater rich in calcium and sulphate in the form of large discrete crystal intergrowths, known as sand roses which can become continuous in the form of a gypcrete. The gypsum grows by fragmentation and displacement of the adjacent soil material. Calcite depleted features can be observed in horizons enriched in gypsum (Plate 4). In waterlogged soils, gypsum precipitates as soon as evapotranspiration exceeds precipitation. This accumulation ceases when the groundwater table is lowered. However this groundwater lowering is not directly related to aridity increase because groundwater in deserts is commonly artesian.

Accretion of Soluble Salts and their Impact on the Soil Structure

Soluble salts accumulate essentially in waterlogged soils, named solonchaks (Salorthids in Soil Taxonomy) in which a water table progressively supplies salts to the soil by evaporation of its capillary fringe (Fig. 4). Uncontrolled irrigation frequently raises the water table to the point where its capillary fringe intersects the soil surface. However, soluble salts can be also present in freely drained soils in smaller amounts which can be derived from the parental material, or supplied by salted dust or spray. The most common salt in solonchaks is sodium chloride, but salts composed of other cations, such as calcium, magnesium, potassium and sodium and anions like sulphate, carbonate, nitrate and borate are present in various amounts. Crystallization and distribution of salts in the soil profile occur according to their solubility. During the rainy season, salts are dissolved and lixiviated downwards while during the dry season, evaporation induces a capillary rise of the soil water together with the ions. Consequently, salts crystallize in the soil subsurface and at its surface in the form of a salt crust. When sodium becomes dominant in the soil solution, it first replaces calcium and magnesium in the soil due to cation exchange capacity. In a second stage, if its abundance increases, $NaCO_3$ is crystallized, while the organic matter solubilizes and migrates to the soil surface and stains it black. The soil becomes alkaline, with pH above 9, resulting in soils known as solonetz (Natrargids in the Soil Taxonomy).

Forces of crystal growth from solution promote the desintegration of the host ground mass. The pressure exerted on the adjacent material by growing crystals depends mainly on the degree of supersaturation. Salt crystals can continue to grow against a confining pressure when a film of solution is maintained at the salt/adjacent material. Maintenance of this solution depends on the interfacial tensions at the boundary of salt with the adjacent material and at the solution/adjacent material boundary. When the sum of the latter two is smaller than the first one, the solution can penetrate between the growing crystals and the adjacent material (Correns, 1949). Disruptive stresses may also be exerted by anhydrous salts which become hydrated from time to time, eg. anhydrite to gypsum. The effect of salts on the adjacent material, when heated, must also be mentioned.

Salts in deserts can be held responsible for, (1) rock alterations such as rock flaking and, rock and alluvial gravel desegregation which produces sand and silt size grains, and (2) loose salted mulch at the surface of sebkhas eventually associated with mounds and polygonal patterns. The loose and salted mulch contains variable percentages of non-saline soil ground mass in the form of subangular aggregates.

Exchangeable sodium (SAR, sodium absorption ratio) disperses clays. When the soil is water saturated, this dispersion induces, a structure collapse characterized by small, closed polyconcave vughs and a subsequent hardsetting in which various textured intercalations can be recognized. Hydraulic conductivity of sodium hardsetted soil ground mass is close to zero which induces, (1) surface runoff, if hardsetting affects the soil subsurface associated with a high surface micro-erosion because of the sensitivity of clays to dispersion, and (2) perched water table if the hardsetted layer is a B-horizon in which vertical fissures and the overlying surficial horizon are characterized by infillings and intercalations of dusty clay to sands.

Chemical Weathering and Clay Neoformation

Pedro (1994) has proposed two processes, salynolysis and alkalinolysis for explaining the weathering of primary minerals in arid environments, especially of silica and silicates in contact with an alkaline soil solution. Unfortunately the theory has not yet been correlated with field observations. According to Singer (1995), expanding smectite, or partly expanding smectite/chlorite mixed layers, constitute the most common weathering product of igneous rocks under the semi-arid to arid conditions of central Negev (Israel). The most typical clay mineral in desert soils is palygorskite, which can be found in soils affected by a rising water table, or in calcretes. The threshold concentration requirements for palygorskite stability are an alkaline pH, and high silica and magnesium activity in the soil solution (Singer, 1989).

Relationships Between Pedological, Erosional and Accretionary Processes under Increasing Aridity

Soil forming processes in deserts are more closely imbricated with erosional and accretionary processes compared to humid climates (Cooke et al., 1993). Some of their relationships are discussed below.

Interactions in Well Drained Soils between Pedological, Erosional and Accretionary Processes in the Sequence from Humid to Hyper-arid Climates

Under humid forests, pedological (including weathering) processes are intense, reach a great depth and as a first approximation can be considered as independent of erosional processes (Fig. 6.1). In areas which are presently under subhumid climates (i.e. 700 to 400 mm of annual precipitation) with a long dry season and/or which have been affected in the past by aridification, evidence of erosion in soils is frequent (Fig. 6.2) (Fedoroff, 1997). Such evidence is in the form of truncations commonly emphasized by stonelines, while pedosediments lie above truncations. A lower rainfall (less than 400-300 mm per annum) leads to an extension of vegetation free zones and regions with bare soils become almost continuous, and surface crusting occurs (Fig. 6.3). At this stage, surface runoff and erosion are at their maximum while pedological processes are at their minimum, being restricted to surface crusting. In the rolling landscape of Liptako (central Niger), in response to the drought which began in 1973, soil biota could not counteract soil crusting (Ousseini, 1987). Consequently, most of the surficial horizons were eroded down to the argillic horizon. The coarser part of the eroded material was deposited near-by in the form of loose sands, while silts and clays were removed. During sand storms, these loose sands were moved by saltation and accumulated around tufts of grass in the form of small dunes, known in the literature as nekhba. Impact of hydric erosion and sedimentation decreases with a further decrease in rainfall (less than 150-100 mm per annum). This climatic zone is characterized by pavement crusts that are stable (Fig. 6.4). In areas with a quasi absence of rains like the southern Egyptian desert, all processes (pedological and accretionary) resulting from water action are inactive, while bedrock erosion by sand grains and dune displacement ensuing from wind action are at their maximum. (Fig. 6.5).

Subhumid to semi-arid zones (annual precipitation between 800 to 300 mm) are characterized by banded vegetation where the cover has been preserved. This banded vegetation consists of a regular alternation of, (1) a crusted denuded band on which runoff is intense while only the soil subsurface is hydrated, and (2) a downslope vegetated band in which litter and strong macro-faunal activity favour infiltration of upslope runoff and consequently, the whole profile is hydrated. The downslope edge of the vegetated band dries out as a consequence of decreased infiltration insufficient to maintain the vegetation cover. This banded vegetation is characterized by the following pedo-sedimentary processes, (1) in the crusted band, soil crusting and sheet erosion occurs, while the soil profile remains almost dry and inactive, "dead", and (2) in the vegetated band, an intense biological activity and particle translocation from runoff suspension occurs. This banded vegetation was first described in the Sahel as "tiger bush" by Clos-Arceduc (1956) and Ambouta (1984). It also covers wide areas in the Chihuahua desert of Mexico (Delhoume et al., 1987, 1988) and in Australia.

Interactions in Depressions and Their Margins Between Pedological, Erosional and Accretionary Processes

The closest relationship between pedological and erosional sedimentary processes probably exists in and around closed basins that can attain high densities in depressions in deserts and their margins (Goudie and Wells, 1995). According to their functioning, two basic types of closed basins can be distinguished, (1) basins occasionally submerged by floods in which salt efflorescence does not appear, known as takyrs, (2) basins characterized by salt efflorescences, named sebkhas or playas.

In takyrs, sedimentation and particle translocation are closely linked (see above). Sand grains transported by saltation on the takyr surface during the dry period are trapped in the cracks. Active takyrs do not

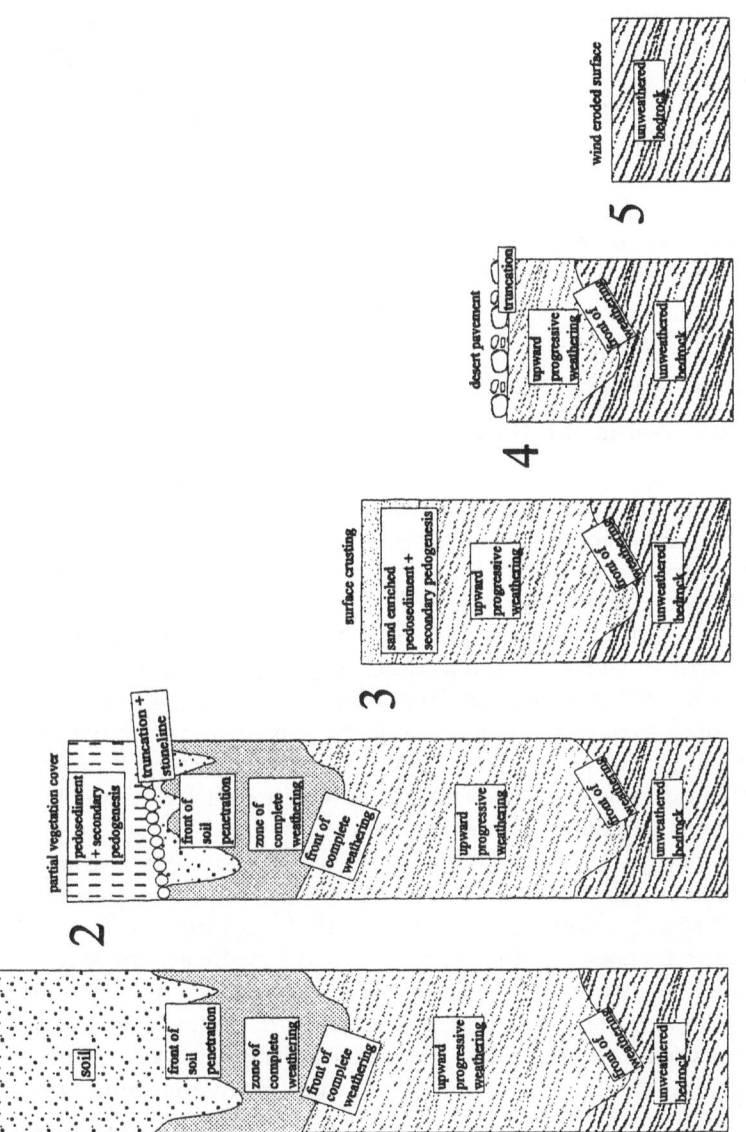

Fig. 6: Interactions of soil forming processes with erosion in a sequence of increasing aridity.
1. Permanent humid climate; 2. Present humid to subhumid climate, with periods of aridity in the past; 3. Present and past (with fluctuations) semi-arid climate; 4. Arid climate with precipitations only affecting the soil surface; 5. Hyper arid climate (almost no precipitations, eg. southern egyptian desert).

exhibit wind erosion signatures while inherited takyrs, like those in the Egyptian desert, are usually wind eroded with relict yardangs.

In sebkhas, the pedological processes of salt crystallization at the soil surface combined with fragmentation of the host ground mass by crystal growth, provide a loose, light, wind erodible mass known as the salt efflorescence. Effectively these efflorescences are wind deflated and are frequently trapped on the lee sides of sebkhas if vegetation is present. The trapped aeolian sediments most frequently are in the form of crescentic aeolian accumulations, called "lunette", however various other forms exist, for example, thin (1-2 m) gypsic crusts of regular thickness, covering a wide area around the sebkha (Coque, 1962). Aeolian deposits around sebkhas have variable grain size distribution and mineralogical composition according to the sebkha functioning, the geological outcrops and the soil cover of the sebkha catchment basin (Goudie and Wells, 1995). Aeolian deposits around sebkhas are affected by a post-depositional pedogenesis and usually most soluble salts are washed away (Hachicha et al., 1987).

Signatures of Past Soil Forming Processes in Deserts and their Palaeoenvironmental Significance

Inherited pedogenic characters are very common in deserts. The soil cover frequently appears as a complex mosaic at all scales of juxtaposed and frequently superimposed soils, horizons, fabrics and features (eg. Courty, 1994). The diversity and great abundance of inherited pedogenic characters results from, (1) the high sensitivity of soils to climatic fluctuations, even minor ones, in arid to semi-arid environments, (2) great abundance of hardened and cemented soil horizons and fabrics such as calcretes and gypcretes resistant to erosion, with fossilized textural features which would otherwise be eroded, and (3) abundance of aeolian deposits of different kinds in deserts and mainly on their margins which contain frequently fossilized palaeosols of various kinds.

Earlier quaternary geologists thought that palaeoenvironmental history of deserts and their margins consisted of an alternation of pluvial periods characterized by soil development and arid periods during which soils were eroded and sediments, dominantly aeolian sands, were accumulated. However, further investigations have shown that the palaeoenvironmental history is far more complex. Past and present records indicate that soil forming and erosional or accretionary processes overlap. However, soils, palaeosols and many pedo-sediments indicate that this overlapping was frequently different from the present day. Moreover, the intensity and the combination of the basic pedological as well as sedimentary processes were also frequently different in the past. Evidence of abrupt events (natural fires, abrupt hydric and aeolian erosion and abrupt severe frost) is also common in deserts.

Significance of Clay Enriched Horizons

Soils with clay-enriched horizons, known as argillic and natric in the Soil Taxonomy extended on large surfaces in arid zones, especially in the deserts of the America (eg. Southard, and Southard, 1985), and also in the Sahel at the southern margin of the Sahara (Fedoroff and Courty, 1987; West et al., 1987). The importance of this type of soil is emphasized by the place given to it in the Soil Taxonomy in which, the Argid suborder, one of the two suborders of Aridisols, deals with soils having an argillic or a natric horizon. Here the genesis of some these Argids is discussed.

Development of Argillic Horizons by Accretion of Aeolian Fine Dust in Aeolian Sands

At the southern margin of the Sahara, sand-dunes have advanced some five hundred kilometres further south during two periods, viz. 35,000-29,000 yrs. BP and 21,000-17,000 yrs BP (Durand and Lang, 1986). Then soils characterized by argillic horizons developed and stabilized these dunes (Fedoroff and Courty, 1987; West et al., 1987). These argillic horizons consist of quartzitic sand grains and clay, the amount of which rarely exceeds 10%. The quartz grains are coated with a thin, reddish, dusty (up to 5 μm), moderately to well oriented clay to a variable thickness of 20-50 μm. Most of these grains are bridged with the same clay. In a few profiles, these bridges are in the form of crescentic microlaminated clay coatings. At the bottom of some channels, the sand grains are embedded in a dusty clayey mass.

Weathering of quartz grains cannot provide the clay of these argillic horizons. The regular distribution of these argillic horizons on the rolling sand-dunes excludes any fluvial origin. The only acceptable origin for these clays is consequently aeolian. The grain size distribution of these dusty clays shows that only fine dust (up to 5 μm) was transported, probably at high altitude and for long distances. The mineralogy of clay (essentially kaolinites mixed with iron oxides) of these argillic horizons indicates that the dust originated from Ultisols and Oxisols. The distribution of clays in the form of grain coatings and grain bridges has a typical morphology which results from clay translocation in suspension from the surficial part of the soil towards the B horizon. Absence of layering, irregular thickness of these clay coatings and bridges as well as moderate orientation of abundant very fine silt must be interpreted as a rapid suspension flow through the soil and a rapid sedimentation of particles. These characters also indicate heavily loaded suspensions and an absence of sorting during translocation. The few observed crescentic microlaminated clay coatings were probably formed in closed voids. The stabilization of sand-dunes by these argillic horizons indicates an increase in precipitation associated with a decrease in wind intensity, however deflation must have occurred in the source areas.

At least two phases of sand-dune stabilization by these argillic horizons occurred during Late Pleistocene and Holocene optimum in the Sahel, (Ousseini, 1987).

Development of Clayey Soils by Accretion of Aeolian Clayey Pseudosands Blown from Sebkhas (Playas)

Saline and eventually alkaline clayey pseudosands can be blown from sebkhas, with a sufficiently wide and erodible catchment basin, and sufficient rainfall to supply a large amount of fine sediments (Hachicha et al., 1987). Relict accumulations of such clayey pseudosands exist around temporary lakes and palaeolakes, although in some cases their source area cannot be ascertained. Rognon et al. (1987) suggested that the clayey aeolian accumulations in the northern Negev were deposited in the form of pseudosands blown from the continental shelf. A similar hypothesis has been proposed by Goodfriend et al. (1996) for explaining the palaeo-vertisols of Madeira.

Significance of Terrestrial Carbonate Accumulations

The most developed carbonate accumulations are observed under semi-arid mediterranean climates, in northern Africa (e.g. Wilbert, 1962; Ruellan, 1970; Elloy and Thomas, 1981), in south-western United states (e.g. Monger et al., 1991) and in southern Africa (Netterberg, 1969). Their frequency decreases progressively towards the heart of deserts. Under the dry tropics, they are apparently less frequent and in hyper-arid zones, e.g. in the southern Egyptian desert, they are absent.

The formation of carbonate enriched horizons and calcretes has been attributed to processes related to, (1) the dissolution in a surficial horizon of carbonate, either from parent material or brought in by dust, or due to the release of calcium from Ca-bearing mineral, and (2) translocation from A to B horizon, and a net accumulation in the B horizon. These carbonate accumulations were considered to be only time dependent. This model was earlier described by Hawker (1927) and then more thoroughly elaborated in the classic paper Gile et al. (1966). Recent investigations, most of them based on micromorphological analysis, e.g. Fedoroff et al. (1994) have shown that Gile's model is an oversimplification of the history of terrestrial carbonate accumulations.

The main characteristic of carbonate accretion during the Pleistocene and Holocene is its discontinuity. Phases of aggradation are separated by periods of gaining maturity i.e., micritization and dissolution which are organized in more or less well expressed cycles. A cycle or more generally a sequence, usually consists of contrasted phases which correspond to varied processes (see examples). During the Late Pleistocene and Holocene, the duration of these phases did not exceed a few thousand years. In most calcretes and other carbonate enriched soils, these phases are juxtaposed. The overall complexity of the carbonate accumulation depends on the overlapping of the successive accretionary phases. Two modes of carbonate accumulations have to be considered (Fig. 7), (1) accretion on parent material, such as loess, which was continuously aggrading and in which each phase of carbonate aggradation had no relationship with the previous, nor with subsequent phase, and (2) accretion on a parent material which was not enriched since carbonate aggradation began and on which a varied number of carbonate accretion phases overlapped. Varied intermediate types exist, in which aggradation of parent material occurred, but was not sufficient to fossilize each phase of carbonate accretion (e.g. in Pampean loess, Argentina; Imbellone and Teruggi, 1987). Overlapping of aggradional phases consequently led soil scientists (who were not able to recognize the successive phases in the field) to conclude that the carbonate accretion was time dependent.

The carbonate accumulations in Chinese loess (Loess Plateau) are a good example of the first mode. The history of carbonate accumulations for the last 135,000 years, based on Guo Zhengtang's (1990) results, is as follows :

Phase 1 which corresponds to the peak of Last Interglacial (isotopic stage 5e) is characterized by a dissolution of detrital carbonate of the loess, synchronous with a clay illuviation;

Phase 2 is characterized by $CaCO_3$ aggradation in the form of large, simple, micritic, impregnative nodules in the heart of the Plateau (Xifeng section) and a continuous, 50 cm thick horizon, with a similar fabric at the southern edge of the Loess Plateau is seen. It is considered that this micrite was supplied by newly deposited aeolian dust from which $CaCO_3$ was dissolved and reprecipitated in a lower B horizon from a temporary perched water table. This phase occurred during the isotopic stage 5d or 5c.

Phase 3 is characterized by a partial dissolution of detrital carbonates and impregnative nodules of phase 2 in the upper B horizon.

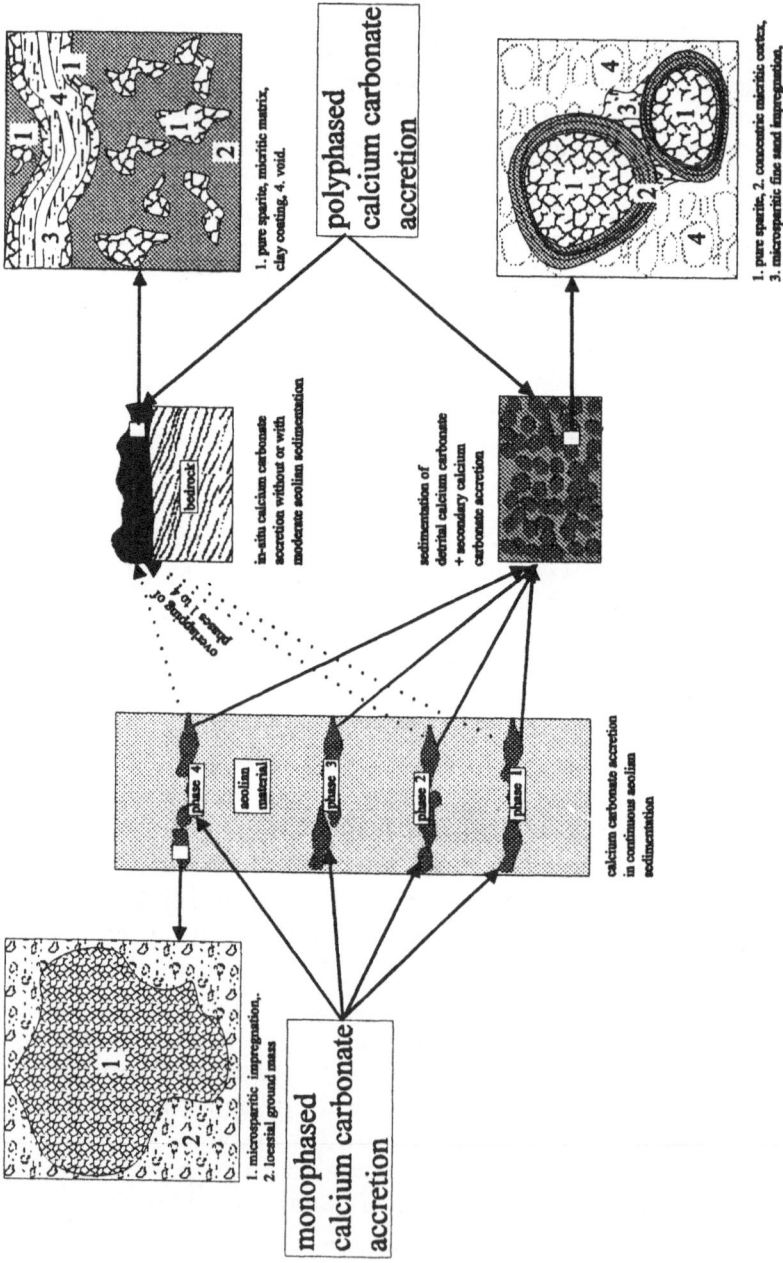

Fig. 7: Comparison of the single mode of calcitic accumulation (monophased accumulation) with the overlaping one (polyphased accumulation).
1. In-situ polyphased accumulative type; 2. Mixed detrital and in-situ accumulative type.

Phase 4 The newly deposited loess around 70,000 years ago induced inside the underlying soil aggradation of biogenic carbonates in the form of needles and micritic coatings and hypocoatings.

Phases 1 to 4 affect the palaeosol S_1, a Kastanozem. The Lower Malan loess which is free of secondary carbonates has however, supplied the biogenic carbonates of phase 4. The intermediate palaeosol, S_0 exhibits features similar to palaeosol S_1 (Phases 2, 4 and probably 3), which are, though, more weakly developed. The Upper Malan loess is similar to the lower Malan Loess in respect of the fact that it is also free of secondary carbonates. The Holocene soil S_0 exhibits a well developed phase 4 at the base of the lower sequence and a weaker phase 4 at the base of upper sequence.

Overlapping carbonate accretions (second mode of accretion) frequently become sealed by a lithified lamellar surface crust. This crusting occurred most commonly when surficial, soft horizons were eroded, bringing to the surface a harder, $CaCO_3$ cemented horizon which rapidly became stabilized by a "microphytic crust". Subsequently, dust trapped by the "microphytic crust" induced growth of successive juxtaposed crusts which progressively became lithified, and produced a lamellar crust (Fig. 8). Lithified crusts also occur inside the soil at the contact between the surficial, non cemented horizon and the $CaCO_3$ cemented horizon. These crusts are a result of juxtaposition of successive root mats, pseudomorphed by carbonates. Lichen, algae and fungi also contribute to these internal lithified crusts. This crusting was probably active during periods of accretion of carbonate rich dust. A lamellar-crust sealed calcrete is fossilized, except for a few vertical and horizontal large cracks in which roots and soil biota remain active. Lefévre (1985) has described calcretes of similar development on a sequence of terraces of increasing age from middle Moulouya valley (Morocco). Each calcrete developed during a single cycle of terrace formation, then became fossilized by the lamellar crust and therefore remained inactive during subsequent cycles.

Two historical examples of reconstruction of complex calcrete (or overlapping carbonate accretions) can be given. The first is a very old (Plio-Villafranchian, Conrad, 1969), complex calcrete (Plate 5) from the Sahara, presently under less than 100 mm of rainfall (Callot, 1987) whereas the second is a recent (probably Late Pleistocene), rather simple, calcitic accretion in the Mexican Altiplano under a subhumid monsoonal climate (Fedoroff et al., 1994).

The following phases, from oldest to youngest, have been distinguished in the Saharan calcrete (Callot, 1987).

Phase 1 is characterized by a porous sparitic fabric which appears in the field as a chalky, white soft crust which hardens at the top. These fabrics are generated in the fluctuating zone of the water table (Fig. 5.3) in depressions under subhumid climates.

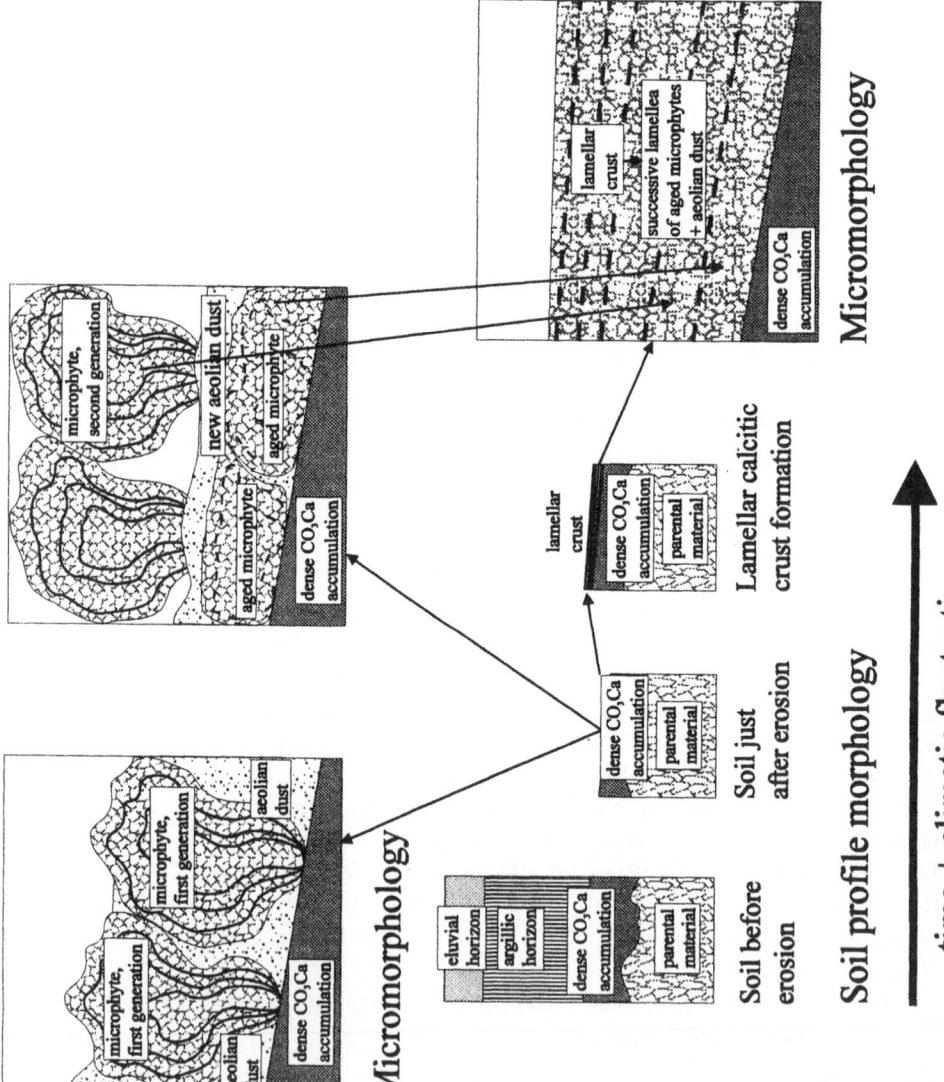

Fig. 8: Development of a calcitic lamellar crust.

Phase 2 Sparite is progressively replaced by a micritic dense fabric while biogenic calcitic features appear (channel hypocoatings, some infilled by calcitic root pseudomorphs and acicular calcite with locally, loose packing of sub-rounded micritic aggregates). During this phase, the soil was well drained. Drainage was slightly negative favouring the sparite micritization and the biogenic accretion of calcite. The depression of phase 1 had dried out, because of its incision by a valley, or due to a drier climate.

Phase 3 Large fissures of the micritic fabric are infilled by sparite. A perched water table developed during the rainy season, or the ground-water level came up. In both cases, the rainfall increased.

Phase 4 Karstic hollows (Plate 6), a few metres deep, which dot this calcrete, give evidence of periods of partial dissolution of calcrete which assumes a positive drainage and consequently indicates a humid to sub-humid climate.

Alternation of carbonate and dust accretion was followed by a partial dissolution of carbonate which lasted long probably from phase 4 to the last humid period. This alternation is registered in the surface lamellar crust and also in the walls of karstic hollows. The surface lamellar crust is, 6-10 cm thick, disjointed in the form of horizontal slabs and are locally split. In thin section, it appears as highly lithified, in which phases of accretion and dissolution cannot be precisely recognized. However, the individual layers in the walls (50 cm thick) of karstic hollows are rather easily recognizable. Approximately fifty layers have been identified. Each layer corresponds to a phase of carbonate accretion followed by its partial dissolution. Sealing of the carbonate accumulation was consequently reinforced through time. These crusts correspond to a long period during which climate fluctuated from semi-arid conditions which favoured crusting to more humid ones during which some carbonate dissolution occurred.

Phase of fossilization of karstic hollows—During this phase, karstic hollows were denuded of the original carbonate free material, which was replaced by a dense packing of rounded micritic nodules and soil micro-aggregates embedded in a tri-modal sandy matrix. Nodules and micro-aggregates have to be considered as pseudosands, and the whole infilling as an aeolian material brought into the karstic hollows by saltation from, eroding desiccated carbonated soils. Hyper-arid conditions in the Sahara were probably established during this period. A final calcitic crusting occurred which also sealed the infilled karstic hollows.

Present evolution—The hard surface crust is presently almost non-functional. It is covered by loose sands with small crust fragments which are clay, which is enriched in the form of dusty clay, poorly oriented coatings and infillings at the contact with the crust, the thickness of which does not exceed 10 cm. Laterally, this calcrete is progressively eroded, mainly by wind action and runoff.

Only three phases of carbonate aggradation could be recognized on tepetates in the Mexican Altiplano. These are,

Phase 1 Characterized by biogenic forms of carbonate - acicular calcite in deeper horizons and successive calcified root mats in the subsurface. Juxtaposition of calcified root mats indicate an increasing calcitization. Acicular calcite and root pseudomorphs were progressively replaced by micrite, i.e., the process of micritization. Carbonate was probably supplied in the form of dust from temporary lakes which partially dried out during a dry period.

Phase 2 Top of these biogenic carbonates, a polyphased lamellar crust developed (Fedoroff et al., 1994) which corresponds to a dry period which favoured development of surface crusts.

Phase 3 Corresponds to present day functioning. Occurrence of calcitic depleted features at the bottom of the Phaeozem which overlies the crusted horizon indicates that leaching of $CaCO_3$ is active under present day conditions, whereas the leached carbonates are probably fixed by algae in the "microphytic crust" present on top of the calcitic crust.

Global correlation of carbonate aggradation and degradation phases has so far not been attempted. These correlations have been established only in some areas at a regional scale, for example, the Loess plateau in China. This is due to difficulties in obtaining high resolution chronology of carbonate accumulations based on radiometric numerical dates. During the last 135,000 years, the following phases of carbonate aggradation and degradation on loess have occurred,

(1) the peak of the Last Interglacial was characterized in the Loess Plateau (as well as in the loess of western and central Europe) by a severe dissolution of carbonates,

(2) accretion of impregnative sparite or micrite took place due to perched water tables or raised groundwater during isotopic phases 5d or 5c on the Loess Plateau as well as in central Europe, but not in western Europe,

(3) during loess deposition, in China and Europe, biogenic calcitic forms were frequent,

(4) interstadial soils in China are characterized by accretion of impregnative micrite which is absent in western Europe,

(5) Holocene optimum in China is characterized by biogenic calcite forms which are absent in the western Europe. The Holocene optimum was a period favourable to carbonate aggradation in the desert margins. This aggradation was induced by a rise in groundwater in intra-dunal area and in valley sediments. It has been reported from India (Courty, 1990) and from north-east Syria (Courty, 1994).

Signature of Abrupt Events (Natural Fires, Droughts, Heavy Rain Spells) in Soils and Palaeosols of Deserts

Inherited black soils as well as black sediments have been described in the Sahara. Most of these are located in and around Holocene palaeolakes (Dutil, 1971; Hugot, 1977; Riser et al., 1983; Callot, 1987). These authors consider that the black colour resulted from humification of organic matter which occurred in poorly drained soils. Meanwhile Lefévre (1985) and Fedoroff and Courty (1989) based on micromorphology have shown that this black colour results from packed micro-charcoal fragments, while very abundant phytoliths and micritic crystals are typical for plant ashes (Wattez and Courty, 1987). Fedoroff and Courty (1989) conclude that the so-called black soils and sediments are, in fact, most frequently in-situ, sometimes wind-reworked, and rarely water-transported natural fires residues (charcoal and ashes). A precise chronology of these natural fires in the Sahara has not yet been established because of the lack of systematic radiocarbon dates. Did they occur in the Sahara in relation, either with recurring droughts or with some intense climatic event? However, in the Thar desert, Courty (1990) recognized two distinct periods of natural fires radiocarbon dated to 12,800 yrs. BP and 8,900 yrs. BP.

Weiss et al. (1993) in Djezireh (north-east Syria) and Fedoroff and Courty (1995) in the Vera basin (south-east Spain) have recognized signatures of an abrupt event in soils which occurred during the Early Bronze Age (2,300-1,960 yrs. 300 Cal BC). This event started with natural fires characterized by burnt soil matrix and charcoal, and strong winds, the signature of which is in the form of dust accumulation and pellety fabric. Then, heavy rain spells occurred which resulted in soil surface crusts and in the collapse of the soil structure due to water saturation. Simultaneously, landslides occurred in the sierras which lie above the Vera basin, while tephra and gypsum have been found in sediments related to this event in Djezireh. For the same period, Demkin (1995) describes soil alkalinization in the arid steppe of southern Russia.

In the Chinese loess, Porter and An Zhisheng (1995) have correlated grain-size maxima with ages that match those of the last six Henrich events (Henrich, 1988) while Guo Zhengtang (1990) has shown that these grain-size maxima are synchronous to, (1) very strong winds which have eroded the underlying soil when loess sedimentation was resumed - the grain-size maximum appears in thin-sections as a pellety fabric (pellets are rounded fragments of earlier soil surface) embedded in a fresh loess matrix, (2) heavy rain spells characterized by fragments of surface crusts and coarse coatings and infillings, and (3) severe frost which has produced well developed frost related features.

Conclusions

In this chapter an effort is made to demonstrate the potential of studies of soil in deserts and their margins to provide a record of environmental fluctuations. Soils react sensitively to an increase of aridity as well as to an increase in rainfall. Various climatic components can be inferred from the soil characters, e.g. yearly rain distribution, rain spells, wind intensities and evapotranspiration. Soils register climatic trends as well as short and abrupt climate changes. Moreover, palaeopedologists are able to reconstruct the palaeo-environmental history of deserts for a much longer time interval than geomorphologists and palynologists because of the existence of lithified, very old inherited soils such as calcretes, that exist in deserts. Past environmental fluctuations of deserts can be consequently reconstructed from the character of inherited soils, palaeosols and from an analysis of the present soil cover.

Most of basic soil forming processes in deserts are reasonably well understood. Of the poorly understood phenomena are the processes of silicate weathering and clay neoformation in highly saline soil solution and under high pH (~10). Geomorphologists and pedologists need to develop a joint research to better evaluate the past and present role of geomorphic and pedogenic processes in the evolution of desert landscapes. Progress has also to be accomplished in the following areas:

1. A clear separation of functional characters from inherited ones has to be made, for example, textural surface crusts are undoubtedly functional, whereas the conditions under which root calcitic pseudomorphosis as well as secondary micritization occur are not precisely known. Moreover, we have to find whether it is possible to use present functional characters as analogues for the past?

2. Chronology of pedo-sedimentary events. Pedologists have so far not achieved full use of the possibilities of radiometric dates. The time intervals in the reconstruction of pedo-sedimentary sequences must be reduced, for example, from 10^4 and even 10^5 years, as they are presently for Late Pleistocene, to 100 years and even less for the Holocene. This can be achieved for this period through close co-operation with archaeologists (Fedoroff and Courty, 1995).

3. Further search for signatures of abrupt events. This is a new research priority. One priority could be a better understanding of the frequency and intensity of past natural fires. The second priority could be research on the causes of these natural fires on one hand, and their impact on climate on the other.

References

Alaily, F. (1987). Genesis of cracks in sandy soils of the extreme arid part of the Sahara: a hypothesis. *Catena* 14 : 345-357.

Amit, R., and Harrison, J.B.J. (1995). Biogenic calcic horizon development under extremely arid conditions, Nizzana region, Negev, Israel. In: *Advances in Geoecology*. H.P. Blume and S.M. Berkowitz (ed.), 28 : 65-88.

Ambouta, K. (1984). *Contribution à l'édaphologie de la brousse tigrée de l'ouest nigérien*. Univ. Nancy I, Thèse, France,

Blank, R.R. and Fosberg, M.A. (1991). Duripans in Idaho, USA: in-situ alteration of aeolian dust (loess) to an opal-A-X-ray amorphous phase. *Geoderma* 48:131-149.

Blume, H.P., Yair, A and Yaalon, D.H. (1995). An initial study of pedogenic features along a transect across longitudinal dunes and interdunes area. Nizzana region, Negev, Israel. In: *Arid Ecosystems* H.P. Blume and S.M. Berkowitz (ed.), *Advances in Geoecology*, 28 : 51-64.

Bresson, L.M. and Boiffin, J. (1990). Morphological characterization of soil crust development stages on an experimental field. *Geoderma*, 47: 293-325.

Bullock, P., Fedoroff, N., Jongerius, A., Stoops, G., and Tursina, T., (1985). *Handbook for soil thin section description*. Waine Research Publications, Wolverhampton, United Kingdom.

Casenave, A. and Valentin, C. (1989). *Les états de surface de la zone sahélienne. Influence sur l'infiltration*. Didactiques. ORSTOM, Paris.

Callot, Y. (1987). *Géomorphologie et paléoenvironnements de l'Atlas saharien.au Grand Erg occidental : dynamique éolienne et paléolacs holocénes*. Univ. P. et M. Curie, Thèse, Paris.

Clos-Arceduc, M. (1956). Etude sur photographies aériennes d'une formation végétale sahélienne: la brousse tigrée. *Bull. IFAN*, sérieA, 7(3): 677-684.

Conrad, G. (1969). *L'évolution continentale post-hercynienne du Sahara algérien (Saoura, erg Chech-Tanezrouft, Ahnet-Mouydir)*. Centre de recherches sur les zones arides, série géologie, n° 10, CNRS; Paris.

Cooke, R., Warren, A. and Goudie, A., (1993). *Desert Geomorphology*, UCL Press, London.

Coque, R. (1962). *La Tunisie présaharienne: étude géomorphologique*. Colin, Paris.

Courty, M.A., (1990). *Environnements géologiques dans le nord-ouest de l'Inde. Contraintes géodynamiques au peuplement protohistorique (Bassin de la Ghaggar-Saraswati-Chautang)*. Thèse Univ. Bordeaux I.

Courty, M.A., (1994). Le cadre paléogéographique des occupations humaines dans le bassin du Haut-Khabbur (Syrie du nord-est). Premiers résultats. *Paléorient*, V. 20/1 : 21-59.

Courty, M.A., and Fedoroff, N. (1985). Micromorphology of recent and buried soils in Northwestern India. *Geoderma*, 35 : 287-332.

Correns, C.A. (1949). Growth and dissolution of crystals under linear pressure. *Discussions of the Faraday Society*, 5 : 267-271.

Dan, J. (1983). Soil chronosequence in Israel. *Catena* 10: 287-319.

Dan, J., Yaalon, D.H., Moshe, R and Nissim, S. (1982). Evolution of reg soils in southern Israel and Sinai. *Geoderma*, 28 : 173-202.

Durand, A. and Lang, J. (1986). Approche critique des methodes de reconstitution paléoclimatique : le Sahel nigéro-tchadien depuis 40.000 ans. *Bull. Soc. Géol. Fr.*, (8), II,2 : 267-278.

Delhoume, J.P., Montana, C. and Cornet, A. (1987). Vegetation pattern and soils in the Mapimi bolson. Part I: Vegetation arcs. *A series of the Chihuahuan desert research institute*, 13: 1-19.

Delhoume, J.P., Montana, C. and Cornet, A. (1988). Vegetation pattern and soils in the

Mapimi bolson. Part II: Polygonal patterns. *A series of the Chihuahuan desert research institute*, 14 : 1-15.

Demkin, V.A. (1995). Temporal and spatial regularities governing the development of solonetz process in dry and desertic steppes soils (in russian). *Pochvovédinié*, 5 : 533-540.

Dutil, P. (1971). Contribution à l'étude des sols et des paléosols du Sahara. Thése Science, Univ. Strasbourg.

Eisenberg, J., Dan, J. and Koyumdjisky, H. (1982). Relationships between moisture penetration and salinity in soils of the northern Negev (Israel). *Geoderma* 28: 313-344.

Elloy, R. and Thomas, G. (1981). Dynamique de la genèse des croûtes calcaires (calcrétes) développées sur séries rouges pléistocènes en Algérie nord-occidentale. Contexte geomorphologique et climatique. Pétrographie et géochimie. *Bull. Centre Rech. Explor.-Prod. Elf-Aquitaine*, 5 : 53-112.

Fedoroff, N. (1991). Possibilities of paleopedology for paléoenvironmental reconstructions. Special Proceedings. *Review reports. For symposia of the XIIIth INQUA Inter. Congress,* Beijing, China, : 117-120.

Fedoroff, N., and Courty, M.A. (1987). Morphology and distribution of textural features in arid and semi-arid regions. In: *Micromorphologie des sols/ Soil Micromorphology,* N. Fedoroff, L.M. Bresson et M.A. Courty (ed.): 213-220. Proceedings of VIIth Inter. Working Meeting Soil Micromorphology, AFES, Plaisir.

Fedoroff, N., and Courty, M.A. (1989). Indicateurs pédologiques d'aridification. Exemples du Sahara. *Bull. Soc. géol. France*, 8, T.V, n° 1: 43-53.

Fedoroff, N., and Courty, M.A. (1994). Organisation du sol aux échelles microscopiques. In: *Pédologie. 2. Constituants et propriétés du sol*, M. Bonneau, B. Souchier (eds.), 2éme édition, Masson, Paris : 349-375.

Fedoroff, N., Courty, M.A., Lacroix, E. and Oleschko, K. (1994). Calcitic accretion on indurated volcanic materials (example from tepetates, Altiplano, Mexico). *15th Inter. Congress of Soil Science*, Acapulco, Mexico, V. 6a: 460-473.

Fedoroff, N. and Courty, M.A. (1995). Le rôle respectif des facteurs antropiques et naturels dans la dynamique actuelle et passée des paysages méditerranéns. Cas du bassin de Vera, sud-est de l'Espagne. In *"L'Homme et la dégradation de l'Environnement"*. XVéme Rencontres Internationales d'Archéologie et d'Histoire d'Antibes. Editions APDCA, Juan-les-Pins: 115-141.

Fedoroff, N., Baroix, I and Ordaz, V. (1997). Multi-disciplinary approach of soil functioning versus genesis under tropical rain forest (a case study in the littoral plain of Tabasco, Mexico). *European Journal of Soil Science* (in press).

Flach, K.W., Nettleton, W.D. and Chadwick, O. (1992). The criteria of duripans in the US soil taxonomy and the contribution of micromorphology to characterize silica indurated soils. In: *Terra. Suelos Volcanicos Endurecidos*. Mexico, 10: 34-45.

Gile, L.H., Petersen, F.F. and Grossman, R.B. (1966). Morphological and genetic sequences of carbonate accumulation in desert soils. *Soil Sci.* 101: 347-360.

Goodfriend, G.A., Cameron, R.A.D., Cook, L.M., Courty, M.A., Fedoroff, N., Livett, E. and Tallis, J. (1996). The Quaternary aeolian sequence of Madeira: stratigraphy, chronology and palaeoenvironmental interpretation. *Palaeogeography, Palaeoclimatology, Palaeoecology*, 120: 195-234.

Goudie, A.S. and Wells, G.L. (1995). The nature, distribution and formation of pans in arid zones. *Earth Science Reviews*, 38: 1-69.

Guo Zhengtang, (1990). *Succession de paléosols et des loess du Centre-Ouest de la Chine. Approche micromorphologique*. Thèse Univ. P. et M. Curie, Paris.

Hachicha, M., Stoops, G. and M'hiri, A. (1987). Aspects micromorphologiques de l'évolution des sols de lunettes argileuses de Tunisie. In: *Micromorphologie des sols/ Soil*

Micromorphology, N. Fedoroff, L.M. Bresson and M.A. Courty (eds.), pp : 193-197, Proceedings of VIIth Inter. Working Meeting Soil Micromorphology, AFES, Plaisir.

Hairsine, P.B. and Hook, R.A. (1994). Relating soil erosion by water to the nature of the soil surface. In: *Sealing, Crusting and Hardsetting soils: Productivity and Conservation*, H.B. So, G.D. Smith, S.R. Raine, B.M. Schafer and R.J. Loch (eds.), pp : 77-91 Australian Soc. Soil Sci. Inc. (Queensland branch), Brisbane, Australia.

Harden, J., Taylor, E.M., Reheis, M.C. and McFadden, L.D. (1991). Calcic, gypsic and siliceous soil chronosequences in arid and semi-arid environments. In: *Occurrence, characteristics and genesis of carbonate, gypsum and silica accumulations in soils*, W.D. Nettleton (ed.) SSA Special publication, Madison, USA, 26 : 1-16.

Hawker, H.W., (1927). A study of the soils of Hidalgo County, Texas and the stages of their lime accumulation. *Soil Sci.* 23 : 475-485.

Henrich, H. (1988). Origin and consequences of cyclic ice rafting in the northeast Atlantic ocean during the past 130,000 years. *Quaternary Research*, 29 : 142-152.

Herrero, J., Porta, J. and Fedoroff, N. (1992). Hypergypsic soil micromorphology and landscape relationships in Northeastern Spain. *Soil Sci. Soc. of America Journal*, pp : 1188-1194.

Hugot, G. (1977). Un secteur du Quaternaire lacustre mauritanien : Tichitt (Aouker). Elements pour servir à une étude géomorphologique. *Mémoire Inst. mauritanien de la Recherche scientifique (section préhistoire)*, Mauritanie.

Imbellone, P.A. and Terrugi, M.E. (1987). Discontinuous calcretes in loessic paléosols near La Plata, Argentina. In: N. Fedoroff, L.M. Bresson et M.A. Courty (eds.) pp : 625-630 *Micromorphologie des sols/Soil Micromorphology*. Proceedings of VIIth Inter. Working Meeting Soil Micromorphology, AFES, Plaisir.

Jones, C.G. and Shachak, M. (1990). Fertilization of the desert by rock eating snails. *Nature* 346 : 839-841.

Lefévre, D. (1985). *Les formations plio-pléistocénes du bassin de Ksabi (moyenne Moulouya, Maroc)*. Thèse, 3ème cycle, Univ. Bordeaux I.

Milnes, A.R., Wright, M.J. and Thiry, M. (1991). Silica accumulations in saprolites and soils in South Australia. In: *Occurrence, characteristics and genesis of carbonate, gypsum and silica accumulations in soils*. W.D. Nettleton (ed.). SSA Special publication, Madison, USA, 26 : 121-149.

Monger, H.C., Daugherty, L.A. and Gile, L.H. (1991). A microscopic examination of pedogenic calcite in an Aridisol of southern New-Mexico. In: *Occurrence, characteristics and genesis of carbonate, gypsum and silica accumulations in soils*. W.D. Nettleton (ed.). SSA Special publication, Madison, USA, 26 : 37-60.

Nettenberg, F. (1969). Ages of calcrete in southern Africa. *South African Archaeological Bull.* 24 : 88-92.

Ousseini, I. (1987). *Etude de la répartition des formations sableuses et interprétation des dépôts éolien sdans le Liptako oriental (République du Niger)*. Univ. P. et M. Curie, Thèse 3ème cycle, Paris.

Pedro, G., (1994). Les conditions de formation des constituants secondaires. In: *Pédologie. 2. Constituants et propriétés du sol*. M. Bonneau et B. Souchier (eds.), 2ème édition, Masson, Paris : 65-78.

Phillips, S.E., Milnes, A.R. and Forster, R.C. (1987). Calcified filaments: an example of biological influences in the formation of calcrete in south Australia. *Aust. J. Soil Sci.* 25 : 405-428.

Porter, S.C. and An Zhisheng. (1995). Correlation between climatic events in the North Altlantic and China during the last glaciation., *Nature* 375 : 305-308.

Riser, J., Hillaire-Marcel, C and Rognon, P. (1983). Les phases lacustres holocènes. In: N. Petit-Maire and J. Riser (eds.), *Sahara ou Sahel. Quaternaire récent du bassin de Taoudenni (Mali)*. Lab. Géologie du Quaternaire du CNRS, Marseille : 65-86.

Rodriguez Hernandez, M.C, Fedoroff, N., Fernandez-Caldas, E. and Quantin, P. (1979). Les Vertisols des Iles Canaries occidentales. Etude physico-chimique, minéralogique et micromorphologique. *Pédologie* 29 : 71-107.

Rognon, P., Coude-gaussen, G., Fedoroff, N. and Goldberg, P. (1987). Micromorphology of loess in the northern Negev (Israel). In: N. Fedoroff, L.M. Bresson M.A. Courty (eds.): 631-638, *Micromorphologie des sols/Soil Micromorphology*. Proceedings of VIIth Inter. Working Meeting Soil Micromorphology, AFES, Plaisir.

Ruellan, A. (1970). *Les sols à profil calcaire différencié des plaines de la Basse Moulouya (Maroc oriental)*. Mémoire ORSTOM 54.

Shachak, M., Jones, C.G. and Granoty, Y. (1987). Herbivory in rocks and the weathering of a desert. *Science* 236 : 1098-1099.

Shachak, M., Jones, C.G. and Brand, S. (1995). The role of animals in an arid ecosystem: snails and isopods as controllers of soil formation, erosion and desalinization. In: *Arid Ecosystems*, H.P. Blume and S.M. Berkowitz (eds.) *Advances in Geoecology* 28 : 37-50.

Singer, A. (1989). Palygorskite and sepiolite group minerals. In: *Mineral in Soil Environments*. J.B. Dixon and S.B. Weeb (eds.) pp : 829-872, Soil Sci. Soc. Am., Madison, U.S.A.

Singer, A. (1995). The mineral composition of hot and cold deserts. In: *Arid Ecosystems*. H.P. Blume and S.M. Berkowitz (eds.), *Advances in Geoecology* 28 : 13-28.

Soil Survey Staff (1992). *Keys to Soil Taxonomy*. SMSS Technical Monograph n° 19. Fifth edition. Pocahontas Press Inc. Blacksburg, Virgina, USA.

Southard, R.J. and Southard, A.R. (1985). Genesis of cambic and argillic horizons in two northern Utah aridisols. *Soil Sci. Soc. Am. J.* 49 : 167-171.

Steinberger, Y. (1995). Soil fauna in arid ecosystems: their role and functions in organic matter cycling. In: *Arid Ecosystems*. H.P. Blume and S.M. Berkowitz (eds.), *Advances in Geoecology*. 28 : 29-36.

Thiry, M. (1992). Pedogenic, silicifications: structures, micromorphology, mineralogy and thier interpretation. In: *Terra. Suelos Volcanicos Endurecidos*. Mexico, 10: 46-59.

Valentin, C. (1986). Surface crusting of arid sandy soils. In: *Assessment of soil surface sealing and crusting*. F. Callebaut, D. Gabriels, and M. De Boodt (eds.) pp: 40-47, Research Centre for Soil Erosion and Soil Conservation, Gent, Belgium

Valentin, C. (1994). Sealing, crusting and hardsetting soils in sahelian agriculture. In: *Sealing, Crusting and Hardsetting soils: Productivity and Conservation*, H.B. So, G.D. Smith, S.R. Raine, B.M. Schafer and R.J. Loch (eds.) pp. 53-76, Australian Soc. Soil Sci. Inc. (Queensland branch), Brisbane, Australia.

Wattez, J., and Courty, M.A. (1987). Morphology of ash of some plant materials. In: *Micromorphologie des sols/ Soil Micromorphology*. N. Fedoroff, L.M. Bresson and M.A. Courty (eds.) pp. 677-683: Proceedings of VIIth Inter. Working Meeting Soil Micromorphology, AFES, Plaisir.

Weiss, H., Courty, M.A., Wetterstrom, W., Guichard, F. Senior, L. Meadow, R. and Curnow, A. (1993). The genesis and collapse of third millenium north Mesopotamian civilization. *Science* 261 : 995-1004.

West, L.T., Wilding L.P., and Calhoun, F.G. (1987). Argillic horizons in sandy soils of the Sahel, West Africa. In: *Micromorphologie des sols/Soil Micromorphology*. N. Fedoroff, L.M. Bresson and M.A. Courty (eds.) pp : 221-225 Proceedings of VIIth Inter. Working Meeting Soil Micromorphology, AFES, Plaisir.

West, N.E. (1990). Structure and function of microphytic soil crusts in wildland ecosystems of arid to semi-arid regions. *Advances in Ecological Research*, Academic press, 20 : 179-222.

Wilbert, J. (1962). Croûtes et encroûtements calcaires au Maroc. *Al Awamia*, 3 : 175-192.

Wright, V.P. (1990). A micromorphological classification of fossil and recent calcic and petrocalcic microstructures. In: *Soil Micromorphology. A basic and applied science*. L.A. Douglas (ed.) *Developments in Soil Science*, 19 : 401-407.

5 Semi-Arid/Arid Zone Calcretes: A Review

S.K. Tandon[*] and Sushil Kumar[**]

ABSTRACT

The methods and concepts required for understanding the properties and genesis of semi-arid/arid zone calcretes are described. Calcretes are near-surface terrestrial, secondary calcium carbonate accumulations in soil profiles, bedrocks, and sediment. They commonly occur in both Quaternary and pre-Quaternary continental and marginal marine sequences and are of value in palaeoenvironmental and palaeoclimatic interpretation.

Without pretending to be comprehensive, the present review provides a brief coverage of the aspects of calcrete classification, profiles, maturity, hydrological setting, micromorphology, cathodoluminescence signatures, mineralogy and chemistry, mode of introduction of carbonates, and the environments and rates of calcrete formation. Additional aspects include the climatic/palaeoclimatic implications of calcretes, the use of stable isotopic compositions of calcretes for palaeoenvironmental interpretation and calcrete dating.

Through its interdisciplinary character, the study of calcretes improves our understanding of the evolution of Quaternary landscapes and can be potentially applied to the stratigraphic record from the Early Proterozoic onwards—the time of the first occurrence of calcretes.

Introduction

Calcrete is a term widely used for near-surface terrestrial, secondary calcium carbonate accumulation in soil profiles, bedrocks and sediment introduced by displacive/replacive and passive modes of precipitation (Goudie, 1983; Wright and Tucker, 1991). Caliche, kunkar, capstone and nari etc. are other names used for these carbonate accumulations. The term calcrete excludes tufas, travertines, beachrocks, lake carbonates and aeolianites.

Although calcretes may occur in subhumid, temperate and polar regions, they are widespread only in semi-arid regions (rainfall zone of

---

[*]Department of Geology, Delhi University, Delhi, India.
[**]Present Address: Atomic Minerals Division, Shillong, India.

400-600 mm), where $CaCO_3$ precipitates under conditions of net annual moisture deficit (Goudie, 1983). Groundwater calcretes may, however, occur in annual rainfall zones of 1000-1500 mm (Seminuik and Searle, 1985).

These secondary accumulations of calcium carbonate cover an estimated 20 million square kilometres, that is, about 13% of the earth's present-day land surface (Yaalon, 1988). So calcretes constitute a sharp geochemical tool that shapes the landscape in semi-arid areas (Nahon, 1976). Calcretes are preserved in various stages of formation, and retain a range of physical and chemical features (Harden and Taylor, 1983; Machette, 1985; Mcfadden and Tinsley, 1985). Calcretes also occur commonly in Quaternary and ancient continental and marginal marine sequences (Goudie, 1973; Reeves, 1976; Allen, 1974, 1986; Wright and Tucker, 1991; Wright, 1994, Tandon et al., 1998).

The common occurrence of calcretes in the geological record and their use for palaeoenvironmental reconstruction, both in Quaternary and pre-Quaternary sequences, has evoked considerable interest from geomorphologists, stratigraphers, sedimentologists, pedologists, archaeologists, and palaeoclimatologists. This interest has been enhanced by the value of calcretes, as non-traditional constructional and road material, in cement and sugar industries, and in agricultural development and mineral exploration (Anand et. al., 1997).

Previous accounts of the genesis and properties of calcretes are given by Reeves (1976), Braithwaite (1983), Goudie (1983, 1996), Esteban and Klappa (1983), Wright and Tucker (1991), and Milnes (1992). These reviews are quite comprehensive, yet the present review is offered because the subject has grown from the diverse contributions of stratigraphers, sedimentologists, pedologists, geochemists, geomorphologists, and Quaternary scientists. Calcrete study is a subject which is actively pursued both in Quaternary and pre-Quaternary sequences, and the literature from both these major streams is surveyed. Interpretations of calcretes have recently been refined by determining their carbon and oxygen isotopic composition, and luminescence characteristics. A summary of calcrete classification schemes, including that on calcareous palaeosols (Mack et al., 1993) is also given.

Classification

One of the simplest means of classification is based on the morphological features of calcretes. Terms such as laminar, platy, blocky, bedded and massive were used (Gile, 1961) to describe the calcic horizons. Reeves (1976) provided a detailed list of morphological features observed in calcretes of South-western USA. These are breccia conglomerate, buckle cracks, concretions, fractured cobbles, glaebules, honeycomb, irregular masses, laminae, nodules, papules, pedodes, pipes, pisolites, plates, septaria, and slickensides.

Netterberg (1967, 1969, 1980) proposed a classification for South African calcretes which has been used in most other semi-arid regions of the world. Netterberg's studies included both geotechnical and genetic characteristics. He classified these carbonate accumulations into calcareous soils, calcified soils, powder calcretes, glaebular calcretes, cutans, pedotubules, honeycomb calcretes, hardpan calcretes, calcrete boulders and calcrete cobbles. In cases where roots are the main factor causing calcification, Wright et al. (1995) proposed a category termed rhizogenic calcretes.

In view of the extensive literature on the morphological attributes of calcretes and their analysis (Netterberg, 1980; Goudie, 1983; Esteban and Klappa, 1983; Wright and Tucker, 1991) no further comments are required. However, some common macromorphological features are illustrated in Figures 1a and 1b.

Profiles

The vertical succession of morphologically distinct layers or horizons is termed as a caliche profile (Esteban and Klappa, 1983). An idealised calcrete profile (Fig. 2) consists of host rock—transitional zone—chalky-nodular-platy—hardpan morphologies overlaid by an active soil. Generally, calcrete profiles are incomplete. Stacks of multiple calcrete profiles (Fig. 3) are recognised in the geologic record.

Boundaries between horizons of a profile tend to be gradual rather than abrupt although the main elements of a profile are quite distinct. Due to the irregularity of surfaces and lateral impersistence of some of the morphological layers, upper layers may be directly juxtaposed on lower ones with omission of intervening layers.

Read (1974) and Arakel (1982) emphasized the profile approach and recognised specific horizons such as pisolitic soils at the land surface, and laminar, massive and mottled calcretes lower down in the profile. Milnes (1992) outlined the profile characteristics of the calcretes of the Neogene coasts of Australia. Typically, calcretes in these sequences constitute a cómplex assemblage of horizons and are up to 5 m thick. They consist of a partly indurated, reddened palaeosol sequence at the base and indurated, clast-rich slope deposits above. Thin layers of lithoclasts occur in the palaeosol and the upper part of the palaeosol sequence consists of scattered lithoclasts and rhizoliths. Upper parts of some calcrete profiles are considerably disrupted through dissolution and

Fig. 1a: Outcrop photograph of mature calcrete profiles (Quaternary?) near Ladnu, Rajastan, India; a) note the development of a profile from powdery dissemination of calcrete to nodular calcrete and a prominent hardpan with pronounced prismatic structure; these mature profiles possibly represent durations of the order of 10^5 years; note also the well-developed buckle cracks.

Fig. 1b: Outcrop photograph of nodular and rhizocretionary calcrete overlying massive groundwater calcrete (Quaternary sequence near Jodhpur, Rajasthan). Courtesy Dr. J.E. Andrews.

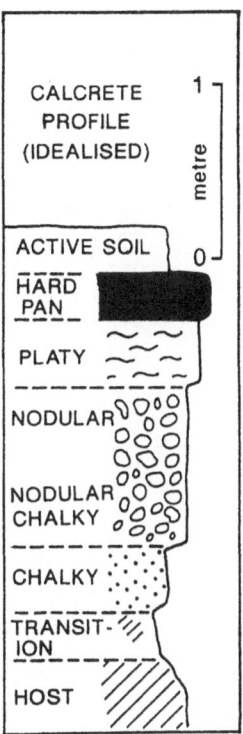

Fig. 2: Idealised calcrete profile consisting of a vertical succession of morphologically distinct horizons (after Estaban and Klappa, 1983).

the effects of plant roots (Klappa, 1980 b). The main horizons (Milnes, 1992) are carbonate mottled horizons, laminar structured zones, hardpan, brecciated, and laminar calcrete.

In calcretes without horizon development, vertical variations in attributes can still be examined. In thick nodular calcretes, variations in the size and density of nodules, nodule-host sediment ratio and the degree of coalescence of nodules give rise to zones without the formation of distinct horizons (Fig. 4).

Maturity

Closely connected to the profile aspects is the concept of development of calcretes in a series of stages. Progressive stages are based on increasing amounts of calcium carbonate due to increase in the size and density of carbonate nodules through continued replacement or displacement of the non-carbonate host sediments. Maturity-related classifications, constituting

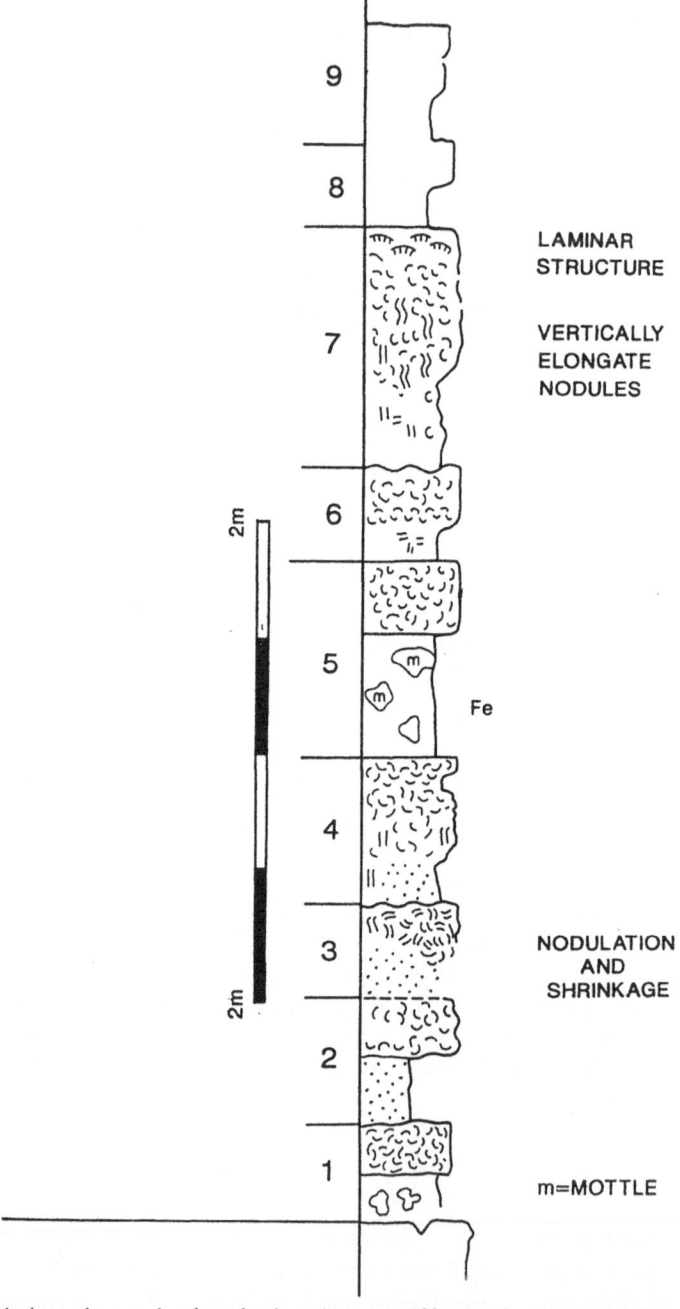

Fig. 3: Litholog of a stack of multiple calcrete profiles in the Maastrichtian Mottled Nodular Beds, central India; superimposition of profiles implies alternating periods of aggradation and pedogenesis (after Sood, 1995).

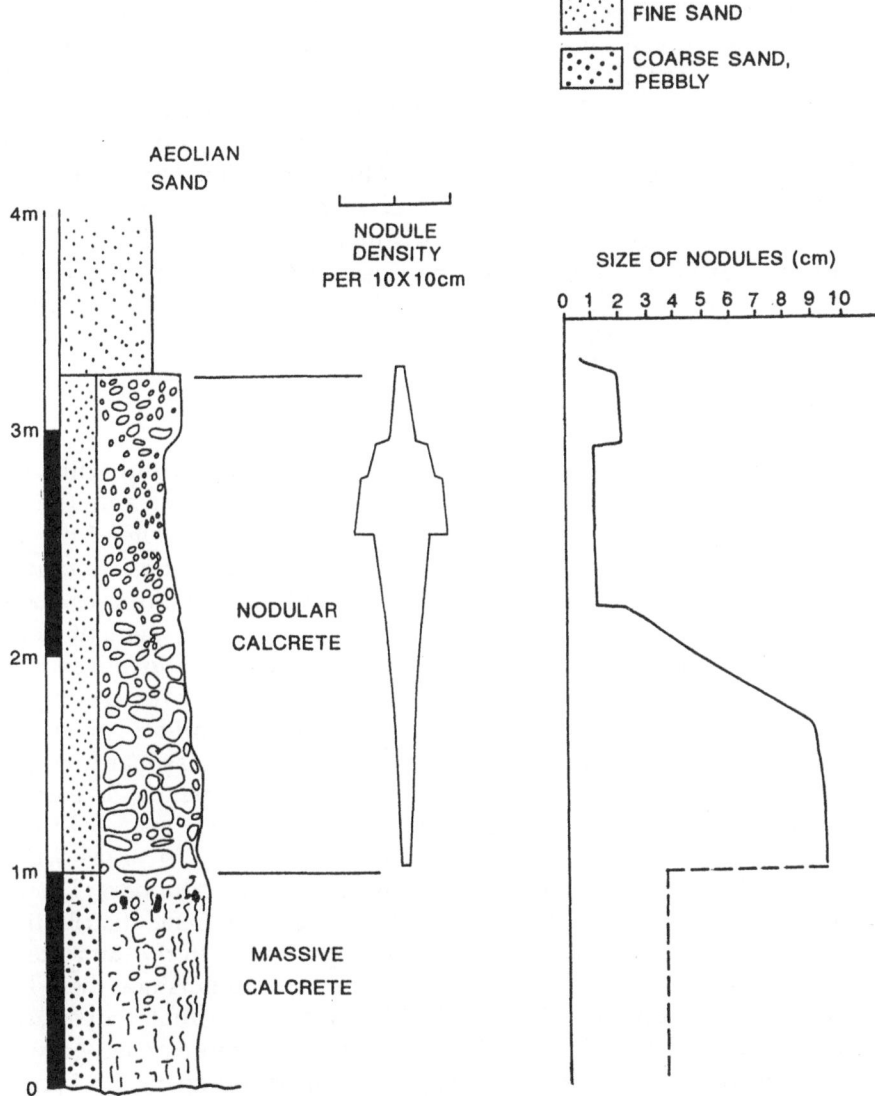

FINE SAND

COARSE SAND, PEBBLY

Fig. 4: Section of nodular calcretes showing vertical variations in nodule properties such as nodule size and density; subdivision into laterally impersistent zones is possible (Late Quaternary, Jodhpur, India).

a chronosequence, have been suggested for both Quaternary and pre-Quaternary sequences (Gile et al.,1966; Allen, 1974; Steel, 1974, Machette,1985). Gile et al. (1966) classified calcretes of desert soils into four stages of morphogenetic sequences on the basis of carbonate morphologies associated with increasing amounts of authigenic carbonate.

Machette (1985) proposed a more comprehensive classification (Table 1) with two additional morphological stages. These advanced stages are commonly formed in the Middle Pleistocene and older soils in the calcic soils of the south-western USA. Stage 5 morphology is characterized by thick laminae and incipient pisolites; stage 6 morphology reveals products of 'multiple cycles of brecciation, pisolite formation, and wholesale relamination and recementation of breccia fragments' (Machette,1985). Care must, however, be exercised in distinguishing between fine-grained sand and carbonates overprinted by desiccation-related processes and mature calcretes, a point emphasised by Esteban and Klappa (1983), Platt (1989) and Wright and Tucker (1991).

Classification: Hydrological Setting

Carlisle (1980, 1983) classified calcretes by their relationship with soil moisture in the vadose and phreatic zones. Pedogenic calcretes are formed by the precipitation of carbonates leached by meteoric water from the upper soil horizon (cf. Goudie,1983). Non-pedogenic calcretes are formed by the precipitation of carbonates in the host soil, sediment or rock above a perched or permanent groundwater table, carbonate cement being introduced to the host soil/sediment by absolute accumulation. In contrast to pedogenic calcretes, capillary fringe and groundwater non-pedogenic calcretes may form at depths of several metres. In semi-arid tracts, for example in parts of India (Dhir et al., 1992) a strong seasonality with a prolonged dry season drives formation of non-pedogenic calcretes at depths of up to tens of metres. These groundwater calcretes are difficult to distinguish from pedogenic calcretes, particularly in areas where fluctuating water tables result in a moving soil/subsoil boundary. Despite some overlapping attributes, vertical trends in carbonate distribution, carbonate-parent rock relationships, petrographic criteria, and stable isotopic composition of carbonates are useful in distinguishing these two broad calcrete categories. Slate et al. (1996) recognised that vadose zone palaeosols show a morphology and texture similar to those of non-gravelly well-drained Quaternary soils developed in semi-arid regions. Carbonate-enriched zones occur beneath clay-rich, carbonate depleted zones. Palaeosols, formed within the influence of the groundwater table (hydromorphic), show large 2-10 cm size carbonate segregations in a non- calcareous to weakly calcareous matrix. Such vadose zone non-pedogenic calcretes are termed hydromorphic calcretes (Slate et al., 1996) and show 'pedogenic' features such as a massive to nodular carbonate horizon, mottling and gleying. Further, Slate et al. (1996) argued that interpretation of drier climates and slower sedimentation based on the model of Gile et al. (1966) should be made only for those palaeosols

Table 1: Stages of Calcium Carbonate Morphology in Calcic Soils and Pedogenic Calcretes (from Machette, 1985)

Stage	Gravel Content	Diagnostic Morphological Features	CaCO$_3$ Distribution	Maximum CaCO$_3$ Content	Remarks
1	High (> 50%)	Thin discontinuous coatings on pebbles, usually on underside	Coatings sparse to common	Trace to 2%	
	Low (< 50%)	Few filaments in soil or faint coatings on ped surfaces	Filaments sparse to common	Trace to 2%	
2	High	Continuous, thin to thick coatings on pebbles	Some carbonate in matrix	2-10%	
	Low	Nodules, Soft, 0.5 cm to 4 cm in diameter	Nodules common, matrix noncalcareous to slightly calcareous	4-20%	
3	High	Massive accumulations between clasts, cemented in advanced forms	Continuous in matrix (k fabric)	10-25%	K is a carbonate soil horizon
	Low	Many coalesced nodules, matrix is firmly to moderately cemented	Continuous in matrix (k fabric)	20-60%	
4	Any	Thin (<0.2 cm) to thick (1 cm) laminae in upper part of Km horizon	Cemented, platy to weak tabular structure, Km horizon is 0.5 - 1 m thick	>25 in high gravel content; >60 in low gravel content	
5	Any	Thick laminae (> 1 cm); small to large pisoids; laminated carbonate coats soil surfaces	Indurated, dense, strong platy to tabular structure, Km horizons 1-2 m thick	>50 in high gravel content; >60 in low gravel content	
6	Any	Multiple generations of laminae, breccia and pisolites, recemented	Indurated, dense, thick strong tabular structure, Km horizon >2 m thick	>75 in all gravels	

where the evidence indicates vadose-soil processes (Slate et al., 1996), and must not be extended to vadose zone non-pedogenic calcretes.

Micromorphological Classification

A wide variety of microfabrics occur in calcretes, with two end-member types. Alpha-calcretes consist of floating, etched, exploded siliciclastic grains, spar filled (crystallaria) cracks—planar, wedge, circumgranular to partly circumgranular, and large euhedral crystals in a dense micritic to microsparitic groundmass. Beta-calcretes have microfabrics consisting of calcified filaments, needle fibre calcretes, alveolar and alveolar-septal fabrics, microbial tubes, rhizocretions, root mats and calcified faecal pellets (Bal, 1975 a,b; Klappa, 1978, 1979; Cohen, 1982; Mount and Cohen, 1984, Wright and Tucker, 1991)

Calcic features associated with arid zone soils/palaeosols (aridisols/calcic vertisols) mostly exhibit alpha fabrics. The nature of the substrate—carbonate or non-carbonate—on which the calcretes develop, and the latitudinal-climatic setting appear to exercise some control on the distribution of beta-calcretes. In the geological record, beta-calcretes occur most commonly in subaerial exposure surfaces within shallow marine carbonate sequences (cf. Wright,1994).

Palaeosol Classification

Calcretes and calcrete nodules occur commonly in palaeosols (Sehgal and Stoops, 1972; Allen, 1974; Steel, 1974; Hubert, 1978; Blodgett, 1988; Mack et al., 1992; Driese et al., 1992; Wright et al., 1995; Tandon et al., 1995; Tandon and Gibling, 1997). Mack et al. (1993), therefore, introduced a new order—Calcisol—in their palaeosol taxonomy. According to them, 'any palaeosol in which a calcic horizon is a prominent horizon may be classified as calcisol. A calcic horizon is defined as a subsurface horizon which is calcium carbonate or dolomite rich. If a calcic horizon is not the most prominent feature, then the term calcic should be used as subordinate modifier (eg. calcic vertisol)'.

In addition to these, a classification of calcretes and dolocretes based on dolomite content was proposed by Netterberg (1980).

Micromorphology

Micromorphologic (petrographic) studies are important in understanding the genesis of calcretes and for their recognition in the geologic record. A variety of microfabrics occur in calcretes of all ages (Precambrian to Recent). Pedogenic calcite nodules, calcans, neocalcans, and crystallaria show micromorphological features such as fine-grained micrite glaebules, calcans-neocalcans-quasi-calcans, acicular calcite needles, coarse-grained spar crystallaria, poikilotopic crystals and sparry blocky calcite (Drees and Wilding,1987). Following initial efforts (Brewer, 1964; Nagtegaal, 1969; Sehgal and Stoops, 1971; Bal, 1975a,b), several significant contributions to calcrete petrographic and micromorphological studies (Wright and Tucker, 1991 and references therein) have been made.

Alpha-calcretes, which are predominantly physicochemical precipitates, are more common in aridisols and in non-carbonate host rocks. In beta-calcretes, carbonate accumulation is aided by biological activities of remains of plants, insects and micro-organisms (Esteban, 1974; Kahle, 1977; Klappa, 1978, 1980; Esteban and Klappa, 1983; Phillips et al., 1987).

Alpha-calcretes

These show the K fabric, that is dense, continuous, micritic to microsparitic groundmass (Fig. 5a), crystallaria (including circumgranular forms) (Fig. 5b), floating and etched skeleton grains (Fig. 5c), large intercalary crystals, rhombic crystals and mottled fabrics (due to areas with different densities or sizes of calcite crystals).

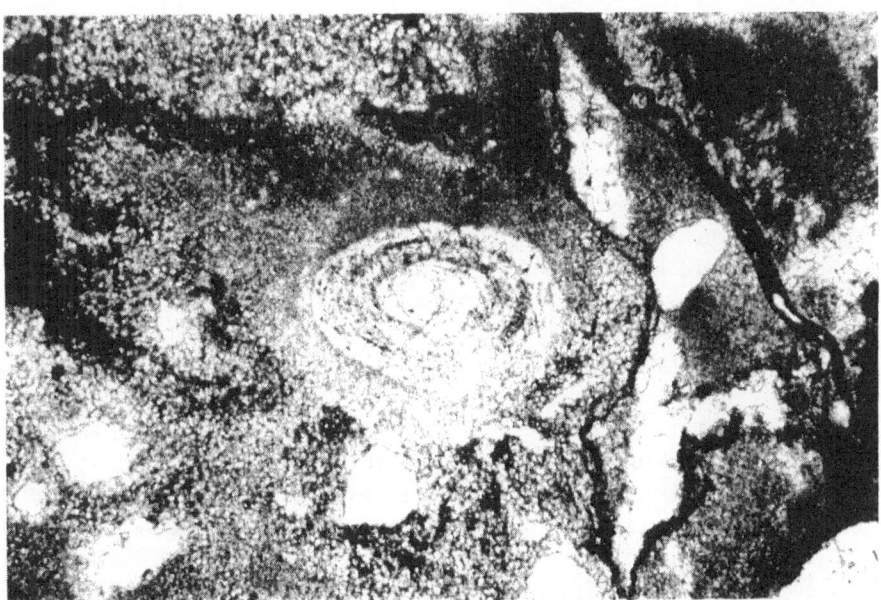

Fig. 5(a): Photomicrograph of dense micrite groundmass of an alpha-calcrete from the Late Devonian-Early Carboniferous Arran Cornstone formation, Scotland. Note also the floating quartz grains, the disorthic nodule with well-defined concentric laminae, spar rims around the quartz grains and the zones of crystal size mottling. Disorthic nodule ~250μm.

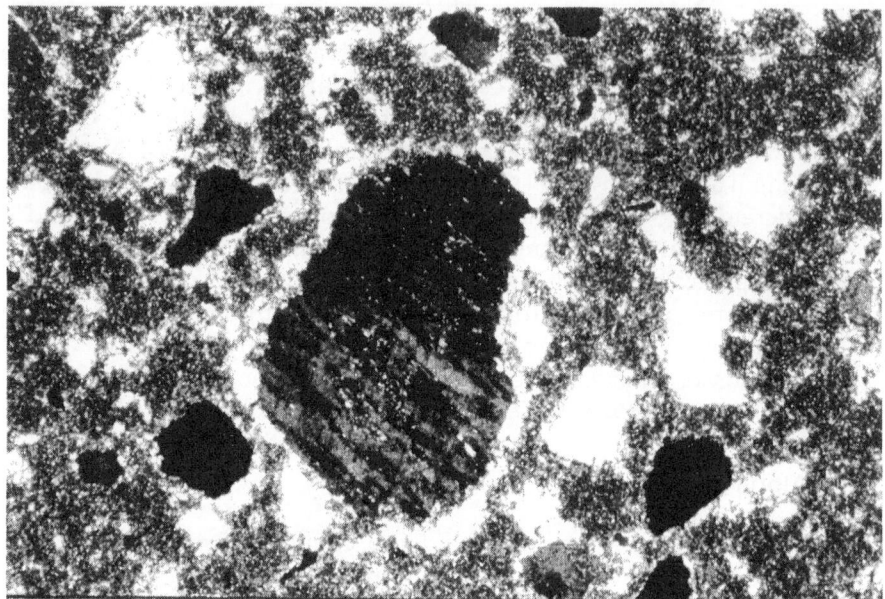

Fig. 5(b): Photomicrograph of floating fabric in alpha-calcretes (Maastrichtian calcretes, Jabalpur, Central India). Note the corroded and frayed margins of the feldspar and quartz grains; calcite spar fills the shrinkage spaces around grains. Feldspar grain ~200 μm.

They show plasmic fabrics, that is a spatial arrangement of plasma (groundmass < 2 μm) or skeletal grains depending on grain size and genesis (Brewer, 1976).

Two types of fabrics are recognised on the basis of intercrystal distance and relation of crystals to the soil matter:

1. Crystic plasmic fabric — crystals are arranged very close to each other.
2. Intercalary crystals — single large crystals are set in the soil materials at rather large mutual distances as compared to crystic plasmic fabric (Brewer, 1964; Bal, 1975 a, b).

Plasmic fabrics are of two types — Simple fabric or Fibrous plasmic fabric. Simple fabrics are further divided into Crystic or Calcic depending on the distance between crystals. Fibrous plasmic fabric is mainly composed of needle-shaped calcite (Lublinite).

Alpha-fabrics are developed in non-carbonate host rock in which calcite forms an adhesive bond with non-carbonate or silicate grains, causing displacement during crystallization (Chadwick and Nettleton, 1990). Displacive growth is a major process in calcrete formation. Floating grains show evidence of grain expansion (Braithwaite, 1989), exploded micas (Saigal and Walton, 1978), fracturing (Buczynski and Chafetz, 1987), and etching reflecting grain replacement (Nagtegaal, 1969).

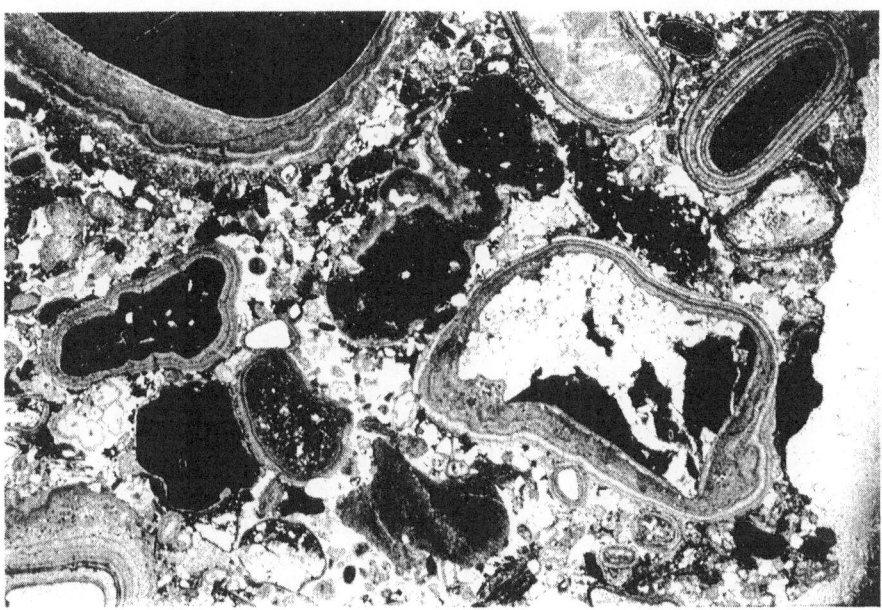

Fig. 5(c): Photomicrograph illustrating grain coatings from Neogene groundwater calcretes (Siwalik Group, Punjab Sub-Himalaya, India); note the well-developed laminated envelopes consisting of dark micritic laminae and spar; shape of the envelopes follows the shape of the cortex. Coated fragments ~200μm – 400 μm.

Alpha-calcretes formed under the influence of fluctuating water tables also contain coated grains and gravity cements (Fig. 5c) (Tandon and Narayan, 1981). From soil environments, Chadwick et al. (1989) described the morphology of calcite crystals in clast coatings and indicated that even after the soil matrix is eliminated as a dominant control on calcite crystal morphology (Wieder and Yaalon, 1974, 1982), other factors interact to create a complex precipitation environment. They include (a) length of time the crystallisation surface is wet, (b) rate of precipitation, (c) ionic strength of the soil solution, and (d) crystal surface interactions with organic and inorganic particulate matter.

Beta-calcretes

This type shows biogenically formed features such as needle fibre calcite, microbial tubes, alveolar and alveolar-septal fabric, *Microcodium* and calcified faecal pellets etc. (Esteban, 1974; Esteban and Klappa, 1983). Other biogenically related features such as rhizocretions (Klappa, 1980 a, b; Mount and Cohen, 1984; Purvis and Wright, 1991), rhizolites and calcified root mats also occur. Monger and Daugherty (1991) emphasised the role of soil micro-organisms in calcrete formation through controlled experiments.

Needle-shaped calcite has been variously termed needle fibres (James, 1972), lublinite (Stoops, 1976), whisker crystals (Calvet, 1982) pseudomy-celia (Ducloux et al., 1984). It is a common habit of secondary calcium carbonates in vadose carbonate soils and in beta-calcretes. More recently, Verrecchia et al. (1994) suggested that some habits of needle fibre calcites are related to biomineralisation of specific fungal hyphae that upon decomposition release the needles to the medium.

Cathodoluminescence Studies

Application of cathodoluminescence (CL)—the emission of light resulting from the bombardment of phosphors by electrons or cathode rays (Sommer, 1972)—has helped in the elucidation of calcrete microfabrics. Sites of imperfection become luminescence centres and these preferentially trap energy from the cathode beam. On subsequent decay to the ground state, photons are emitted and luminescence occurs.

Carbonates viewed under CL show varied luminescence signatures ranging from non-luminescent to dull luminescent, to moderately luminescent, to bright luminescent in orange, brick red, and yellow colours. CL studies of calcretes help in recognition of different events and stages of carbonate precipitation and dissolution. In a Lower Carboniferous calcrete profile, Solomon and Walkden (1985) established a spar cement stratigraphy of subaerial vadose cements and provided evidence of recurrent periods of fabric dissolution. They recognised that the dissolution textures are confined to the walls of the larger pores and early brecciation fractures. Tandon and Friend (1989) used CL to document the effects of the replacive and displacive growth of the pedogenic carbonate as well as neomorphism and dissolution in alpha fabrics (Fig. 6a) of Late Devonian to Early Carboniferous calcretes in Arran, Scotland. Whilst these studies pertain to alpha-calcretes developed in a siliciclastic host sediment, Wright and Peeters (1989) examined calcretes from the Lower Carboniferous marginal marine limestones of South Wales, Germany and Belgium. Their studies revealed that the fabrics resembling recrystallization microspars with irregular crystal forms and non-planar crystal boundaries result from multiple phases of precipitation and dissolution. Tandon and Friend (1989) noted the presence of relict cathodoluminescence patterns within the neomorphic rhombic calcretes and suggested that initial replacement took place by carbonate growing about closely spaced nuclei. Introduction of calcite also involved fracturing and other forms of displacement (Fig. 6b). Luminescence work produced the evidence of continuing replacement, particularly by carbonate spar, of material around the floating particles (Fig. 6c) and of open space filling of voids and cracks by spar (Tandon and Friend, 1989).

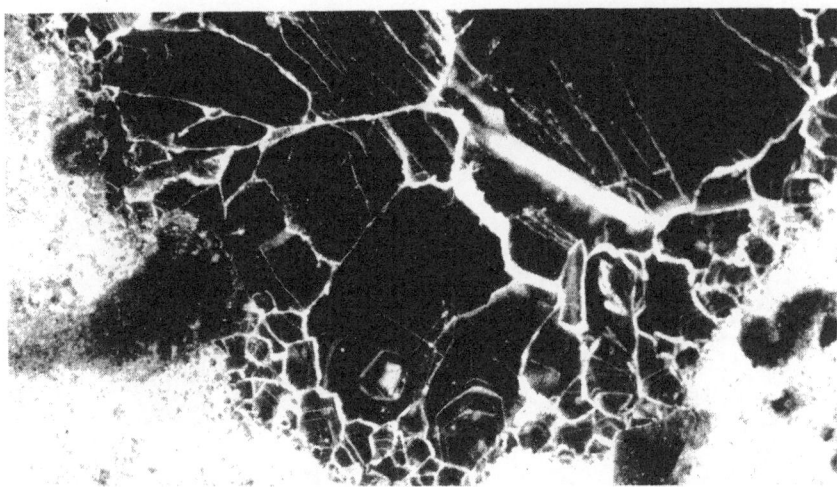

Fig. 6(a): Photomicrograph under cathodoluminescence illustrating complex calcite growth patterns in shrinkage spaces (vughs). The dense micrite groundmass shows an orange luminescence signature; the vugh is characterized by (a) increase in size of crystals from the floor of the vugh towards the centre of the vugh; (b) zoned crystals in the proximity of the vugh floor; (c) bands of bright luminescing spar along the intercrystalline boundaries, their irregular character and band width variation is suggestive of neoformed spar (brightly luminescing) along the dissolutional surfaces related to intercrystalline boundaries (Late Devonian-Early Carboniferous alpha-calcretes, Fallen Rocks, Arran, Scotland). Vugh ~1 mm across.

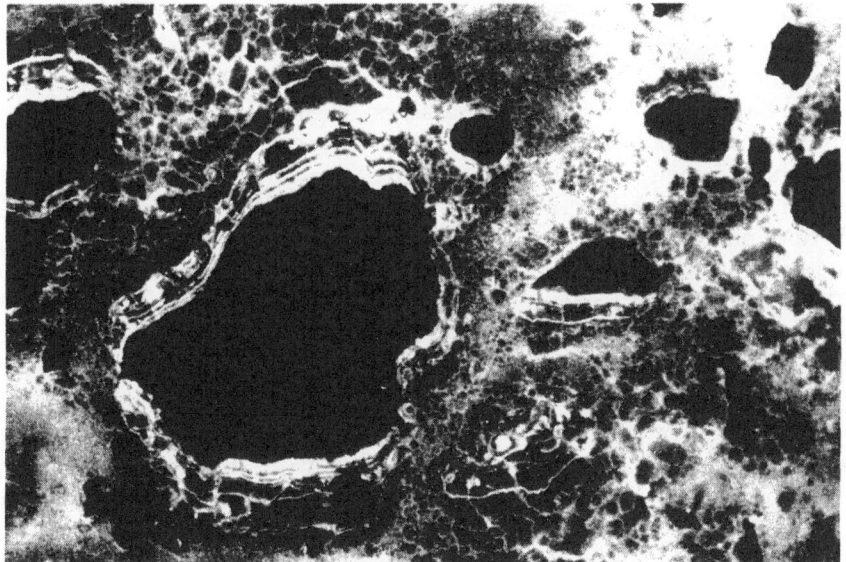

Fig. 6(b): Photomicrograph showing orange luminescing dense micrite, microspar fabrics, and prismatic elongate spar crystals at periphery of the dark clay-rich area; such prismatic crystals commonly occur along the carbonate- clay interface and may be of displacive habit. Prismatic crystals ~50 µm – 100 µm. (Late Devonian-Early Carboniferous alpha-calcretes, Fallen Rocks, Arran, Scotland).

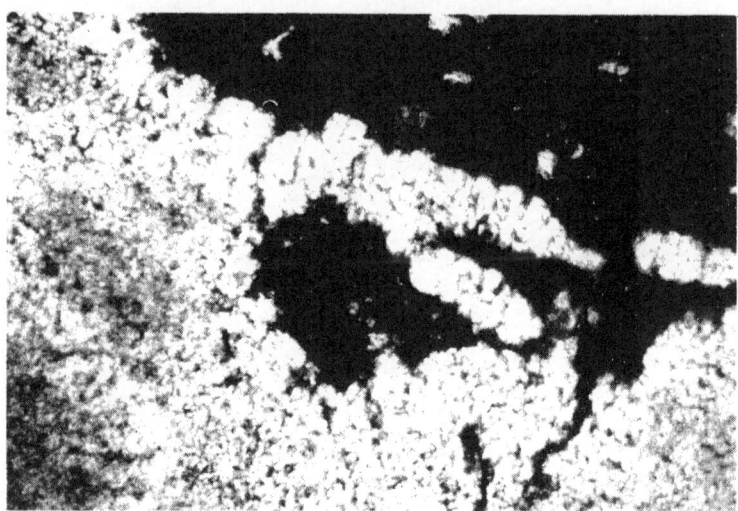

Fig. 6(c): Photomicrograph to illustrate bright orange luminescent bands around the quartz grains; growth patterns show that the spar has most likely grown by replacement of the silicate grains. (Maastrichtian alpha- calcretes from Jabalpur, Central India). Quartz grain ~400 μm.

Most ancient calcretes exhibit a luminescence sequence of uniform dull-red-brown luminescence in the micrite, early non-luminescent to dull luminescent calcite in septarian voids and shrinkage spaces, followed by a bright yellow-orange luminescent spar cement (Tandon and Friend, 1989). Driese and Mora (1993), based on a study of pedogenic carbonate nodules of Devonian vertic palaeosols in the central Applachians, showed that the 'CL petrographic components define chemically distinct fields, which only slightly overlap, indicating that each formed in a chemically distinct environment'. They showed, using microprobe data, that dull-red-brown luminescent micrite typically contains Mn concentrations of 500-2000 ppm and Fe concentrations less than 2000 ppm. Bright yellow-orange luminescent, late calcite spar cements have extremely high Mn concentrations (up to 34,000 ppm) and Fe concentrations below detection limit (Driese and Mora, 1993). Higher concentrations of Mn^{2+} into successively younger calcite spar cements without an attendant increase in Fe^{2+} suggests an increased mobility of Mn^{2+} due to decreasing Eh resulting from oxidation of organic matter after soil burial (Driese and Mora, 1993).

All of the above luminescence data are from ancient calcretes; little or no data are available from the extensive Quaternary duricrusts of semi-arid regions. Even in the case of ancient calcretes, data mainly pertain to alpha-calcretes; beta-calcretes have generally not been examined.

Mineralogy and Chemistry

Calcretes consist dominantly of authigenic carbonates, Mg clay and chert, plus relict minerals of parent rocks (Wang et al., 1994). Most calcretes consist of low-magnesian calcite (LMC); however, high-magnesian calcite and aragonite as well as dolomite have been recorded (for example, from South Africa; Watts, 1980). High magnesian calcite can form under conditions of rapid evaporation despite low Mg/Fe ratios in the solution (Watts, 1980). Many Australian calcretes in the inland regions contain high concentrations of dolomites, especially in the lower parts of profiles (Milnes,1992). The earliest carbonates may be disordered, hydrated, or organic-matter bearing calcite (Ducloux et al., 1984; Dupius et al., 1984)

Although aragonite has been recorded in calcretes of Southern Africa, West Africa and the Mediterranean region (Goudie, 1973; Nahon et al., 1980), its occurrence is uncommon because it gradually recrystallizes into ordered calcite. In beta-calcretes, precursor calcium oxalate may be important for later carbonate formation.

Clay minerals commonly occur in calcretes and include smectite, illite, sepiolite, and palygorskite. Concentrations of illite may occur in the upper horizons of semi-arid soils. Pedogenic illitization may be difficult to recognise in ancient calcareous palaeosols (Wright and Tucker, 1991), although Robinson and Wright (1987) recognised this process in Carboniferous calcrete-bearing vertisols. In the Kalahari calcretes, Watts (1980) reported clays dominated by palygorskite and sepiolite, with subsidiary montmorillonite, mixed-layer clays, illite, glauconite, chlorite and kaolinite. Growth of calcite is accompanied by increasing palygorskite which reaches its maximum towards the top of the nodular horizon (Goudie, 1983). In more humid areas, smectite is formed rather than palygorskite (Nahon et al., 1975). Sepiolite is often associated with palygorskite in Kalahari calcretes and predominates in mature calcretes, often associated with dolomite (Watts, 1980). Montmorillonite mainly occurs in mature calcretes in the lower portions of calcrete profiles (Watts, 1980).

Goudie (1973), on the basis of around 300 analyses of calcretes (Table 2), found that the mean percentages of the main constituents were as follows:
$CaCO_3$ - 79.28%, SiO_2 - 12.3% , Al_2O_3 - 2.12% , MgO- 3.05%, CaO- 42.62%. Based on the composition of kunkar (pedogenic carbonate) nodules in the older alluvium of the Indo-Gangetic Plains in North India, Singh and Lal (1946) concluded that the kankar consists of 50-70% $CaCO_3$, 25-45% host sediment, and a small percentage of iron oxide, magnesium oxide, and phosphate. These nodules consist of 45-73% $CaCO_3$, 1.84-4.08% Fe_2O_3,

1.22-5.9 % Al_2O_3, 0.03- 0.08% P_2O_5, trace to 0.23% Mn_2O_3, 1.3-10.5% MgO, 0.05-0.47% K_2O, and 0.05-0.11% Na_2O.

An important theme that has been followed is to examine the changes in composition in calcrete profiles. Aristarain (1970), on the basis of chemical analysis of calcrete profiles from New Mexico, showed that where aluminium was held constant, the following compositional changes occurred :

1. Ca^{2+}, C, O_2 and H were added relative to the upper parts of the calcrete profile, the abundance decreasing downwards;
2. Mg, S, Fe^{3+} were added in decreasing amount from top to bottom;
3. Si and Ti increased slightly in the upper parts of the profile but considerably in the intermediate zone;
4. K and P remained constant;
5. Na and Fe^{2+} showed no clear trend in variation.

Hay and Reeder (1978) observed an Si/Al and K/Al ratio decrease with increasing replacement of clays by micrite in Olduvai massive calcretes and suggested that this is due to dissolution or leaching of phengitic illite, and formation of clay approaching the composition of halloysite or kaolinite. Si, Al, K and Mg are lost in replacement of clay and are precipitated in the form of dolomite, zeolite and dawsonite. In the calcrete profiles of inland South Australia, Hutton and Dixon (1981) recognised a progressive decrease in Ca/Mg ratio, which relates to a corresponding increase in dolomite to calcite at depth.

Milnes and Hutton (1983) recognised, on the basis of Si/Ti ratio, that non-carbonate components of hardpan and nodular calcretes have a different provenance from underlying calcareous sediments. Using this data, they argued that carbonate hardpan and calcrete nodules in many profiles marked disconformities that represent periods of subaerial exposure and diagenesis with associated erosion and transport.

Because of variable conditions of the host materials, the composition of pedogenic calcretes may be quite variable (Wieder and Yaalon, 1974). Singh and Singh (1972) pointed out that a Ca-rich soil produced a Ca-rich kankar. A compilation of some of the data on the chemical composition of pedogenic calcretes is given in Table 2.

Mode of Introduction of Carbonates in Calcretes

In calcretes there has been growth, in the host rock, of carbonate of greater volume than that admissible by original porosity. Besides the passive filling by carbonate cement, carbonates are precipitated in shrinkage-displacive spaces (Tandon and Friend,1989). Carbonate in excess

Table 2: Chemical Composition of Some Pedogenic Calcretes

Authors	Locality	Profile Depth (cm)	wt. %												ppm					
			CaO	SiO₂	Fe₂O₃	MnO	MgO	Na₂O	K₂O	Al₂O₃	TiO₂	P₂O₅	FeO	Mn₂O₃	Nb	Y	Zr	Sr	Rb	Ba
Singh and Lal (1946)	South eastern Punjab, India	35-75	31.38	30.2	2.32		1.61	0.35	0.25	3.33		0.07		0.1						
Singh and Singh (1972)	Vindhyan range of Mirzapur, India	160-175	29.99	28.5	2.72		2.01		0.22	2.69		0.06		0.1						
		60-90	38.64		1.19		2.42	0.28	0.17	2.17		0.061		0.213						
Aristarain (1970)	High Plains, New Mexico	90-120	37.52		1.57		2.42	0.23	0.39	2.8		0.77		0.213						
		120-150	36.4		1.6		2.9	0.5	0.28	2.5		0.77		0.228						
		85	45.56	14.36	0.6		0.34	0.13	0.13	1.04	0.06	0.02	0.08							
Hay and Reeder (1978)	Northern Tanzania	150	37.68	26.92	0.68		0.49	0.11	0.23	1.59	0.11	0.02								
		225	18.7	57.46	1.27		0.27	0.39	0.57	4.46	0.14	0.05								
			44.15	7.1	3.41	0.09	1.71	0.66	0.69	2.85	0.43				70	20	160	4120	40	400
Patil and Surana (1992)	Western Maharashtra (India)		42.14	15.59		1.39			2.93											
Goudie (1973)	Average chemical composition of Indian calcretes		40.19	17.92	3.62		1.7			3.1										
Goudie (1973)	Average calcrete composition for 300 samples from all over the world		42.62	12.3	2.12		3.1			2.1										

of that allowed by original porosity must be introduced by displacive and replacive precipitation (Fig. 7). Textural studies of these latter types result in better understanding of the conditions of precipitation. Displacive crystals grow from supersaturated solutions (Taber, 1916). Chadwick and Nettleton (1992) suggested that the cohesive bonds due to ionically bonded calcite play a dominant role in displacive growth in calcite. The petrocalcic horizons in soils are formed by growth and interlocking of authigenic crystals that have little affinity for the surface reactive groups on the S-matrix. The O^2 portion of the CO_3^{2-} is more reactive with a strong electron acceptor Ca^{2+}. Strong cementation occurs as crystals cohere to each other through chemical intergrowth and physical interlocking. The S-matrix is not held in place by adhesion and is displaced by the increasing volume of crystic plasma.

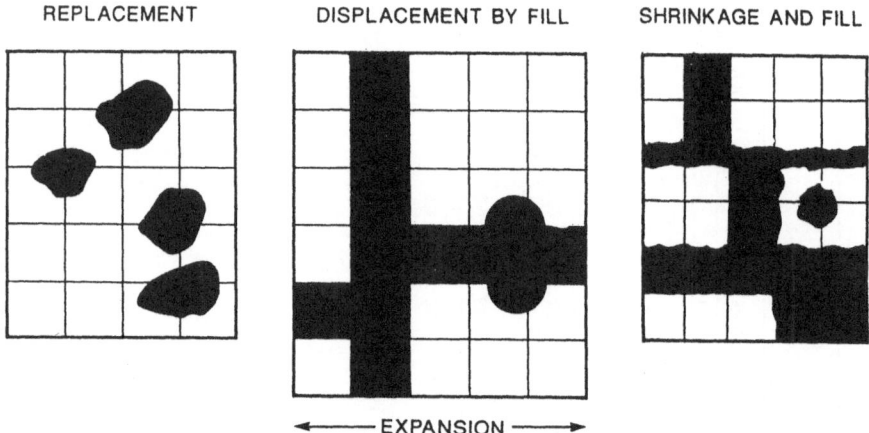

Fig. 7: Diagram illustrating the various modes of introduction of carbonates in calcretes.

Macroscopic evidence in favour of displacive calcite include displacive nodular growth (Wieder and Yaalon, 1974), brecciation of bedrock by calcite (Reeves, 1976), internal buckling and pseudo-anticlines in mature calcrete profiles (Watts, 1977).

Thin sections reveal evidence favouring displacive introduction of calcite such as exploded detrital fragments, expanded micaceous grains, fragmented quartz grains, floating fabrics (Watts, 1978; Saigal and Walton, 1987; Buczynski and Chafetz, 1989; Braithwaite, 1989). Other evidence includes fibrous calcite spar filling irregular cracks with the sides of the cracks showing perfect visual matching. Fibres are parallel to each other and elongated in the direction of crack opening (Watts, 1978, and Fig. 6b)

Braithwaite (1989) used CL to recognise two styles of growth-driven single crystal displacement. Linear growth is elongate with growth zones at right angles to the direction of elongation. Island growth results in equidimensional crystals whose shapes are determined both by the contact with the surfaces of original clastic grains and by compromise with adjacent calcite crystals.

Many calcretes in semi-arid areas display replacement of parent silicate minerals by calcite (Nagtegaal, 1969; Nahon et al., 1975; Monger and Daugherty, 1991b; Tandon et al., 1998). Nagtegaal (1969) suggested that equigranular crystalline calcite mosaics form by replacement. The degree of replacement may vary from nil, with calcite mainly occurring as an inert cement, to strongly corroded detrital silicate grains floating in aggregates of calcites, to stages where almost nothing is left of the original constituents of parent materials.

Abundant crusts or rinds of calcite enveloping single grains or aggregates are a characteristic of calcites of replacement origin. Tandon and Friend (1989) used CL to observe evidence of continuing replacement by carbonate spar of material around the floating exotic particles.

All microscopic and macroscopic observations show conclusively that replacement in calcretes is pseudomorphic regardless of the nature of the parent rocks (Wang et al., 1994).

Reaction : Transport Model of Replacement in Calcrete Formation

Wang et al. (1994) suggested a dynamic model for calcrete genesis based on certain characteristic features, such as:
— their typical occurrence in semi-arid regions with alternating dry/ humid seasons;
— pseudomorphic replacement of parent material, involving a process of import of a large amount of $CaCO_3$ and simultaneous removal of a large amount of silicates;
— replacement of silicate minerals by attapulgite (or other Mg-clays) towards the bottom of the profiles;
— precipitated amorphous silica next to half-dissolved quartz grains.

In the dry season, water evaporation causes precipitation of $CaCO_3$. If groundwater contains enough Mg, precipitation of sepiolite (attapulgite) also takes place. The precipitation of sepiolite causes depletion of aqueous silica resulting in dissolution of parent silicates. In the following wet season, sepiolite, silica and cations are flushed away by rain-water, making space for $CaCO_3$ precipitation in the next dry season. As climate cycles repeat, $CaCO_3$ is accumulated and silicates are removed. If groundwater contains too little Mg(aq.), then the model predicts growth of $CaCO_3$ without removal of silicates, thus producing void-filling and/or displacive texture instead of replacement of detrital silicate grains.

Sources of Calcium

Sources of calcium include carbonate bedrocks, shells, vegetation litter, volcanic emissions, atmospheric dust, rainfall, lakes, surface runoff and groundwater (Goudie, 1983). The relative importance of these sources varies for different types of calcretes. Machette (1985) arrived at this conclusion as many calcic soils occur in non-calcareous sediments well above present and former levels of groundwater. Recent dust fall in Las Cruces, New Mexico, has averaged about 0.2 gm of $CaCO_3$ per cm^2 per 1000 years (Gile et al., 1979). In the south-western USA also, concentration of Ca^{++} in rainfall is high and may exceed 5 mg of Ca^{++} per litre of water (Jinge and Werby, 1958).

Groundwater may constitute an important source of calcium carbonate in the case of non-pedogenic calcretes. Cementation zones in capillary fringe zones (Semenuik and Meagher, 1981) of the water tables (perched and permanent) are controlled by the grain size and texture of the host sediments. Such authigenic carbonates reflect the composition of the groundwater.

In calcretes hosted in limestones, for example in the Edwards Plateau, Texas, these form through in situ dissolution and reprecipitation of carbonate under soil conditions (Rabenhorst and Wilding, 1986). If the carbonate is derived from a marine limestone, wide variation in the stable isotopic composition of the pedogenic carbonate from that of the parental carbonate helps in distinguishing them (Rabenhorst et al., 1984). Where the parental rock is non-marine carbonate, particularly a palustrine carbonate, the distinction between nodular calcrete and parent carbonate can be difficult (Tandon and Gibling, 1997)

Environments/Mode of Calcrete Formation

Calcretes form in semi-arid areas where leaching is inadequate for carbonates to be removed from the system. In siliciclastic host sediment, translocation and accumulation predominate. Dissolution and subsequent precipitation of carbonates are involved in calcrete formation and take place by the following reaction:

$$CO_2 + H_2O + CaCO_3 = Ca^{2+} + 2HCO_3^-$$

Increased CO_2 partial pressure, decreased temperature, and low pH all cause dissolution whereas precipitation is favoured by decrease in CO_2 partial pressure, evaporation, common ion effect, biological degassing, or by freezing. Such conditions aid multiple precipitation-dissolution events in calcretes, and thereby lead to complex histories under near-surface conditions from the soil zone through to the groundwater zone.

According to Goudie (1973), formation of duricrust in different environments is possible as follows:

(a) *Fluvial environment.* Calcretes occur along stream valleys and channels as sinuous wedges of local and regional extent. Calcretes may also form as a result of massive sheetfloods across semi-arid flat lands (Goudie, 1973, Khadkikar et al., 1998, Tandon et al., 1998). Nickel (1985) has dealt with accumulation and distribution of terrestrial carbonates, including calcretes, in these semi-arid alluvial settings. The type, composition and distribution of calcretes is controlled by physical (transportation energy), morphological (slope gradient), chemical (precipitation and evaporation) processes, and by biochemical interactions.

(b) *Lacustrine environment.* Calcretes form in marginal and emergent areas, particularly during periods of low-stand.

(c) *In-situ transformation model.* Formation by redistribution of pre-existing carbonates.

(d) *Detrital model.* Calcrete formation by lateral transportation and redeposition of scattered fragments of calcretes which are recalcretized by in situ processes.

(e) *Pedogenic model.* Calcretes formed due to soil processes because of concentration of authigenic carbonates vertically within the soil profile (*per descensum* model of Goudie, 1983)

(f) *Groundwater model.* Non-pedogenic calcrete is formed by the concentration of authigenic carbonate externally derived from laterally moving groundwater. Accumulation in the vadose zone may take place because of evaporation at the capillary fringe (Goudie, 1983).

Mann and Horwitz (1979) suggested a model of calcrete formation in the phreatic zone through four stages of development :

1. Occurrence of a broad drainage channel with an alluvial valley fill and a shallow groundwater system;
2. Precipitation of $CaCO_3$ at or below the groundwater table, where the solubility of calcite is exceeded;
3. Formation of pools and domes of carbonate creating surface mounds as carbonate is pushed upwards, displacing the overlying colluvium;
4. Maturation in which older carbonate lifted above the water table is displaced upwards (where it may be dissolved) and is replaced by the production of younger calcretes.

It is useful to discriminate between pedogenic and groundwater calcretes (Tandon and Narayan, 1981) inspite of difficulties of overlapping fabrics in some instances. Mora et al. (1993) tabulated the general stratigraphic, geometric, and petrographic criteria for discriminating between these two types. Further data have been compiled (Table 3) to distinguish between pedogenic (vadose), hydromorphic and phreatic calcretes.

Table 3: General Comparison of the Attributes of Pedogenic, Hydromorphic and Phreatic Calcretes

	Pedogenic (Vadose) Calcretes	Hydromorphic Calcretes	Non-Pedogenic (Phreatic) Calcretes
Subsurface water zone	Soil moisture zone	Capillary fringe zone	Phreatic zone
Process involved	Soil forming process	Capillary transport due to fluctuating groundwater table	Precipitation from groundwater in phreatic zone
Predominant transport in soils	Vertical redistribution (relative accumulation in soil profile) *per descensum*	Absolute accumulation *per ascensum*	Absolute accumulation (lateral transport)
Stratigraphic and geometric character	Display profiles with orderly set of horizons; sharp top gradational base	Carbonate segregations, 2-10 cm glaebules distributed throughout horizon with non-calcareous or weakly calcareous matrix	Uniformly massive, single horizon, gradational top and base
	Generally thin, 1-2m; varies laterally over shorter distance		Generally thick (up to 10 m or more)
Sedimentological occurrence	Typically associated with more stable surfaces or floodplains	Axial floodplain and lower piedmont deposits	Associated with drainage channels, playas and lake deposits
Macrostructures	Include nodular, massive, laminar, pisolitic forms	Glaebular (spherical and massive)	Rarely display laminar horizons, never-pisolitic or prismatic unless at the top of unit
Microfabrics	Exhibit beta-fabrics (biogenic microbial/root related carbonates such as alveolar-septal structure, needle fibre calcite) and alpha fabrics	Predominantly alpha-fabrics; some overlapping of fabrics may occur	Typically alpha- fabric (densely crystalline)
	Vadose fabrics such as meniscus and pendant cements		May be present, towards top
	In situ brecciation (autobrecciation)		Not common, but may be present towards top

(Contd.)

Table (*Contd.*)

	Replacive and displacive micrite and or microspar nodules	May be present
Mineralogical and geochemical	Rarely display lateral changes as calcite to dolocrete to gypcrete Low Mg-calcite constitutes higher proportion (76-78%)	Shows regular mineralogical changes reflecting salinity gradients

Although known for several decades, calcareous palaeosols and caliche profiles (Mülter and Hoffmeister, 1968; James, 1972; Read, 1974; Knox, 1977; Harrison, 1977; Harrison and Steinen, 1978; Rao and Thamban, 1997) in shallow marine carbonate sequences have received less attention. Wright (1994) aptly stated '...the need to be able to recognise such exposure surface features has arguably increased with the advent of sequence stratigraphy'. Palaeosols not only provide the means of correctly identifying exposure surfaces, but also allow the differentiation of major ones such as major low-stands in the sense of Van Wagoner et al. (1988), from less important ones, such as those associated with small parasequences in high-stand system tracts, or those formed during the retrogradational phases found in transgressive system tracts. For example, Tandon and Gibling (1997) recognised that nodular and associated groundwater calcretes in coal bearing cyclothems of the Sydney basin mark the boundary between underlying marginal-marine and overlying alluvial deposits. The calcretes are inferred to represent low-stand surfaces (sequence boundaries). The host sediment for calcretes includes sand/siltstone and palustrine limestone (Fig. 8). Palaeosols identified at regressive, low-stand surfaces include calcretes (Hanneman and Widerman, 1991), vertisol-like palaeosols (Joeckel, 1994, 1995) and hydromorphic palaeosols (Aitken and Flint, 1995). Calcretes at sequence boundaries also help in determining low-stand palaeoclimates (Tandon and Gibling, 1994). Further the maturity of these palaeosols—calcareous or others—provides insights to changes in rates of accommodation space created during sea level rise and fall.

Palustrine carbonates are shallow freshwater deposits which show evidence of subaqueous deposition and subaerial exposure (Freytet, 1973; Platt and Wright, 1992). They possess a lacustrine biota but display microfabrics similar to those of calcretes. Many workers, for example Platt(1989), have noted that palustrine limestones form through pedogenic

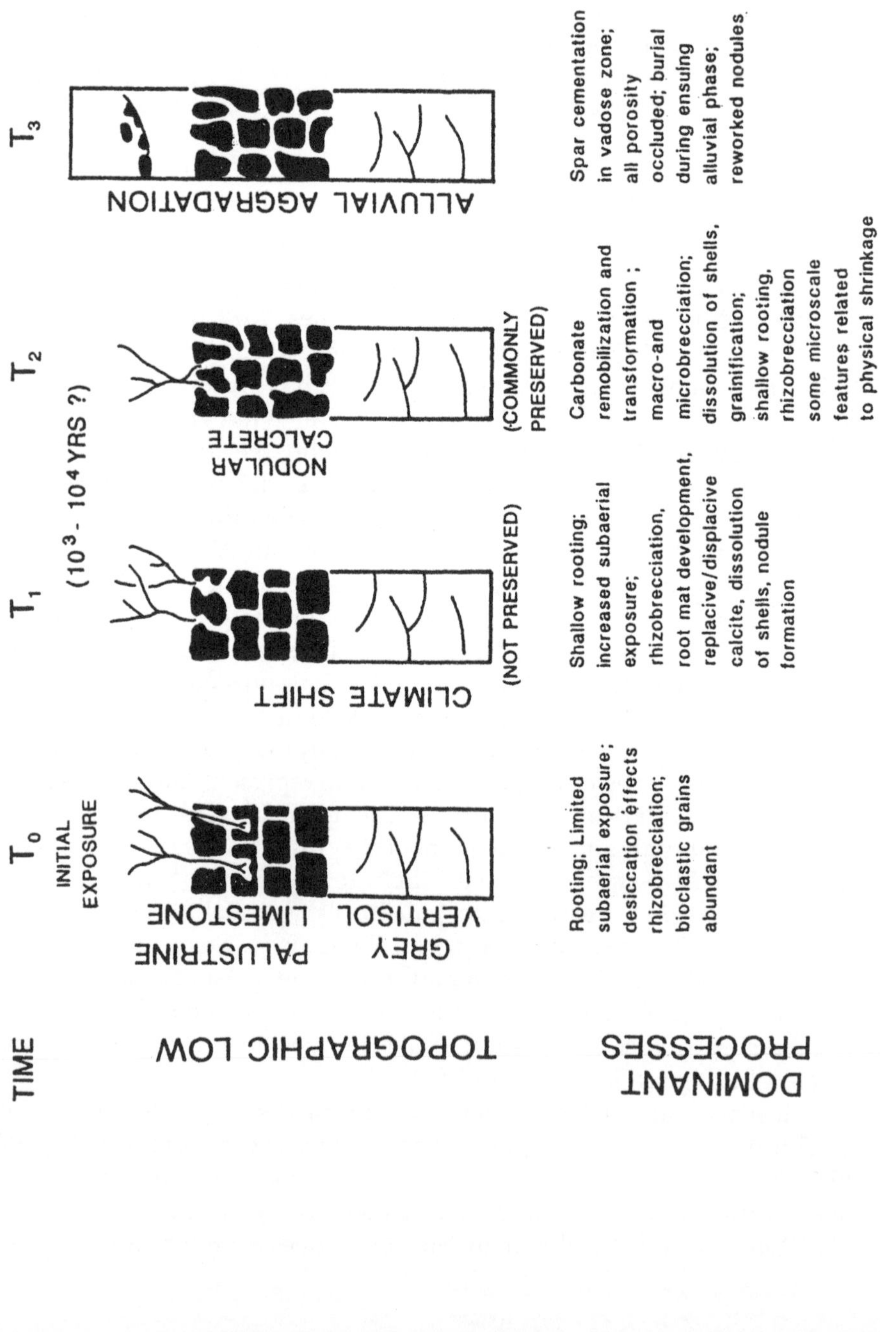

TIME

T_0
INITIAL
EXPOSURE

T_1
(10^3 - 10^4 YRS ?)

T_2

T_3

TOPOGRAPHIC LOW

PALUSTRINE
LIMESTONE

GREY
VERTISOL

CLIMATE SHIFT

(NOT PRESERVED)

NODULAR
CALCRETE

(COMMONLY
PRESERVED)

ALLUVIAL AGGRADATION

DOMINANT
PROCESSES

Rooting; Limited
subaerial exposure;
desiccation effects
rhizobrecciation;
bioclastic grains
abundant

Shallow rooting;
increased subaerial
exposure;
rhizobrecciation,
root mat development,
replacive/displacive
calcite, dissolution
of shells, nodule
formation

Carbonate
remobilization and
transformation ;
macro-and
microbrecciation;
dissolution of shells,
granification;
shallow rooting,
rhizobrecciation
some microscale
features related
to physical shrinkage

Spar cementation
in vadose zone;
all porosity
occluded; burial
during ensuing
alluvial phase;
reworked nodules

Fig. 8: Diagram illustrating calcrete profile development in carbonate substrates (after Tandon and Gibling, 1997).

modification of lake carbonates. Where palustrine limestones are subjected to prolonged exposure, especially on low-stand surfaces, they are transformed to calcretes through dissolution and reprecipitation (Tandon and Gibling, 1997). Hence, it is useful to evolve criteria to discriminate between palustrine limestone and calcretes (Table 4).

Table 4: Broad Criteria for Distinguishing Between Calcretes and Palustrine Limestones

	Palustrine Limestone	Calcrete
1	Subaqueous precipitation affected by subaerial exposure	Secondary accumulation in host sediment or transformation of pre-existing carbonates through soil CO_2
2	Rooting and shrinkage	Physical shrinkage more common in alpha-calcretes; beta-calcretes show both rooting and shrinkage
3	Occurs in areas of raised water tables and basinal lows	Occurs commonly in upland locations
4	Fossils: aquatic community present	Generally absent
5	More positive $\delta^{13}C$ values relative to pedogenic calcretes	Depleted $\delta^{13}C$ values because of influence of soil CO_2 in pedogenic calcretes; groundwater calcretes similar to palustrine limestone

Rates of Calcrete Formation

Rates of calcrete formation are a useful input in quantitative stratigraphic studies. Considering the wide variation in calcrete types and the multiple processes involved in their formation, generalised rates cannot be simply arrived at. However, a time significance has been attached to calcretes because they develop in easily identifiable stages (Gile et al., 1961, 1966; Allen, 1974; Machette, 1985).

Leeder (1975) developed a model for pedogenic carbonate development in alluvial sediments based upon rates of flood sediment accretion versus time taken for carbonate profile development. The stage of calcrete maturity depends upon its residence time within that part of the soil where precipitation has occurred. If the accretion rate is high, immature stage 1 to stage 2 carbonate nodules would develop. If, however, the rate of sediment accretion is low, mature profiles would develop. Using the then—available data on the time intervals of formation of profiles occurring on distinct geomorphic levels, Leeder (1975) took minimum and maximum ages of formation of various profiles as follows:

Stage 1 : min. 1000, max. 4500 years
Stage 2 : min. 3500, max. 7000 years

Stage 3 : min. 6000, max. 10,000 years

Stage 4 : min. 10,000 years

Later, Wright (1990a) pointed out that geochronological data on the geomorphic surfaces of the south-western USA have been revised. In most cases, the estimates of Leeder (1975) must be increased by one order of magnitude. Using the data of Hay and Reeder (1978) on the rapid development to stage 4 calcretes in East Africa, Wright (1990b) also pointed out that time estimates for a mature calcrete profile may range from 3 k yrs to 1 m yrs—a variation of over two orders of magnitude.

Radiocarbon and ^{230}Th - ^{234}U dates of calcic horizons from calciorthid horizons have been used to calculate a rate of deposition of 1.0 to 3.5 g $CaCO_3/m^2/yr$ during soil formation (Schlesinger, 1985). This rate is consistent with present-day rates assuming that the atmospheric deposition of Ca limits the process (Schlesinger, 1985). Further, by simulating stochastic precipitation, evapotranspiration, thermodynamic relationships, soil parameters and soil water and $CaCO_3$ flux, Marion et al. (1985) developed a regional model for soil $CaCO_3$ deposition in south-western USA. The range in predicted $CaCO_3$ deposition rates agreed with those of most field studies, that is 1 to 5 g/m^2/yr. One of the characteristics of old $CaCO_3$ is the plugging of the $CaCO_3$ horizon. For some plugged horizons in the south-western USA, model estimates of 47 k yrs, 55 k yrs, and 100 k yrs were obtained by Marion et al. (1985). Gile et al. (1981) estimated that it took a minimum of 25 k yrs to 75 k yrs for plugging to occur in gravelly soils near Las Cruces, New Mexico. The model estimates and the field observations are in broad agreement.

Soil carbonate accumulation rates are variable in both time and space (Machette, 1985). cS has been defined by Machette (1985) as the weight of $CaCO_3$ in a 1m^2 vertical soil column (g/cm). In the Las Cruces area, uranium trend ages and cS contents of relict soils have shown that $CaCO_3$ accumulated twice as fast during the past 50 k yrs as against the 100 k yrs - 500 k yrs interval.

McFadden and Tinsley (1985) used a vertical sequence of 1cm^2 area compartments, each with specific properties of texture, bulk density, water-holding, lithologic and mineralogic composition, soil-air capacity pCO$_2$, ionic strength and temperature, to determine their respective rates of carbonate solubility and dissolution. This modelling indicates that with a mean carbonate influx rate of 1×10^{-4} g/cm^2/yr in a semi-arid thermic climate, the maximum depression of the top of the Cca horizon is attained within only a few thousand years.

Climatic Implications and Palaeoclimates

Pedogenic calcretes commonly occur extensively in semi-arid regions. They occur typically in the rainfall zone of 400-600 mm/year where net moisture deficit and seasonal conditions prevail (Blumel, 1982; Goudie, 1983). Under such conditions, the carbonate is mobilised, translocated, and then reprecipitated. In zones of higher annual rainfall, carbonates are flushed out of the system. However, calcretes may also form locally because of favourable local relations among precipitation, temperature, runoff and relief (Reeves, 1976, 1983; Retallack, 1990). Uneven distribution of rainfall through the year favours the formation of calcretes in drier periods.

Semenuik and Searle (1985) showed that groundwater calcretes are well developed in humid regions of south-western Australia whereas calcrete is less common or rare to absent in the semi-arid zone. The limits of occurrence of these calcretes correlate with a rainfall of approximately 800 mm/year, evaporation 1900 mm/yr, and mean summer temperature of 23°C. In the pedogenic calcretes of the Edwards Plateau, Texas, Rabenhorst and Wilding (1986) noted that petrocalcic horizons are common and most strongly expressed in the dry western parts. Calcretes are reported even from Subarctic environments with occurrences in northwestern Alaska (Dijkmans et al., 1986) and Central East Greenland (Swett, 1974). Possibly these thin crusts are caused by permafrost, which favours calcrete precipitation because of a perched water table and inhibition of groundwater percolation. Dijkmans et al. (1986) noted that these calcretes are cemented by low magnesian calcite derived from carbonate grains of dune sands. Stable isotope analysis of these calcretes indicated $CaCO_3$ precipitation because of the loss of CO_2 in groundwater without significant evaporation.

Because of reports of occurrences of calcretes in various climatic zones, their value as palaeoclimatic indicators is questionable. Nonetheless they mainly belong to the semi-arid zones with net moisture-deficient and seasonal conditions. In studying calcretes in the geologic record, it is prudent to interpret semi-arid seasonal climates only after eliminating the possibilities of calcretes having formed purely because of local basinal factors (Tandon and Gibling, 1994).

Empirical relationships between the annual precipitation and the depth of calcareous accumulation have been derived for modern calcareous soils (Jenny, 1941; 1963). Retallack (1992) derived an empirical regression equation between these two attributes to enable estimation of palaeoprecipitation.

In recent years, ancient calcretes have been used for understanding palaeoclimates (Goodfriend and Magaritz,1988; Mack et al., 1991; Mack et al., 1994; Ghosh et al., 1995; Tandon et al.,1995; Tandon and Gibling, 1997; Tandon et al., 1998; Alam et al., 1997). For example, Mack et al. (1991) used the oxygen isotopic composition of calcretes in the Permian Abo Formation to interpret an upsection warming trend (approximately 15°C to nearly 30°C), and a decrease in precipitation (rainfall). In the Late

Cretaceous palaeosol carbonates in central India, Ghosh et al. (1995) explained a lighter oxygen isotopic composition in terms of the cumulative effects of a highly seasonal (monsoon-like) climatic regime and a more pronounced 'continental effect' due to the larger size of Cretaceous India. Tandon et al. (1995) suggested the prevalence of semi-arid conditions during the Maastrichtian in central India based on calcrete facies of the dinosaur nest-bearing Lameta Beds. In Late Carboniferous cyclothems in Atlantic Canada, Tandon and Gibling (1994, 1997) used the systematic alternation of coals (histosols) and other hydromorphic soils with calcretes and calcic vertisols as indicators of strong variation in seasonality during the duration of a cyclothem, estimated at 200 k yrs. In at least one cyclothem, calcic palaeosols formed on an interfluve adjacent to a palaeovalley cut through marine strata (Tandon and Gibling, 1994). More seasonal and possibly drier climates were interpreted by them during low-stands.

Quade et al. (1989) commented on the development of the Asian monsoon on the basis of the carbon isotopic composition of the 18 m yrs calcareous palaeosol record of the Siwalik Group. A dramatic ecological shift from C_3 to C_4 dominated floodplain biomass began approximately 7.4 - 7.0 m yrs ago. Oxygen isotopes also exhibit a shift in the latest Miocene. Although, such changes have now been recognised in other continents, Quade et al. (1989) originally suggested that this dramatic ecological shift marked the inception or strengthening of the Asian monsoon.

Undoubtedly, calcretes host important palaeoclimatic information that needs to be understood more fully. The variability in calcretes and their properties (textures, chemical and isotopic composition) should be related to specific environmental and climatic conditions. Such data collected across climatic zones in the Holocene and Pleistocene should enhance their value as palaeoclimatic indicators.

Carbon and Oxygen Isotope Composition

Determinations of carbon and oxygen isotopes of calcretes and related facies in both Quaternary and pre-Quaternary sequences are available (for reviews see Talma and Netterberg, 1983; Wright and Tucker, 1991; Mora et al., 1993). The carbon and oxygen isotopes of calcretes provide useful indications of climatic influences on soil-forming processes (Magaritz et al., 1981; Amunsdon et al., 1988, 1989; Andrews et al., 1998). Also they provide insights into conditions of carbonate precipitation (Magaritz et al., 1981; Pendall and Harden, 1994) and serve as indicators of the composition of the vegetation biomass (Magaritz, 1986 and others), palaeotemperature and CO_2 concentration in palaeoatmospheres (Cerling, 1991; Cerling, 1992; Mora et al., 1991). Cerling developed a model, using standard diffusion reaction equations, for concentrations and carbon isotopic composition of soil CO_2, and demonstrated the potential of $\delta^{13}C$ of pedogenic carbonates as a proxy for CO_2 concentrations of the Phanerozoic palaeoatmospheres. In a recent application on Late Cretaceous palaeosols in central India, Andrews et al. (1995) showed that the Maastrichtian atmosphere is unlikely to have contained more than about 1300 ppm by volume of CO_2. This value of CO_2 agrees with an independently modelled value of CO_2 in the Late Cretaceous atmosphere (Berner, 1994) providing a verification from the stratigraphic record via pedogenic carbonate nodules. Also, Mora et al. (1996) showed that atmospheric CO_2 levels decreased by a factor of 10 during the Middle to Late Palaeozoic. In applying this technique care must be exercised to use only Stage I, II nodules, ensure absence of groundwater influences, and diagenetic alteration. Wright and Vanstone (1993) also pointed out difficulties in proving the clear coexistence of a C_3 biomass.

$\delta^{18}O$ values may be used to obtain palaeotemperatures and assess the role of evaporation but interpretations must consider effects of elevation, latitude, seasonality, proximity to sea, monsoonal effects, seasonal variations in rainfall and infiltration (Wright and Tucker, 1991). The stable isotope composition of calcretes also provides insights for assessing driving mechanisms of precipitation, and may help in distinguishing between pedogenic and non-pedogenic calcretes, although data on the isotopic composition of the latter are limited (Jacobson et al., 1988; Slate et al., 1996).

Mature calcretes are the products of multiple processes resulting in carbonate dissolution and precipitation events. Such calcretes may have varied isotopic compositions; also calcretes with variable mineralogy (high magnesium calcite, dolomite) may have isotopic compositions which reflect a later diagenetic overprint (Wright and Tucker, 1991). However, careful facies controlled studies of the stable isotopic compositions of ancient calcareous regoliths containing a variety of calcareous soils/calcretes and palustrine limestones may provide clear palaeoenvironmental signatures (Fig. 9). Tandon et al. (1995) showed from the carbon and oxygen isotopes of calcareous components in a Maastrichtian palaeoregolith that the bulk of the calcretes formed in soil-zone environments. The $\delta^{13}C$ values are low, typically –8 to –9 per mil. PDB, demonstrating a strong input of carbon from the decay of terrestrial land plants. Calcrete $\delta^{18}O$ values are variable, –5 to –10 per mil. PDB consistent with precipitation from meteoric water, some of which was evaporatively modified in pools on the alluvial/palustrine flat. The palustrine limestone data have $\delta^{18}O$ values between –4.5 to –7 per mil PDB. The calcrete values have generally similar carbon isotopic composition. In general, more positive $\delta^{18}O$ values correspond with more positive $\delta^{13}C$ values (Fig. 9). Covariation of this kind has been linked to evaporative processes (Salomons et al., 1978).

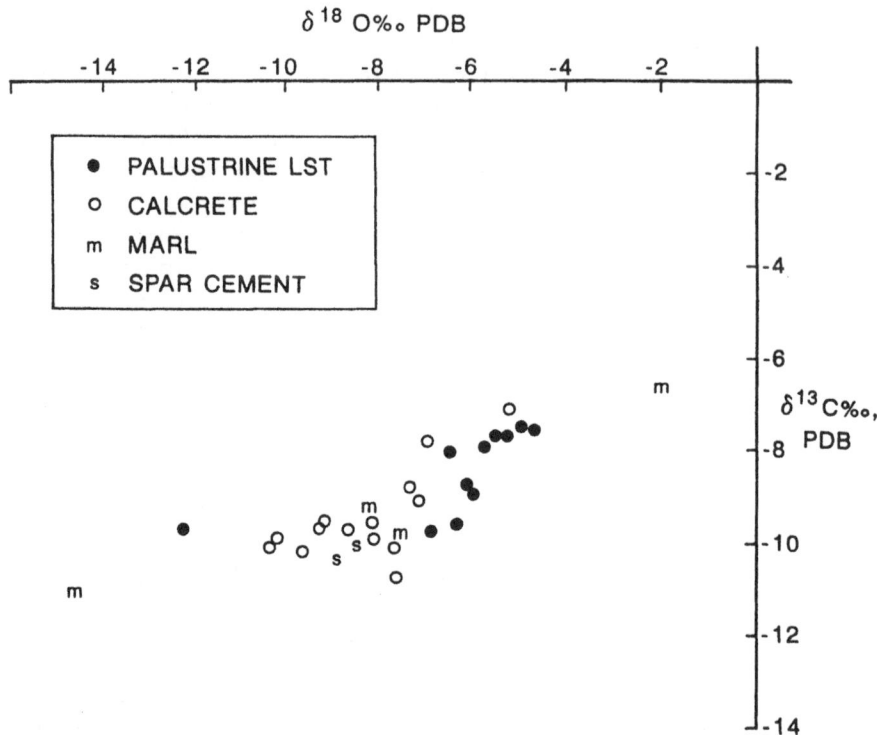

Fig. 9: Diagram illustrating the variable isotopic compositions of calcareous soils, calcretes and palustrine limestones from a Maastrichtian palaeoregolith in central India (after Tandon et al., 1995).

From the earlier literature on the isotopic composition of calcretes it might be useful to recapitulate the following points:

1. The carbonate $\delta^{13}C$ content of calcretes on a world-wide basis ranges from –12 to +4 per mil. with a mean of –4 per mil. The carbonate $\delta^{18}O$ content of calcretes on a world-wide basis range between –9 and +3 per mil. with a mean of –5 per mil. (Talma and Netterberg, 1983).

2. The possible end member compositions of soil carbonate of ^{13}C are –14 per mil. for biologically active soil with 100% C_3 biomass to around +8 per mil. for soil carbonate precipitated from freezing water (Andrews, pers. comm.).

3. Carbon system can be modelled using the diffusion theory and the carbon isotope composition of pedogenic carbonate correlates well with the proportion of C_3 and C_4 photosynthesis in the local ecosystem (Cerling, 1984). $\delta^{18}O$ values of pedogenic carbonates correlate well with the isotopic composition of local meteoric water, despite modification by differential filtration and evaporation (Cerling and Quade, 1992).

Data regarding microbial influence on the isotopic composition of pedogenic carbonate (Mora et al., 1993) are lacking. Highly depleted values of $\delta^{13}C$ −17.6 per mil. are reported for *Microcodium* (Bodergat, 1974). Isotopic compositions of the fabric elements of beta-calcretes should be explored, a plea already made by Wright and Tucker (1991).

Dating of Calcretes

Dating of calcretes and impure carbonates is a difficult task and has defied satisfactory solution. Many investigators have attempted dating of calcretes by the radiocarbon method (Gile et al., 1966, New Mexico; William and Polach, 1969; 1971, Australia). The basic assumptions in this method are:-

(1) $^{14}C/^{13}C$ ratio in the past was the same as that obtaining now,
(2) carbonates of interest have remained as closed systems.
Limitations of this method are:
— ^{14}C fluctuations have taken place during the last few thousand years and have occurred throughout the geologic record;
— difficulties in determination of actual initial $^{14}C/^{13}C$ ratio;
— contamination of calcretes by addition of more recent carbonates may take place so that the obtained absolute date underestimates the actual age. Broecker (1965) determined that 1% contamination of contemporary carbon in a sample of 100,000 years would yield a radiocarbon age of only 30,000 years;
— Soil and atmospheric CO_2 may contaminate calcretes i.e. they may cause enrichment of ^{14}C in calcretes.

Accurate radiocarbon ages of calcretes depend on the source and amount of the original carbon in the calcretes. Gile et al. (1966) determined the ages of desert soil carbonates occurring in Rio Grande desert of southern Mexico by radiocarbon method. He determined the age of carbonates of plugged horizons, which he found to be the oldest, the lower hard laminae somewhat younger, and the carbonate soft laminae to be the youngest. The organic material entrapped in hard laminae was found to be younger than the inorganic carbonates. They also reported two inversions of whole soil dates and attributed them to wide variability in amounts of different carbonate forms within certain horizons. Because of the possibility of contamination by the younger carbonates, the reliability of radiocarbon dates is questionable. The dates serve the limited purpose of providing relative ages and minimum age estimates for calcretes.

Absolute Dating of Pedogenic Carbonates by Thermoluminescence

Singhvi et al. (1996) described a luminescence method for dating 'dirty' pedogenic carbonates. This approach exploits subtle changes in natural radiation field of a mineral in a sediment consequent to its being enclosed in a carbonate precipitate (Fig.10).

The total luminescence of a sediment is the sum of predepositional optically bleached luminescence (I_0) and a fresh acquisition of luminescence initiated due to irradiation from the radioactivity (^{238}U, ^{232}Th, ^{40}K) in the ambient strata.

In the case of carbonate formation, some of the mineral grains that are trapped in the carbonate matrix suffer a change in their dose rate due to dilution (enrichment) of the radioactivity of the sediment matrix by carbonate.

Singhvi et al. (1996) developed an equation for age of the carbonate precipitation:

$$Tc = \frac{D(I_d)_s - D(I_d)_c}{D_s - D_c}$$

where $D(I_d)_s$ = TL level of a mineral grain from a host sediment;
          $D(I_d)_c$ = TL level of mineral grain from carbonate;
          D_s = dose rate in host sediment;
          D_c = dose rate in carbonate matrix.

Dating of carbonates requires estimation of differences in

(1) accumulated radiation of quartz mineral separated from both the carbonate matrix and host sediment;

(2) annual radiation dose (i.e., the difference in the net annual dose rate to a quartz grain in a sand matrix and carbonate nodule).

Uranium Series Dating

Uranium series dating methods are generally applicable over a range of a few thousand to one million years (Schwarz, 1989). Disequilibrium relationships between ^{230}Th, ^{234}U and ^{238}U are used to assess the ages of pedogenic carbonates. Ku et al. (1979) first suggested that calcretes could be dated by U-series if correction was made for the Th and U contributed by the detrital component of the soil leached out during the dissolution of calcite.

Samples consisting of dense carbonate rind around pebbles from soil horizons were chosen by Ku et al. (1979). Analytically, they used the method of leaching the samples with dilute HCl and measuring the ^{238}U, ^{230}Th and ^{232}Th activity in both leachate and residue fractions.

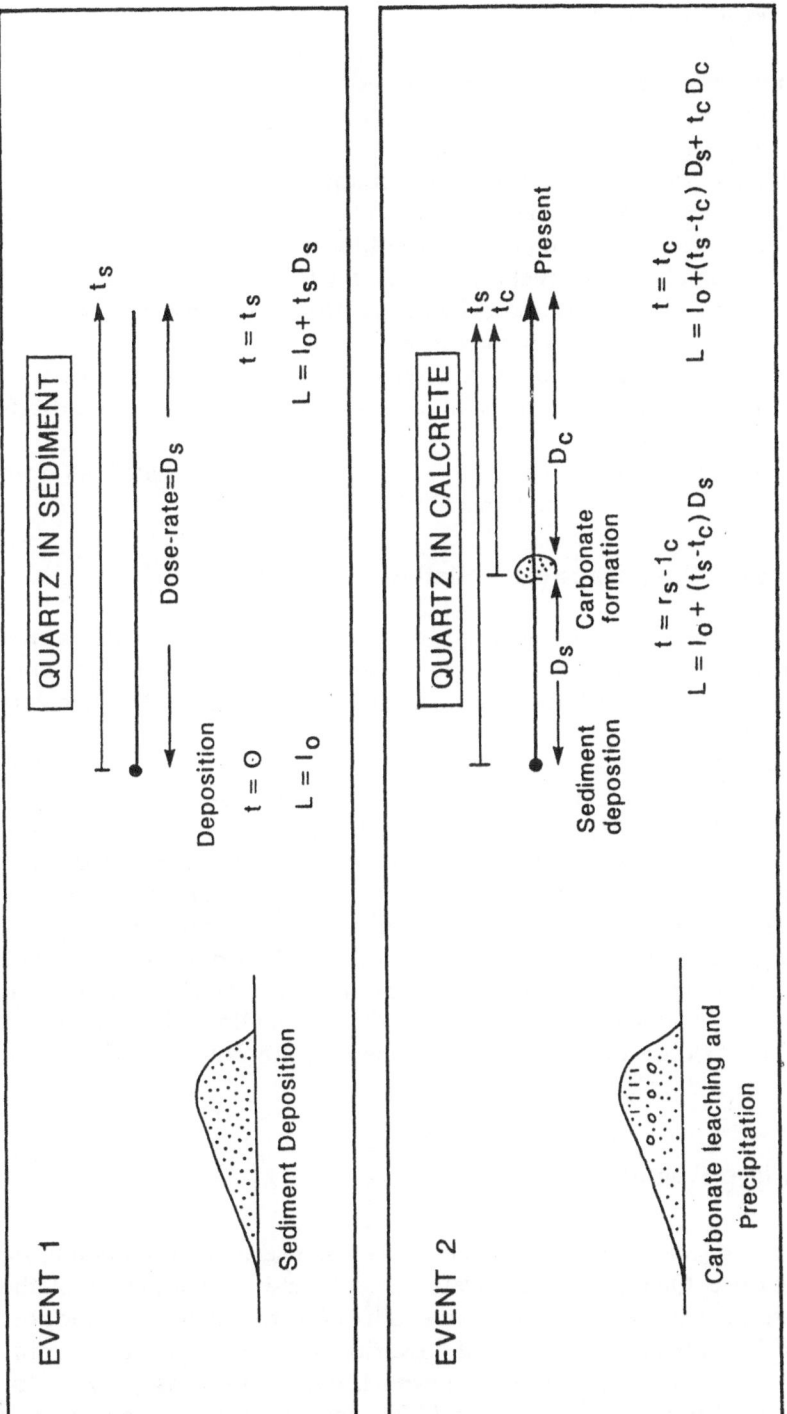

Fig. 10: Diagram illustrating the basic principles involved in the luminescence dating of calcrete nodules which enclose clastic mineral grains (quartz, feldspar) (after Singhvi et al.,1996).

These workers assumed that :
(1) carbonates, when formed, incorporate a few parts per million of uranium but a negligible amount of thorium isotopes;
(2) subsequent to its formation, the carbonate acts as a closed system with respect to U and Th isotopes.
The problems and limitations of the technique are as follows:
(1) Detrital contaminants may contain a measurable amount of Th and U. Since these detrital materials are older than the carbonate matrix, their elimination or correction is necessary to avoid overestimation of ages. Kaufmann (1971) and Schwarz (1989) have suggested methods for detrital contamination correction.
(2) Open system behaviour of calcretes poses a great problem since formation of mature calcretes involves durations of several hundreds and thousands of years.
Schlesinger (1985) correlated the C^{14} and Th/U ages of calcretes from Mojave desert and found the latter to be uniformly older.

Electron Spin Resonance (ESR) Dating

The Electron spin resonance technique determines the accumulated natural radiation dose received by a sample since deposition and annual dose rate at the site from which the sample is taken. These are used to calculate the ESR ages
A(yr) = AD/D
where AD is accumulated dose and D is annual dose.
Jacobson et al. (1988) used the ESR technique to date calcretes from Central Australia. The samples were taken from playas near Curting and were dated by ESR calibrated to ^{14}C dates. They obtained ESR dates in the range 22,000 - 27,000 yr. BP for the calcretes in the vadose zone, suggesting episodes of high rainfall during that period. The phreatic calcrete, below the water table, with ages in the range 34,000–75,000 yr. BP, provides evidence of older episodes of calcretization. The older dates obtained by Jacobson et al. (1988) demonstrate the potential of the ESR technique as a dating tool.

Concluding Remarks

Calcretes and calcareous soils/palaeosols are useful in understanding semi-arid/arid Quaternary landscapes and in the interpretation of the Phanerozoic stratigraphic record. In the latter case, they serve as indicators of subaerial exposure events and lowering of sea level, geomorphic stability, short and intermediate term climatic shifts, as proxies for vegetation biomass composition, and CO_2 concentration in palaeoatmospheres. Future work on macro- and micromorphology, luminescence signatures and stable isotopic composition of young calcretes (Late Pleistocene/Holocene) from various climatic zones, particularly semi-arid/ arid, should continue to seek insights into relationships between petrographic/chemical/isotopic signatures and climatic parameters (precipitation/temperature/seasonality). Besides adding to our understanding of the Quaternary landscapes, these data are potentially applicable to the stratigraphic record from the Early Proterozoic onwards— the time of the first occurrence of calcretes.

Acknowledgement

The authors are grateful to Sri C. Pattnaik who helped in the early stages of preparation of this manuscript. Dr. Nigel Woodcock, University of Cambridge, Cambridge (U.K.) kindly read the manuscript and offered several suggestions for improvement.

The persuasion of Prof. A.K. Singhvi, PRL, Ahmedabad and, the support of Dr. J.E. Andrews (University of East Anglia, Norwich, U.K.) and Prof. E. Derbyshire, Royal Holloway, London) helped us in completing the manuscript within the deadline. We express our sincere gratitude for the help given in preparation of the manuscript by Sri Mayank Jain, Dr. S.C. Bhatt and Sri S. Rana, Department of Geology, University of Delhi.

We thank Rosie Cullington, Phil Judges and Shiela Davies, University of East Anglia, Norwich, for assistance in secretarial, reprographic, and photographic work.

References

Aitken, J.F. and Flint, S.S. (1995). The application of high-resolution sequence stratigraphy to fluvial systems: a case study from the Upper Carboniferous Breathitt Group, Eastern Kentucky, USA. *Sedimentology* 42: 3-30.

Alam, M.S., Keppens, E. and Paepe, R. (1997). The use of oxygen and carbon isotope composition of pedogenic carbonates from Pleistocene palaeosols in NW Bangladesh, as palaeoclimatic indicators. *Quat. Sci. Rev.* 16: 161-168.

Allen, J.R.L. (1974). Studies in fluviatile sedimentation; implications of pedogenic carbonate units, Lower Old Red Sandstone, Anglo-Welsh outcrop. *Geol. J.* 11: 181-208.

Allen, J.R.L. (1986). Pedogenic carbonates in the Old Red Sandstone facies (Late Silurian-Early Carboniferous) of the Anglo-Welsh area, Southern Britain. In: *Palaeosols: Their Recognition and Interpretation*. Wright V.P. (ed.), 58-86. Blackwell Scientific, Oxford.

Amundson, R.G., Chadwick, O.A. Sowers, J.M., and Doner, H.E. (1988). Relationship between climate and vegetation and the stable carbon isotope chemistry of soils in the eastern Mojave Desert, Nevada. *Quat. Res.* 29: 245-254.

Amundson, R.G., Chadwick, O.A., Sowers, J.M. and Doner, H.E. (1989). The stable isotope geochemistry of pedogenic carbonates at Kyle Canyon, Nevada. *Soil Sci. Soc. Amer. J.* 53: 201-210.

Anand, R.R., Phang, C., Wildman, J.E. and Lintein, M.J. (1997). Genesis of some calcretes in the southern Yilgarn Craton, Western Australia—Implications for mineral exploration. *Aust. J. Earth. Sci.* 44: 87-103

Andrews, J.E., Tandon, S.K. and Dennis, P.F. (1995). Concentration of carbon dioxide in the Late Cretaceous atmosphere. *J. Geol. Soc. London* 152: 1-3.

Andrews, J.E., Singhvi, A.K., Kailath, A.J., Kuhn, R., Dennis, P.F., Tandon, S.K., Dhir, R.P. (1998) Do stable isotope data from calcrete record Late Pleistocene Monsoonal Climate Variation in the Thar Desert of India. *Quat. Res.* 50: 240-251.

Arakel, A.V. (1982). Genesis of calcrete in Quaternary soil profiles, Hutt and Leeman lagoons, western Australia. *J. Sed. Pet.* **52**: 109-125.

Aristarian, L.F. (1970). Chemical analyses of calcrete profiles from the High Plains, New Mexico. *J. Geology* 78: 201-212.

Bal, L. (1975a). Carbonate in soils: a theoretical consideration on, and proposal for its fabric analysis. I. *Netherlands J. Agri. Sci.* 23: 18-35.

Bal, L. (1975b). Carbonate in soils: a theoretical consideration on, and proposal for its fabric analysis. II. *Netherlands J. Agri. Sci.* 23: 163-176.

Berner, R.A. (1994). Geocarb 11: a revised model of atmospheric CO_2 over Phanerozoic time. *Amer. J. Sci.* 294: 56-91.

Blodgett, R.H. (1988). Calcareous palaeosols in the Triassic Dolores Formation, south-western Colorado. *Geol. Soc. Amer. Special Paper no.* 216, pp. 103-121.

Blumel, W.W. (1982). Calcretes in Nambia and S.E. Spain: relations to substratum, soil formation and geomorphic factors. In: *Aridic Soils and Geomorphic Processes Catena* (Supplement) D.H. Yaalon, (ed.), pp. 67-82.

Bodergat, A.M. (1974). Les Microcodiums, milieux et modes du dévelopement. Doc. Labu. *Geol. de Fac. Sci. Lyons* 62: 137-235.

Braithwaite, C.J.R. (1983). Calcrete and other soils in Quaternary limestones: structures, processes and applications. *J. Geol. Soc. London* 140: 351-363.

Braithwaite, C.J.R. (1989). Displacive calcite and grain breakage in sandstones. *J. Sed. Pet.* 59, 258-266.

Brewer, R. (1964). *Fabric and Mineral Analysis of Soils.* John Wiley and Sons, New York, pp. 470.

Brewer, R. (1976). *Fabric and Mineral Analysis of Soils.* R.E. Kreiger Publishing, Huntington, New York, 482pp.

Broecker, W.S. (1965). Isotope geochemistry and the Pleistocene climatic record. In: *The Quaternary of the United States*, H.E. Wright, and D. Frey, (eds.), pp. 737-753, Princeton University.

Buczynski, C., and Chafetz, H.S. (1987). Siliciclastic grain breakage and displacement due to carbonate crystal growth: an example from the Lueders Formation (Permian) of north-central Texas, USA. *Sedimentology* 33: 837-843.

Bullock, P., Fedoroff, M., Jongerius, A., Stoops, G. and Tursina, T. (1985). *Handbook for Soil Thin Section Description.* Waine Research, Wolverhampton.

Calvet, F. (1982). Constructive micrite envelops developed in vadose continental environments in Pleistocene eolianites of Mallorca (Spain). *Acta Geologica Hispanaca* 17: 169-178.

Carlisle, D. (1980). Possible variation in the calcrete-gypcrete uranium model. Open File Report US Department of Energy, GJBA-53(80), 38pp.

Carlisle, D. (1983). Concentration of uranium and vanadium in calcretes and gypcretes. In: *Geol. Soc. London Sp. Publ.* R.C.L. Wilson, (ed.), no. 11, pp. 185-195.

Cerling, T.E. (1984). The stable isotopic composition of modern soil carbonate and its relationship to climate. *Earth Planet. Sci. Lett.* 71: 229-240.

Cerling, T.E. (1991). Carbon dioxide in the atmosphere: Evidence from Mesozoic palaeosols. *Amer. J. Sci.* 291: 377-400.

Cerling, T.E. (1992). Use of carbon isotopes in palaeosols as an indicator of the pCO_2 of the palaeo-atmosphere. *Global Biogeochem. Cycles* 6: 307-314.

Cerling, T.E., and Quade, J. (1993). Stable carbon and oxygen isotopes in soil carbonates. In: *Continental Isotopic Indicators of Climate*, P. Swartz, K.C. Lohmann and J.A. McKenzie, (eds.), AGU Monograph.

Chadwick, O.A. and Nettleton, W.D. (1990). Micromorphological evidence of adhesive and cohesive forces in soil concentration. In: *Soil Micromorphology: A Basic and Applied Science*. L.A. Douglas (ed.) pp. 207-212, Elsevier, Amsterdam.

Chadwick, O.A., Sowers, J.M. and Amundson, R.G. (1989). Morphology of calcite crystals in clast coatings from four soils in the Mojave Desert region. *Soil Sci. Soc. Amer. J.* 52: 211-219.

Cohen, A.S. (1982). Palaeoenvironments of root casts from the Koobi Formation, Kenya. *J. Sed. Pet.* 52: 401-414.

Dhir, R.P., Kar, A., Wadhawan, S.K., Rajaguru, S.N., Misra, V.N., Singhvi, A.K., and Sharma, S.B. (1992). *Thar Desert in Rajasthan*. Geol. Soc. India, Bangalore.

Dijkmans, J.W.A., Koster, E.A., Galloway, J.P. and Mook, W.G. (1986). Characteristics and origin of calcretes in a subarctic environment, Great Kobuk sand dunes, Northwestern Alaska, USA. *Arctic and Alpine Research* 18: 377-386.

Drees, L.R. and Wilding L.P. (1987). Micromorphologic record and interpretations of carbonate forms in the Rolling Plains of Texas. *Geoderma* 40: 157-175.

Driese, S.G., Mora, C.I., Cotter, E. and Foreman, J.L. (1992). Palaeopedology and stable isotope geochemistry of Late Silurian vertic palaeosols, Bloomsburg Formation, Central Pennsylvania. *J. Sed. Pet.* 62: 825-841.

Driese, S.G. and Mora, C.I. (1993). Physico-chemical environment of pedogenic carbonate formation in Devonian vertic palaeosols, central Appalachians, USA. *Sedimentology* 40: 199-216.

Ducloux, J., Butel, P. and Dupuis, T. (1984). Microsequence mineralogique des carbonates de calcium dans une accumulation carbonate sous galets calcares, dans l'ouest de la France. *Pedologie* 34: 161-177.

Dupuis, T., Ducloux, J., Butel, P. and Nahon, D. (1984). Etude par spectrographic infrarouge d'un encroutement calcaire sans galet. Mise en evidence et modelisation experimental d'une suite minerale-volutive à partir de carbonate de calcium amorphe. *Clay Mineral* 19: 605-614.

Esteban, M. (1974). Calcite textures and *Microcodium*. *Bull. Soc. Geol. Italy* 92: 105-125.

Esteban, M. and Klappa, C.F. (1983). Subaerial exposure environment. In : *Carbonate Depositional Environments*. P.A. Scholle, D.G. Bebout and C.H. Moore, (eds.) *Amer. Ass. Petr. Geol. Mem.* 33: 1-54.

Freytet, P. (1973). Petrology and palaeoenvironment of carbonate continental deposits with particular reference to the Upper Cretaceous and Lower Eocene of Languedoc (southern France). *Sed. Geol.* 10: 25-60.

Ghosh, P., Bhattacharya, S.K. and Jani, R.A. (1995). Palaeoclimate and palaeovegetation in Central India during the Upper Cretaceous based on stable isotope composition of the palaeosol carbonates. *Palaeogeog., Palaeoclim., Palaeoecol.* 114: 285-296.

Gile, L.H. (1961). A classification of Ca horizons in soils of a desert region, Dona Ana County, New Mexico. *Soil Sci. Soc. Amer. Proc.* 25: 52-61.

Gile, L.H., Peterson, F.F. and Grossman, R.B. (1966). Morphological and genetic sequences of carbonate formation in desert soils. *Soil Science* 101: 347-360.

Gile, L.H., Hawley, J.W., and Grossman, R.B. (1981). Soils and geomorphology in the Basin and Range area of southern New Mexico: *Guidebook to the Desert Project. New Mexico Bureau of Mines and Mineral Resources Memoir* no. 39, 222 p.

Goodfriend, G.A. and Magaritz, M. (1988). Palaeosols and Late Pleistocene rainfall fluctuations in the Negev Desert. *Nature* 322: 144-146.

Goudie, A.S. (1973). *Duricrusts in Tropical and Subtropical Landscapes*. Clarendon, Oxford, 174 pp.

Goudie, A.S. (1983). Calcrete. In: *Chemical Sediments and Geomorphology: Precipitates and Residua in the Near-Surface Environment*. A.S. Goudie and K. Pye, (eds.). 93-131 Academic Press, London.

Goudie, A.S. (1996). Organic agency in calcrete development. *J. Arid Environments* 32: 103-110.

Hanneman, D.L. and Wideman, C.J. (1991). Sequence stratigraphy of Cenozoic continental rocks, southwestern Montana. *Geol. Soc. Amer. Bull.* 103: 1335-1345.

Harden, J.W. and Taylor, E.M. (1983). A qualitative comparison of soil development in four climate regimes. *Quat. Res.* 20: 342-359.

Harrison, R.S. (1977). Caliche profiles: indicators of near-surface subaerial diagenesis, Barbados, West Indies. *Bulletin of Canadian Petroleum Geology* 25: 123-173.

Harrison, R.S. and Steinen, R.P. (1978). Subaerial crusts, caliche profiles and breccia horizons. Comparisons of some Holocene and Mississippian exposure surfaces, Barbados and Kentucky. *Bull. Geol. Soc. Amer.* 89: 385-395.

Hay, R.L. and Reeder, R.J. (1978). Calcretes of Olduvai Gorge and Ndolanya Beds of Northern Tanzania. *Sedimentology* 25: 649-673.

Hubert, J.F. (1978). Palaeosol caliche in the New Haven Arkose, Newark Group, Connecticut. *Palaeogeog. Palaeoclim. Palaeoecol.* 24: 151-168.

Hutton, J.T. and Dixon, J.C. (1981). The chemistry and mineralogy of some South Australian calcretes and associated soft carbonates and their dolomitisation. *J. Geol. Soc. Australia*, 28: 71-79.

Jacobson, G., Arakel, A.V. and Yijian, C. (1988). The Central Australian groundwater discharge zone: evolution of associated calcrete and gypcrete deposits. *Australian J. Earth Sciences* 35: 549-565.

James, N.P. (1972). Holocene and Pleistocene calcareous crust (caliche) profiles: criteria for subaerial exposure. *J. Sed. Pet.* 42: 817-836.

Jenny, H. (1941). Factors of Soil Formation—*A System of Quantitative Pedology*, McGraw-Hill Book Company, New York, 281 pp.

Joeckel, R.M. (1994). Virgilian (Upper Pennsylvanian) palaeosols in the Upper Lawrence Formation (Douglas Group) and in the Snyderville Shale Member (Oread Formation, Shawnee Group) of the northern midcontinent, USA: pedological contrasts in a cyclothem sequence. *J. Sed. Res.* A64: 853-866.

Joeckel, R.M. (1995). Palaeosols below the Ames Marine limit (Upper Pennsylvanian Conemaugh Group) in the Appalachian Basin, USA: Variability on an ancient depositional landscape. *J. Sed. Res.* A65: 393-407.

Junge, C.E. and Werby, R.T. (1958). The concentration of chloride, sodium, potassium, calcium, and sulfate in rainwater over the United States. *J. Meteorology* 15: 417-425.

Kahle, C.F. (1977). Origin of subaerial Holocene calcareous crusts: role of algae, fungi and sparmicritisation. *Sedimentology* 24: 413-435.

Kauffman, A. (1971). U-series dating of Dead Sea basin carbonates. *Geochimica et Cosmochimica Acta* 39: 1269-1281.

Khadkikar, A.S., Merh, S.S., Malik, J.N. and Chamyal, L.S. (1998). Calcretes in Semi-arid alluvial systems—Formative pathways and sinks. *Sed. Geol.* 116: 251-260

Klappa, C.F. (1978). Biolithogenesis of *Microcodium*: elucidation. *Sedimentology* 25: 489-522.

Klappa, C.F. (1979). Calcified filaments in Quaternary calcretes: organo-mineral interactions in the subaerial vadose environment. *J. Sed. Pet.* 49: 955-968.

Klappa, C.F. (1980a). Rhizoliths in terrestrial carbonates: classification, recognition, genesis and significance. *Sedimentology* 27: 613-629.

Klappa, C.F. (1980b). Brecciation textures and tepee structures in Quaternary calcrete (caliche) profiles from eastern Spain: the plant factor in their formation. *Geol. J.* 15: 81-89.

Knox, G.F. (1977). Caliche profile formation, Saldanha Bay (South Africa). *Sedimentology* 24: 657-674.

Ku, T.L., Bull, W.B., Freeman, S.T. and Knauss, K.V. (1979). $Th^{230} - U^{234}$ dating of pedogenic carbonates in gravelly desert soils of Vidal Valley, Southeastern California. *Geol. Soc. Amer. Bull.* 90: 1063-1073.

Leeder, M.R. (1975). Pedogenic carbonates and flood sediment accretion rates: a quantitative model for alluvial arid zone lithofacies. *Geol. Mag* 112: 257-270.

Machette, M.N. (1985). Calcic soils of the south western United States. In: *Soils and Quaternary Geology of the Southwest United States.* D.L. Weide, (ed.) *Geol. Soc. Amer.* Special Paper No. 203: pp. 1-21.

Mack, G.H., Cole, D.R., Giordana T.H., Schaal, W.C. and Barcelos, J.H. (1991). Palaeoclimatic controls on stable oxygen and carbon isotopes in caliche of the Abo Fórmation (Permian), South-Central New Mexico, USA. *J. Sed. Pet.* 61: 458-472.

Mack, G.H. and James, W.C. (1992). Calcic palaeosols of the Plio-Pleistocene Camp Rice and Palomas Formations, southern Rio Grande Rift, USA. *Sed. Geol.* 77: 89-109.

Mack, G.H., James, W. C., and Monger, H.C. (1993). Classification of palaeosols. *Geol. Soc. Amer. Bull.* 105: 129-136.

Mack, G.H., Cole, D.R., James, W.C., Giodano, T.H., and Salgards, S.L. (1994). Stable oxygen and carbon isotopes of pedogenic carbonate as indicators of Plio-Pleistocene palaeoclimate in the southern Rio Grande rift, south-central New Mexico. *Amer. J. Sci.* 294: 621-640.

Magaritz, M. (1986). Environmental changes recorded in the Upper Pleistocene along the desert boundary, Southern Israel. *Palaeogeog., Palaeoclim., Palaeoecol.* 53: 213-229.

Magaritz, M., Kaufman, A., and Yaalon, D. (1981). Calcium carbonate nodules in soils: $^{18}O/^{16}O$ and $^{13}C/^{12}C$ ratios and ^{14}C contents. *Geoderma* 25: 157-172.

Mann, A.W. and Horowitz, R.C. (1979). Groundwater calcrete deposits in Australia: some observations from western Australia. *J. Geol. Soc. Australia* 26: 293-303.

Marion, G.M., Schlesinger, W.H., and Fonteyn, P.J. (1985). Caldep: a regional model for soil $CaCO_3$(caliche) deposition in southwestern deserts. *Soil Science* 139: 468-481.

Marriott, S.B., and Wright, V.P. (1993). Palaeosols as indicators of geomorphic stability in two Old Red Sandstone suites, South Wales. *J. Geol. Soc. London* 150: 1109-1120.

McFadden, L.D., and Tinsley, J.C. (1985). Rate and depth of pedogenic carbonate accumulation in soils: formulation and testing of a compartment model. In: *Soils and Quaternary Geology of the Southwestern United States* D.L. Weide (ed.). *Geol. Soc. Amer D.L. Weide (ed.) Special Paper* 203: 23-42.

Milnes, A.R. (1992). Calcrete. In: Weathering, Soils and Palaeosols. I.P. Martini and W. Chesworth, (eds.), pp. 309-345. Elsevier, Amsterdam.

Milnes, A.R. and Hutton, J.T. (1983). Calcretes in Australia. In: *Soils: An Australian Viewpoint.*, pp. 119-162. CSIRO (Academic Press, London), Melbourne, pp. 119-162.

Monger, H.C., Dougherty, L.A., Lindemann, W.C., and Liddell, C.M. (1991a). Microbial precipitation of pedogenic calcite. *Geology* 19: 997-1000.

Monger, H.C., and Dougherty, L.A. (1991b). Pressure solution: possible mechanism for silicate grain dissolution in a petrocalcic horizon. *Soil Sci. Soc. Amer. J.* 55: 1625-1629.

Mora, C.I., Driese, S.G., and Seager, P.G. (1991). Carbon dioxide in the Paleozoic atmosphere: evidence from C-isotopic compositions of pedogenic carbonate. *Geology* 9: 1017-1020.

Mora, C.I., Driese, S.G., and Fastovsky, D.E. (1993). *Geochemistry and Stable Isotopes of Paleosols. Geol. Soc. Amer.* Continuing Education Short Course.

Mora, C.I., Dreise, S.G., and Colarusso, L.A. (1996). Middle to Late Palaeozoic atmospheric CO_2 levels from soil carbonate and organic matter. *Science* 271: 1105-1107.

Mount, J.F. and Cohen, A.S. (1984). Petrology and geochemistry of rhizoliths from Plio-Pleistocene fluvial and marginal lacustrine deposits, East Lake Turkana, Kenya. *J. Sed. Pet.* 54: 263-275.

Multer, H.G. and Hoffmeister, J.E. (1968). Subaerial laminated crusts of the Florida Keys. *Geol. Soc. Amer. Bull.* 79: 183-192.

Nagtegaal, P.J.C. (1969). Microtextures in recent and fossil caliches. *Leid. Geol. Meded.* 42: 131-142.

Nahon, W. (1976). Curvasses ferrugineuses et encroutements calcaires au Senegal accidental et an Mauritanie. Systémes évolutifs: Géochimie structures, relais et coéxistence. *Sci. Geol. Mem.* 44.

Nahon, D., Paquet, H., Ruellan, A. and Millot, G. (1975). Encroutements calcaires dans les altérations des marines eócénes de la falaise de Thiés (Sénegal): Organisation et morphologie. *Sci. Geol. Bull.* 28: 29-46.

Nahon, D., Ducloux, J., Butel, P., Augas, C. and Paquet, H. (1980). Néoformations d'aragonite, premiére etape d'une suite minéralogique évolutive dans les encroûtements calcaires. *C.R. Acad. Sci. Paris* 291: 725-727.

Netterberg, F. (1967). Some road making properties of South African calcretes. In *Proc. 4th Reg. Conf. African Soil Mechanics and Found. Engg.*, pp. 77-81, Cape Town.

Netterberg, F. (1969). The interpretation of some basic calcrete types. *South African Arch. Bull.* 24: 117-122.

Netterberg, F. (1980). Geology of South African calcretes: I. Terminology, description, macrofeatures and classification. *Trans. Geol. Soc. South Africa* 83: 255-283.

Nickel, E. (1985). Carbonates in alluvial fan systems, an approach to physiography, sedimentology and diagenesis. *Sed. Geol.* 42: 83-104.

Patil, D.N., and Surana, A.P. (1992). Origin of the calcrete deposits of Saswad-Mira area, western Maharashtra, India. *J. Geol. Soc. India* 39: 105-117.

Pendall, E.G. and Harden, J.W. (1994). Isotopic approach to soil carbonate dynamics and implications for palaeoclimatic interpretations. *Quat. Res.* 42, 60-71.

Phillips, S.E., Milnes, A.R. and Foster, R.C. (1987). Calcified filaments: an example of biological influences in the formation of calcrete in South Australia. *Aust. J. Soil Res.* 25: 405-428.

Platt, N.H. (1989). Lacustrine carbonates and pedogenesis: sedimentology and origin of palustrine deposits from the Early Cretaceous Rupelo Formation, West Caverns Basin, North Spain. *Sedimentology* 36: 665-684.

Purvis, K. and Wright, V.P. (1991). Calcretes related to phreatophytic vegetation from the Middle Triassic Otter Sandstone of southwest England. *Sedimentology* 38: 539-551.

Quade, J., Cerling, T.E. and Bowman, J.R. (1989). Development of Asian monsoon revealed by marked ecological shift during the latest Miocene in northern Pakistan. *Nature* 342, 163-166.

Rabenhorst, M.C., Wilding, L.P. and West, L.T. (1984). Identification of pedogenic carbonates using isotope and microfabric analyses. *Soil Sci. Soc. Amer. J.* 48: 125-132.

Rabenhorst, M.C. and Wilding, L.P. (1986). Pedogenesis on the Edwards Plateau, Texas. III. New model for the formation of petrocalcic horizons. *Soil Sci. Soc. Amer. J.* 50: 693-695.

Rao, V.P. and Thamban, M. (1997). Dune associated calcretes rhizoliths and paleosols from the western Continental Shelf of India. *J. Geol. Soc. Ind.* 49: 297-306.

Read, J.F. (1974). Calcrete deposits and Quaternary sediments, Edel Province, Shark Bay, Western Australia. *Mem. Amer. Assoc. Pet. Geol.* 22: 250-282.

Reeves, C.C. (1976). *Caliche: Origin, Classification, Morphology, Land Uses*. Estacado Books, Lubbock, 233 pp.

Reeves, C.C. (1983). Pliocene channel calcrete and suspen-parallel drainage in West Texas and New Mexico. In *Residual Deposits*. R.C.L. Wilson, (ed.), pp. 179-183. *Geol. Soc. London*, Special Publication.

Retallack, G.J. (1990). *Soils of the Past—An Introduction to Palaeopedology*. Unwin Hyman, Boston (Massachussets), 520pp.

Retallack, G.J. (1992). A new compilation of depth to calcic horizons in soils for interpreting former rainfall from palaeosols. *Geol. Soc. Amer. Abstracts with Prog.* 24: A227.

Robinson, D. and Wright, V.P. (1987). Ordered illite/smectite and kaolinite/smectite as possible primary minerals in a Lower Carboniferous palaeosol sequence, South Wales *Clay Minerals* 22: 109-118.

Saigal, G.C. and Walton, E.K. (1988). On the occurrence of displacive calcite in Lower Old Red Sandstone of Carnoustie, eastern Scotland. *J. Sed. Pet.* 58: 131-135.

Salomons, W., Goudie, A.S. and Mook, W.G. (1978). Isotopic composition of calcrete deposits from Europe, Africa and India. *Earth Surface Processes* 3: 43-57.

Schlesinger, W.H. (1985). The formation of calcite in soils of the Mojave Desert, California. *Geochim. Cosmochim. Acta* 49: 57-66.

Schwarz, H.P. (1989). Uranium series dating of Quaternary deposits. *Quat. Internat.* 1: 7-17.

Sehgal, J.L. and Stoops, G. (1972). Pedogenic calcite accumulations in arid and semi-arid regions of the Indo-Gangetic alluvial plain of erstwhile Punjab (India). *Geoderma* 8: 59-72.

Semenuik, V. and Meagher, T.D. (1981). Calcrete in Quaternary coastal dunes in southwestern Australia: a capillary-rise phenomenon associated with plants. *J. Sed. Pet.* 51: 47-68.

Semenuik, V. and Searle, D.J. (1985). Distribution of calcrete in Holocene coastal sands in relationship to climate, southwestern Australia. *J. Sed. Pet.* 55: 86-95.

Singh, D. and Lal, G. (1945). Kankar composition as an index of the nature of soil profile. *Indian J. Agri. Sci.* 16: 328.

Singh, S. and Singh, L. (1972). Chemical and morphological composition of kankar nodules in soils of the Vindhyan region of Mirzapur, India. *Geoderma*, pp. 269-276.

Singhvi, A.K., Banerjee, D., Ramesh, R., Rajaguru, S.N. and Gogte, V. (1996). A luminescence method for dating 'dirty' pedogenic carbonates for palaeoenvironmental reconstruction. *Earth Planet. Sci. Lett.* 139: 321-332.

Slate, J.L., Smith, G.A., Wang, Y. and Cerling, T.E. (1996). Carbonate-Palaeosol genesis in the Plio-Pleistocene St. David Formation, southeastern Arizona. *J. Sed. Res.* 66: 85-94.

Solomon, S.T. and Walkden, G.M. (1985). The application of cathodoluminescence to interpreting the diagenesis of an ancient calcrete profile. *Sedimentology* 32: 877-896.

Sommer, S.E. (1972). Cathodoluminescence of carbonates II. *Geological Applications. Chem. Geol.* 9: 275-284.

Steel, R.J. (1974). Cornstone (fossil calcite) its origin, stratigraphic and sedimentologic importance in the New Red Sandstone, Western Scotland. *J. Geol.* 82: 351-369.

Swett, K. (1974). Calcrete crusts in Arctic permaforst environments. *Amer. J. Sci.* 274: 1050-1063.

Taber, S. (1916). The growth of crystals under external pressure. *Amer. J. Sci.* 41: 532-556.

Talma, A.S. and Netterberg, F. (1983). Stable isotope abundances in calcretes. In: *Residual Deposits.* R.C.L. Wilson, (ed.), pp. 221-233.

Tandon, S.K. and Narayan, D. (1981). Calcrete conglomerate, case hardened conglomerate and cornstone—comparative account of pedogenic and non-pedogenic carbonates from the continental Siwalik Group, Punjab, India. *Sedimentology* 28: 353-367.

Tandon, S.K. and Friend, P.F. (1989). Near-surface shrinkage and carbonate replacement processes, Arran Cornstone Formation, Scotland. *Sedimentology* 36: 1113-1126.

Tandon, S.K., and Gibling, M.R. (1994). Calcrete and coal in Late Carboniferous cyclothems of Nova Scotia, Canada: climate and sea-level changes linked. *Geology* 22: 755-758.

Tandon, S.K., Sood, A., Andrews, J.E., and Dennis, P.F. (1995). Palaeoenvironments of the dinosaur bearing Lameta beds (Maastrichtian), Narmada Valley, Central India. *Palaeogeog., Palaeoclim., Palaeoecol.* 117: 123-154.

Tandon, S.K. and Gibling, M.R. (1997). Calcretes at sequence boundaries in Upper Carboniferous cyclothems of the Sydney Basin, Atlantic Canada. *Sed. Geol.* 112: 43-67.

Tandon, S.K., Andrews, J.E., Sood, A. and Mittal, S. (1998).Shrinkage and sediment supply control on Multiple Calcrete profile development: a case study from the Maastrichtian of Central India. *Soil Geol.* 119: 25-45.

Van Wagoner, J.C., Posamentier, H.W., Mitchum, R.M., Vail, P.R., Sarg, J.F., Loutit, T.S. and Hardenbol, J. (1988). An overview of the fundamentals of sequence stratigraphy and key definitions. In: *Sea Level Changes: An Integrated Approach*. C.K. Wilgus, B.S. Hastings, C.G.S.C. Kendall, H.W. Posamentier, C.A. Ross, and J.C. Van Wagoner, (eds.), SEPM Special Publ. pp. 39-45.

Verrecchia, E.P. and Verrecchia, K.E. (1994). Needle fibre calcite: A critical review and a proposed classification. *J. Sed. Res.* A64: 650-664.

Wang, Y., Nahon, D. and Merino, E. (1994). Dynamic model of the genesis of Calcretes replacing silicate rocks in semi-arid regions. *Geochim. Cosmochim. Acta* 58: 5131-5145.

Watts, N.L. (1977). Pseudo-anticlines and other structures in some calcretes of Botswana and South Africa. *Earth Surface Proc. Landforms* 2: 63-74.

Watts, N.L. (1978). Displacive calcite: evidence from recent and ancient calcretes. *Geology* 6: 699-703.

Watts, N.L. (1980). Quaternary pedogenic calcretes from the Kalahari (southern Africa): mineralogy, genesis and diagenesis. *Sedimentology* 27: 661-686.

Wieder, M. and Yaalon, D. (1974). Effect of matrix composition on carbonate nodule crystallisation. *Geoderma* 11: 95-121.

Wieder, M. and Yaalon, D.H. (1982). Micromorphological fabrics and developmental stages of carbonate nodular forms related to soil characteristics. *Geoderma* 28: 203-220.

Williams, G.E. and Polach, H.A. (1969). The evaluation of ^{14}C ages for soil carbonate from the arid zone. *Earth Planet. Sci. Lett.* 7: 240-242.

Williams, G.E. and Polach, H.A. (1971). Radiocarbon dating of arid-zone calcareous palaeosols. *Geol. Soc. Amer. Bull.* 82: 3069-3086.

Wright, V.P. (1990a). A micromorphological classification of fossil and recent calcic and petrocalcic microstructures. In: *Soil Micromorphology: A Basic and Applied Science* L. A. Douglas, (ed.), Elsevier, Amsterdam vol. 19, pp. 401-407.

Wright, V.P. (1990b). Estimating rates of calcrete formation and sediment accretion in ancient alluvial deposits. *Geol. Mag.* 127: 273-276.

Wright, V.P. (1994). Palaeosols in shallow marine carbonate sequences. *Earth Sci. Rev.* 35: 367-395.

Wright, V.P. and Peeters, C. (1989). Origins of some early Carboniferous calcrete fabrics revealed by cathodoluminescence: implications for interpreting the sites of calcrete formation. *Sed. Geol.* 65: 345-353.

Wright, V.P. and Tucker, M.E. (1991). Calcretes: an introduction. In: *Calcrete. Internat. Assoc. Sedimentologists Reprint Series* 2: 1-22.

Wright, V.P. and Vanstone, S.D. (1991). Assessing the carbon dioxide content of ancient atmospheres using palaeocalcretes: theoretical and empirical constraints. *J. Geol. Soc. London* 148: 945-947.

Wright, V.P., Turner, M.S., Andrews, J.E. and Spiro, B. (1993). Morphology and significance of super-mature calcretes from the Upper Old Red Sandstone of Scotland. *J. Geol. Soc. London* 150: 871-883.

Yaalon, D.H. (1988). Calcic horizon and calcrete in Aridic soils and palaeosols: progress in the last twenty two years. *Soil Sci. Soc. Amer. Agron. Abstracts.*

6 Dunes as Indicators of Climatic Change

K.W. GLENNIE[*]

ABSTRACT

Relatively simple geometrical patterns of sand-dune morphology in modern sand seas can be used to infer past climatic change, including changes in the direction of prevailing winds and to a lesser extent, of long-term humidity. After time spans of several millenia since they were formed, especially the larger (mega) dunes of many areas are out of equilibrium with modern wind systems, and as a result their original morphology has been altered. The reason for these changes in wind direction is associated with the growth and collapse of Northern Hemisphere glaciations, which had a periodicity of about 100 ka during the later Quaternary, and the effect that they had on global wind systems. The Emirates, situated in the path of an important sand-transporting wind, is in an ideal location for showing these effects. The Arabian Gulf dried out during the last and earlier glaciations, and had carbonate-rich dunes migrating over its exposed floor; that supply of sand ceased with each post-Glacial flooding, causing deflation of dunes immediately down wind of the new coast. Gypsum-rich plant-root moulds indicate proximity to a desert water table, which could signal a change from interglacial humidity to a more arid climate.

Introduction

Sand-dunes are built of sand-size grains that are capable of being moved by the wind; and winds form a very important aspect of our global climate. Change the climate, locally or globally, and the wind systems will adjust to the new conditions by altering in both strength and direction of flow. Change the climate, and dunes will also adjust to the new conditions, whether by changing shape under a new wind regime, by becoming vegetated in a more humid climate, or even, in extreme cases, by converting interdune areas into interdune lakes. In each case, the evidence will be there for interpretation, which, essentially, is the topic of

[*]Department of Geology & Petroleum Geology, University of Aberdeen, Aberdeen, Scotland, UK.

this paper. But first, the scene needs to be set by describing how aeolian dunes form and migrate.

Dune Morphology and the Controlling Wind Direction

A sand-dune is an accumulation of wind-blown sand that possesses one or more avalanche slopes. Its size depends on the availability of sand and its height on the ability of the wind to carry sand to the top without removing it again. The finest sand grains are usually found at the dune's crest. A dune migrates by the removal of sand from its windward end (deflation) and redeposition at its downwind margin.

Aeolian dunes have a wide variety of shapes and sizes, but when considering how they form, it is possibly advantageous to describe two relatively simple end members:

1. Transverse Dunes which, as their name implies, have their long axes at right-angles to the dune-forming wind (Fig. 1A). Crescent-shaped barchans form where the supply of sand is limited, the interdune areas are commonly sand free, but large volumes of sand can be trapped where the axes of transverse dunes are relatively continuous.

2. Longitudinal Dunes, whose long axes are essentially parallel to the winds that formed them (Fig. 1B); they grow by adding sand to their down-wind end and, in some circumstances, also at their up-wind end. When sand is in short supply, the longitudinal dune migrates parallel to its axis by losing sand from its up-wind end and adding sand downwind.

Hopefully, an understanding of the origin of these end-member elements will lead to an appreciation of how more complex dunes are built, and of the wind systems that formed them.

Bagnold (1940) made several important observations concerning the interaction between wind and dry sand, which are important for a study of the origins of dunes of different types, and of the modifications they may be subjected to if the climate changes:

1. "a given wind can drive sand over a hard immobile surface at a considerably greater rate than is allowed by the loose sand-covered surface".
2. "a strong wind causes an accretion of sand on an existing sand patch together with an extension up-wind of the border...   ...this action lasts only as long as there is a plentiful supply of sand...".
3. "in a strong sand-laden wind a uniform drift of sand over a uniform rough surface has a transverse instability, so that sand tends

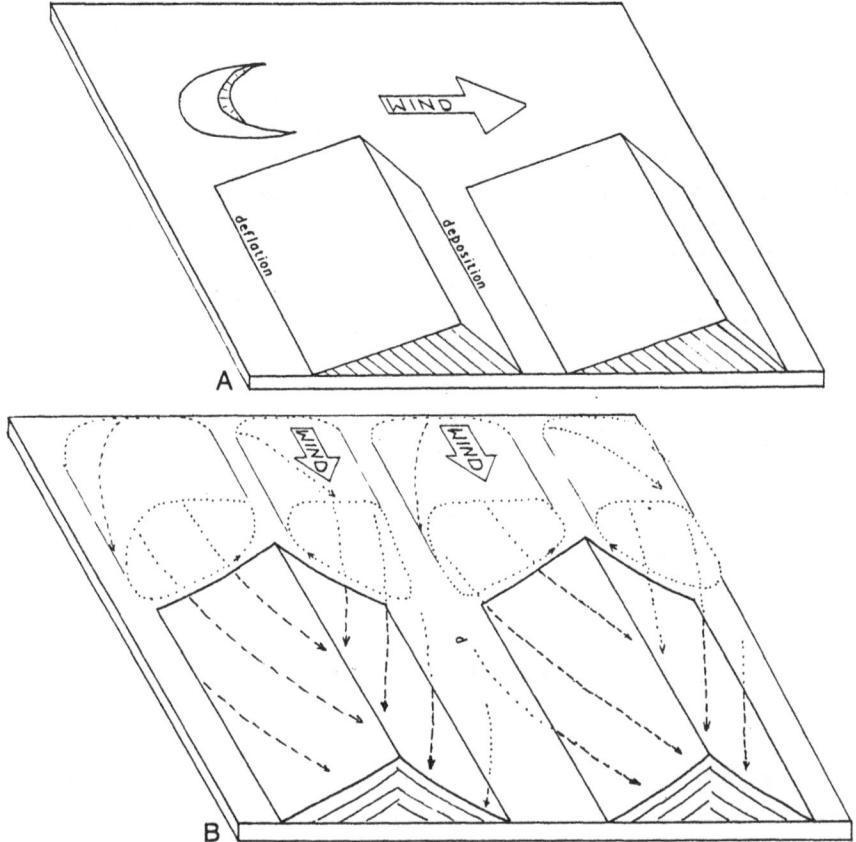

Fig. 1: Isometric sketches of the relationship between dune geometry and wind direction. A. Transverse and Barchan Dunes, which migrate by deflation of sand at the up-wind edge of the dune and re-deposition at the leeward edge. B. Longitudinal Dunes, thought to be formed by contra-rotating helicoidal wind vortices. The up-wind end of these dunes will recede down-wind if there is a shortage of sand.

to deposit in longitudinal strips".

4. "sand movement is proportional to the cube of the excess wind velocity above that at which sand begins to move"

Observation 1 implies that sand will tend to accumulate over areas that already have a cover of sand, and thus leads automatically to observation 2. This is because more energy is absorbed by the loose sand than by the hard surface. Combined with observation 2, it also implies that sand will naturally tend to be swept from hard interdune areas towards existing patches of sand, which eventually grow into dunes. Under conditions of moderate winds and a plentiful supply of sand, the natural dune form has its long axis at right-angles to the wind (Fig. 1A), although, according to Cooke et al. (1993), their outline tends to be linguoid or barchanoid rather than straight. With a reduced supply of sand, the barchanoid dune may be replaced by individual barchans in which the two horns point down-wind. The linear shape of the barchan's horns has affinities with the other end-member, the longitudinal dune. Bagnold's third observation indicates that transverse dunes are much less stable in very strong winds and, via 'blow-outs' and parabolic dune forms, may be re-shaped to longitudinal patterns in which the dune axes are essentially parallel to the prevailing sand-transporting wind (Fig. 1B). Observation 4 lays great stress on the ability of strong winds to move large volumes of sand in a relatively short time; this factor will assume great importance when considering times in the past history of deserts when winds seem to have been generally stronger than at present.

The above outline of the origins of dune types is certainly over simplified, but it contains a strong element of truth. The theme I wish to follow concerns mainly the interpretation of wind directions and their strengths, past and present, as indicators of climatic change, based on the superimposition of simple end-member dune morphologies.

Quaternary Glacial Cyclicity and its Control on Desert Climate

If the axial trend of longitudinal sand dunes is an indication of the flow direction of the winds that formed them, this information can be used to deduce the size and shape of the sub-tropical high-pressure cells that were centred over the Sahara and Arabian deserts, for instance, when the dunes were constructed.

Winds are the outcome of nature trying to even out the differences between areas of high and low atmospheric pressure, the shorter the distance between the two, the higher the wind velocity.

The Earth can be considered as being ringed by four bands of high atmospheric pressure, separated by zones of low pressure, of which the central one is the low-pressure equatorial 'Doldrums' (Fig. 2A). Two important factors control their origin and distribution. The first is because of temperature differences between the Equator and the Poles, and the second is because of the change in the Earth's circumference between the Equator (about 40 000 km) and the Poles (zero).

In an ideal model, hot air rises over the Equator and attempts to reach the Poles at high altitude, with cold air flowing towards the Equator at ground level to replace it. Because of the shorter circumference when the high-altitude air reaches about 30° N or S of the Equator, the air becomes concentrated and thus heavier, and it sinks, warming as it does so, in an

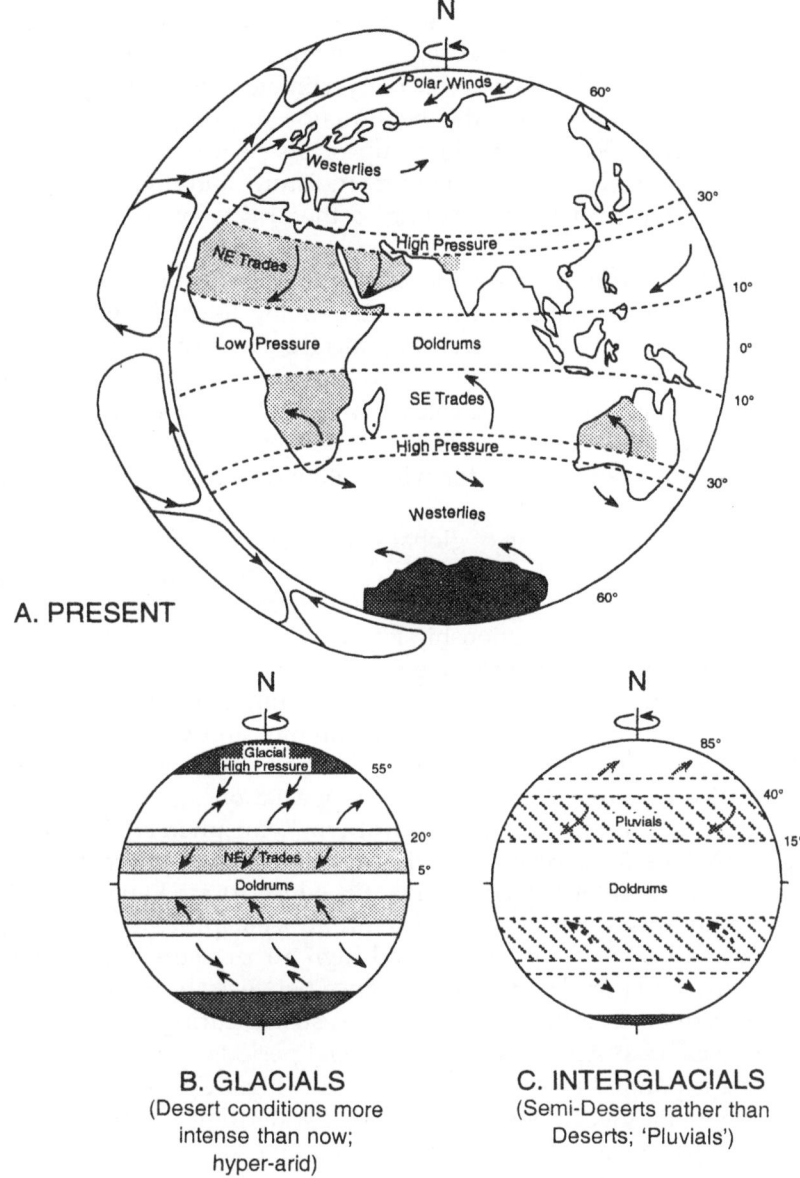

Fig. 2: Cartoon of the Earth showing the pattern of air flow between the Equator and the Poles, and the resulting alternating zones of high and low atmospheric pressure. Under the influence of high-latitude glaciations, the zones are squeezed towards the Equator (Fig. 2B) and in interglacial periods expand towards the Poles (Fig. 2C), with their respective influence on wind velocities.

essentially cloud-free sky. The sinking air divides into two parts. Over the western Mediterranean, for example, the northern part cools as it moves towards the North Pole; since it cannot now retain its content of moisture, it gives NW Europe its rainy climate. The other part of the descending air flows towards the Equator to complete the tropical convection cycle. Because it warms up on its near-surface route to the Equator, it is capable of absorbing more moisture, and thus desiccates the ground over which it flows, leading to the aridity of the Saharan and Arabian deserts.

The final important factor is the Earth's rotation, which, looking down on the North Pole, is anticlockwise. Air moving towards the Equator is crossing land that has an increasing surface velocity, which results in winds that apparently are deflected to the west (NE Trade Winds over the North Atlantic and Pacific Oceans and SE Trade winds south of the Equator, leading to the term 'Trade Wind deserts' for Arabia and the Sahara). The converse is true for winds blowing towards the poles; they veer to the east.

The above simple outline of global wind systems is complicated by the distribution of land and sea, which does not match the outline of the ideal air-pressure cells, and because land heats and cools much more rapidly than water; this relationship leads to the development of monsoon winds in the tropics, which reverse their sense of direction seasonally, as the land is relatively hotter than the sea in summer, and cooler in winter.

Just as climates today vary between summer and winter, particularly in high-latitude areas, so in the past, global climates were controlled by the presence or absence of high-latitude glaciations, especially those of Quaternary age in the sub-polar northern hemisphere. Large ice-caps over North America and Scandinavia, several thousand kilometres across, were overlain by intensely cold dense air, which formed an equally large area of high atmospheric pressure. With an ice-cap of similar size over Antarctica, all other zones of low and high air pressure were squeezed towards the Equator. The smaller distance between the zones is thought to have resulted in strong global wind systems during glaciations (Fig. 2B), and weaker systems during interglacial periods (such as now) when each zone expanded away from the Equator (Fig. 2C).

During major glaciations, the amount of water that went into building large ice-caps caused global sea level to fall by 100m or more (Fig. 3). At the peak of the last glaciation, probably the whole of the Arabian Gulf was dry, the site of sand-dunes migrating over its exposed surface towards the Emirates and on to the Rub'al Khali. Although global climates are known to have been several degrees cooler than those experienced today, even in the desert, the associated strong winds must have had a strong

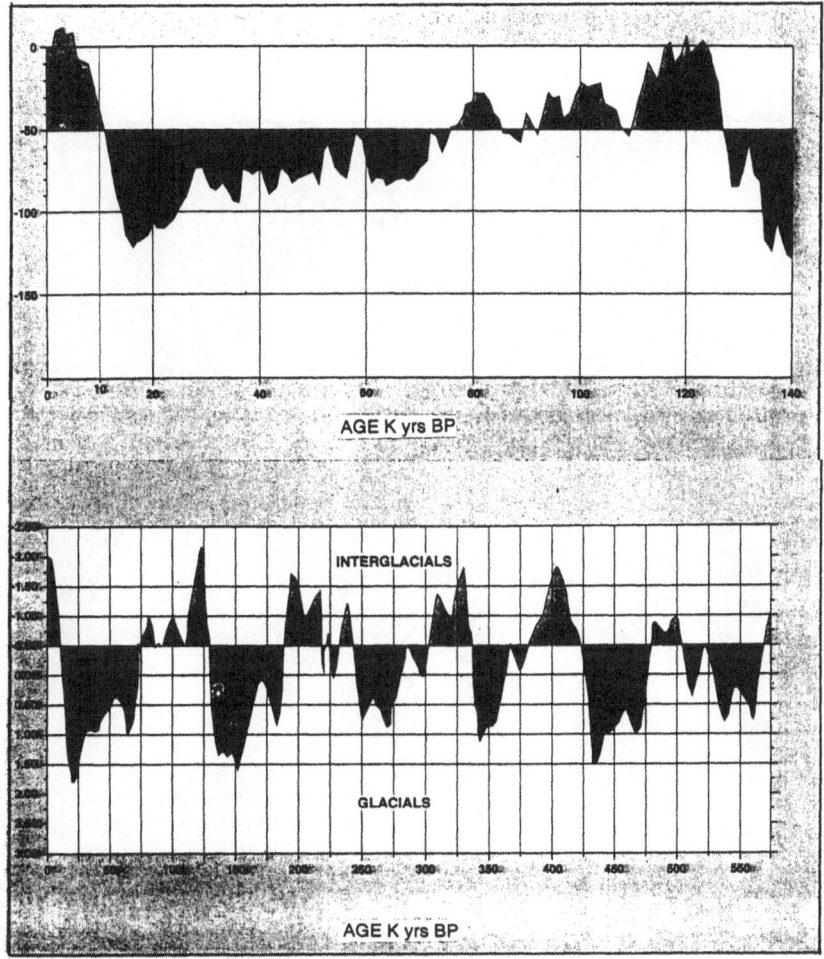

Fig. 3: Sea-level curves based on isotopes of oxygen from deep-sea ocean cores.
A. Last Glacial Cycle., simplified from Shackleton (1987). Depth in metres
B. Last six glaciations and interglacials modified from Boulton (1993).
Horizontal line separating glacials and interglacials is arbitrary.

desiccating effect, leading to a hyperarid climate. At the end of the glaciation, sea level rose again (Fig. 4), the supply of sand from the Gulf was cut off, and near-coastal dunes were deflated down to the level of the water table, which then became the sites of both inland and coastal sabkhas (Figs. 5 and 6). With increasingly weaker wind systems, and a closer proximity to a source of moisture, Arabia's climate became warmer and more humid (the so-called 'Climatic Optimum'), with convection-controlled thunderstorms probably providing much of the rainfall.

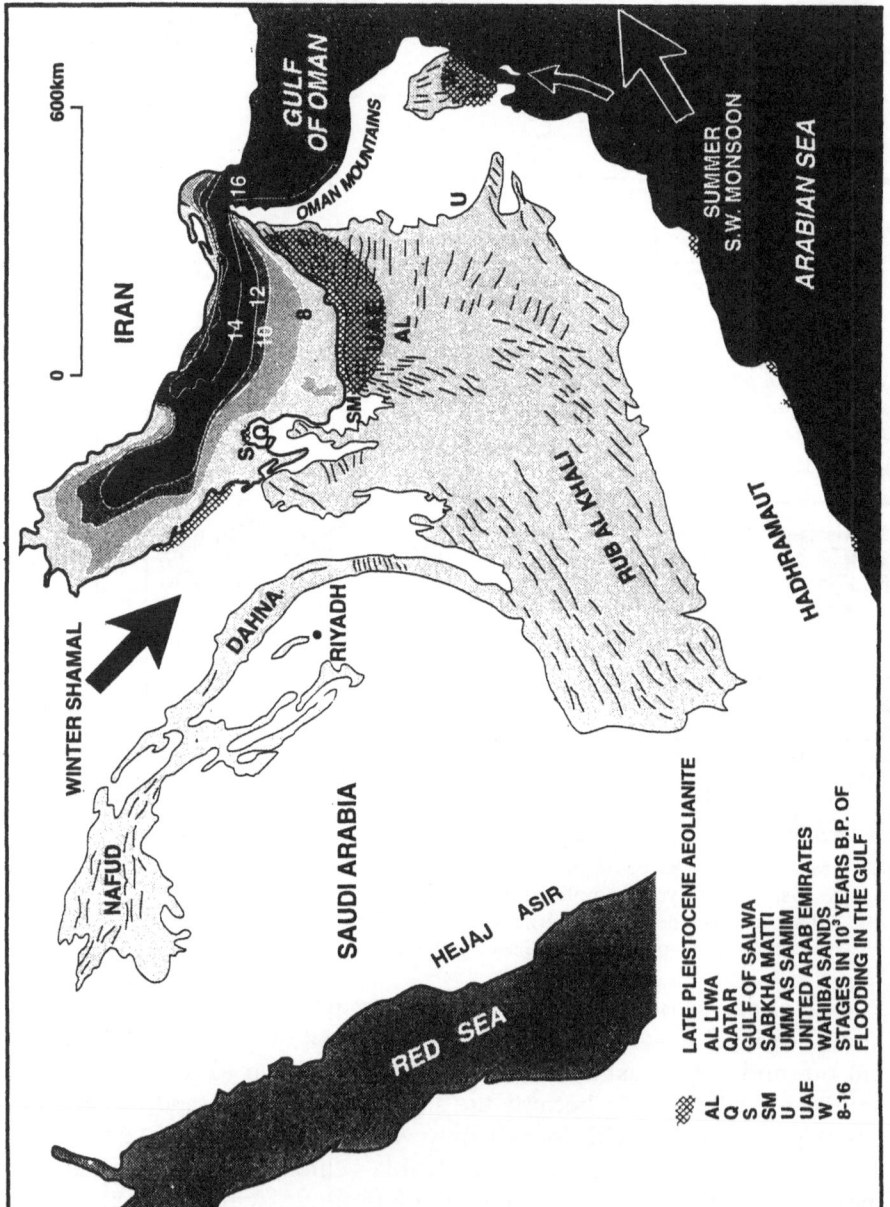

Fig. 4: Map of South Arabia showing the distribution of unconsolidated dune sand and aeolianite, together with lines marking the approximate dates of progressive post-Glacial flooding of the Arabian Gulf in thousands of years ago.

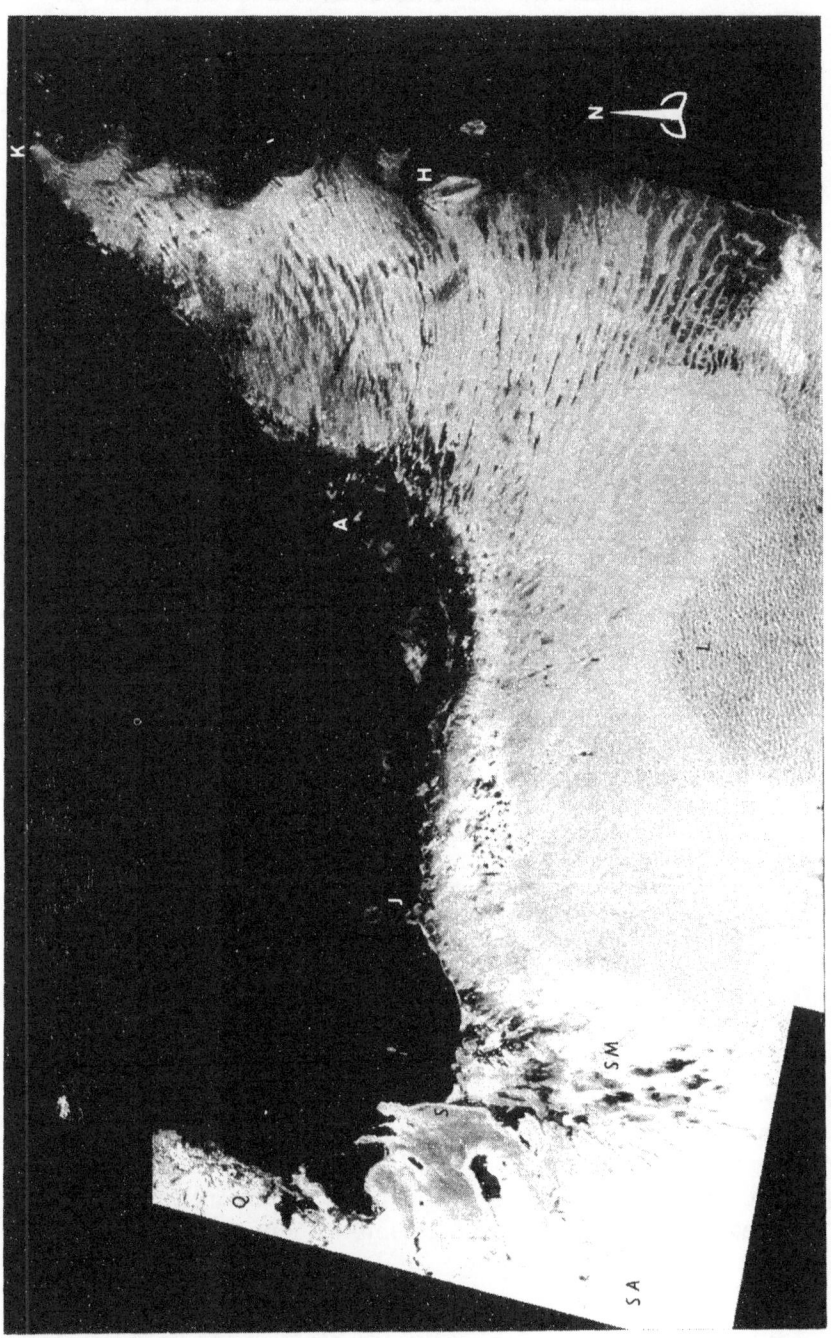

Fig. 5: Landsat image of the Emirates showing the relationship between dune and interdune areas, coastal and inland sabkhas, and the Arabian Gulf. From Sila to Ras al Khaimah is about 500 km. Image by courtesy of ADCO. A Abu Dhabi; H Jebel Hafit; J Jebel Dhanna; K Ras al Khaimah; L Liwa; Q Qatar; S Sila; SA Saudi Arabia; SM Sabkha Matti.

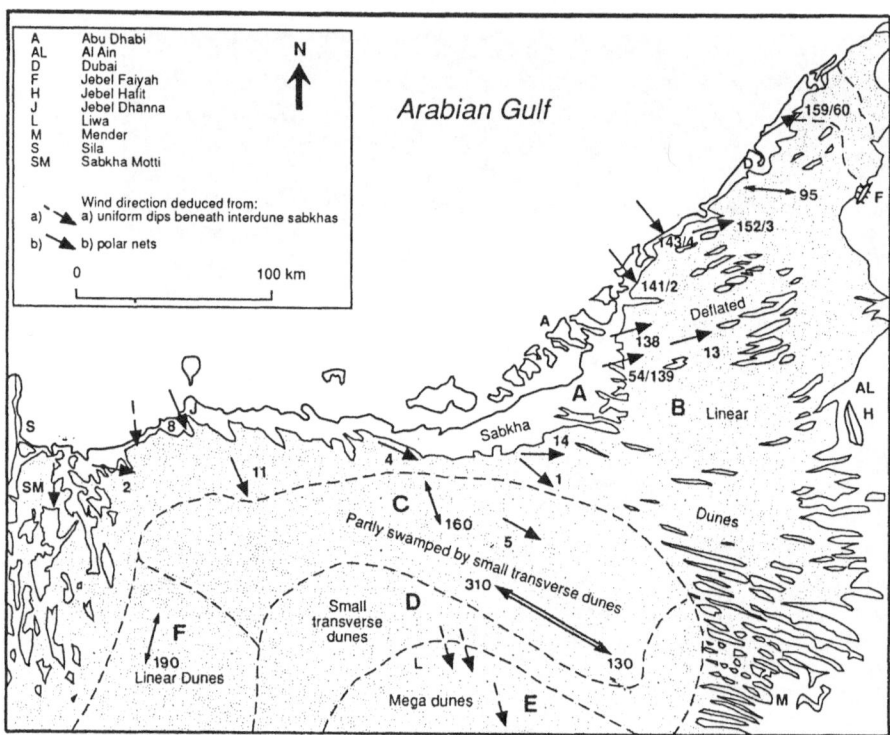

Fig. 6: The Emirates desert can be subdivided into several discrete areas. A. Coastal Sabkha and Sabkha Matti; B. Deflated linear megadunes; C. Dunes of type B largely masked by cover of small transverse dunes; D. Close-knit sand sea of small transverse dunes; E. Mega barchanoid dunes and interdune sabkhas of Al Liwa; F. N-S trending small linear dunes east of Sabkha Matti. Double-headed arrows indicate axial trend of longitudinal dunes (parallel lines with azimuths at both ends = axial orientation of a typical mega dune; short double-headed arrows with one azimuth = small, cross-cutting longitudinal dunes). Solid arrows indicate wind directions deduced from the bedding attitudes of miliolite, which is best developed in near-coastal areas; figures are locality numbers referred to in Glennie (1994). Dashed arrows indicate uniform dip directions beneath some sabkha surfaces.

The above somewhat theoretical approach to Arabia's past changes in climate does not address the role of dunes as indicators of climatic change. But with the ground rules in place, it will be much easier to appreciate the significance of dune morphology, and more importantly, obvious changes to an earlier morphology, and the implications that each bit of evidence has for climatic change.

Dune Distribution in Arabia

Most of the dunes of Arabia, partly transverse, partly longitudinal, are distributed over a giant arc centred SW of Riyadh in Saudi Arabia (Fig. 4). This arc conforms to a general trade-wind pattern, similar to that seen in the Sahara, of sand transport changing from W to E in the north (Nafud), to the SE over the Dahna, and veering to the SW, across the Rub' al Khali towards the mountains of Yemen, where they reach an elevation of some 1200 m above sea level. The dune sands are believed to have been sourced by deflation of the fluvial sediments that, at some time in the past, were transported radially away from the uplifted centre of Saudi Arabia.

If the sand-transporting wind for the dunes of central Arabia was the Shamal, which today blows down the Gulf from the NNW, how is it that the Gulf forms the up-wind limit of the bulk of the quartz-rich dune sands of the Rub' al Khali? There is far too much sand to have been derived solely from the coastal beaches, which in any case consist almost entirely of carbonate grains (Foraminifera and other small bioclastic fragments). The clue to this conundrum is provided both by the dune sands themselves and, just south of the coastline, their absence.

The longitudinal megadunes of the Emirates display excellent evidence of former wind directions that differ from those of today. Their axial orientations are taken to indicate the path the wind followed when each individual dune was formed, possibly at the peak of the last high-latitude glaciation some 15-25 kyrs (thousands of years ago). The regional dune pattern implies a wind that blew across the southern Arabian Gulf towards the SE. The organized nature of the near-coastal longitudinal dunes suggests that, east of Abu Dhabi island, for instance, we now see the relics of a dune system that must have been derived from the west, beneath the waters of the southern Gulf at a time of lower sea level (Figs. 4 and 5), presumably during the last glaciation. This interpretation is supported by a high content of shallow-marine bioclasts in the near-coastal sands, which decreases to zero inland over a distance of up to 80 km. Again east of Abu Dhabi, a northern branch of the main dune system curves to the NE as the Oman Mountains are approached (Figs. 5 and 7), while west of a line between Jebel Hafit and Abu Dhabi city, the wind adopted an increasingly southerly direction (e.g. east of Sabkha Matti), eventually blowing SW across the Rub' al Khali in a typical 'trade wind' direction (Fig. 4).

A little over 100 km SSW of Jebel Hafit, the WNW-ESE trending longitudinal dunes have developed star dunes along their crests, an indication that winds from more than one direction have been blowing in the area. More pertinently in terms of a change in wind direction, is the

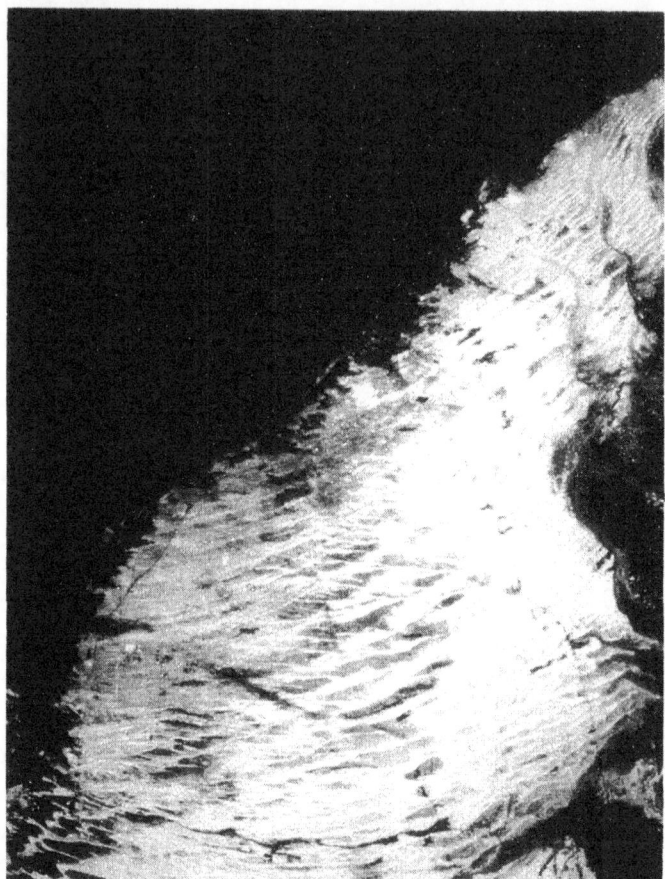

Fig. 7: Landsat image of the northern Emirates showing small, E-W trending longitudinal dunes crossing the interdune areas between Glacial-age mega longitudinal dunes. Near-surface flow by the W-to-E winter jet stream is possibly the reason for erosion of the crests of the megadunes, whose axes are not in equilibrium with such a wind direction.

development of NS trending subsidiary dunes that eventually contribute to a rectilinear boxwork pattern of new dunes and relics of the old dune pattern over the deflation plain (Fig. 8).

Superimposed on this pattern of older dunes are other, much smaller (commonly 3-5m high), widely spaced longitudinal dunes whose axes are mildly to strongly oblique to the mega dunes. In the Northern Emirates, for example, east of Abu Dhabi and Dubai, the axes of these small dunes trend almost W-E across the interdune areas (Fig. 7), the sands in each case having been deflated from the adjacent mega dune to its west; some indeed, possibly because modern winds are relatively weak, fail to extend to the next dune to the east. In this respect, it is pertinent to note that in

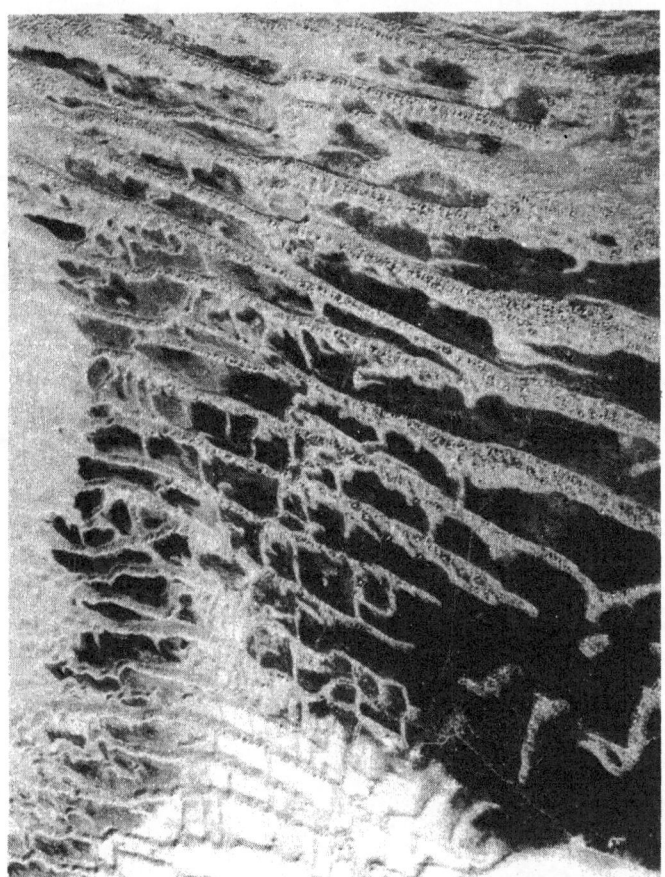

Fig. 8: Landsat image of WNW-ESE trending longitudinal dunes whose crests are capped by star dune. The longitudinal dunes are not in equilibrium with the modern Shamal (north) wind, have developed south extending 'arms' to create a rectilinear boxwork pattern.

the month of January, the sub-tropical jet stream typically flows from west to east across the southern Gulf (United Arab Emirates University, 1993), and can be close enough to the ground for smoke from the refinery near Jebel Dhanna to blow in the same direction.

In marked contrast to the northern Emirates, north of the Liwa (Fig. 9), the NW-SE trending megadunes have small, closely spaced, SSE-trending (approx. N160°E) linear dunes superimposed upon them; and to their south is a system of small, closely spaced transverse dunes that are encroaching over the NW margin of the Liwa barchanoid megadunes (Figs. 6 and 9). In the Emirates, there were clearly at least two phases of building the unconsolidated dunes of the area; first creation of the megadunes, and then their partial destruction by deflation and the development of a partial cover of much smaller new dunes.

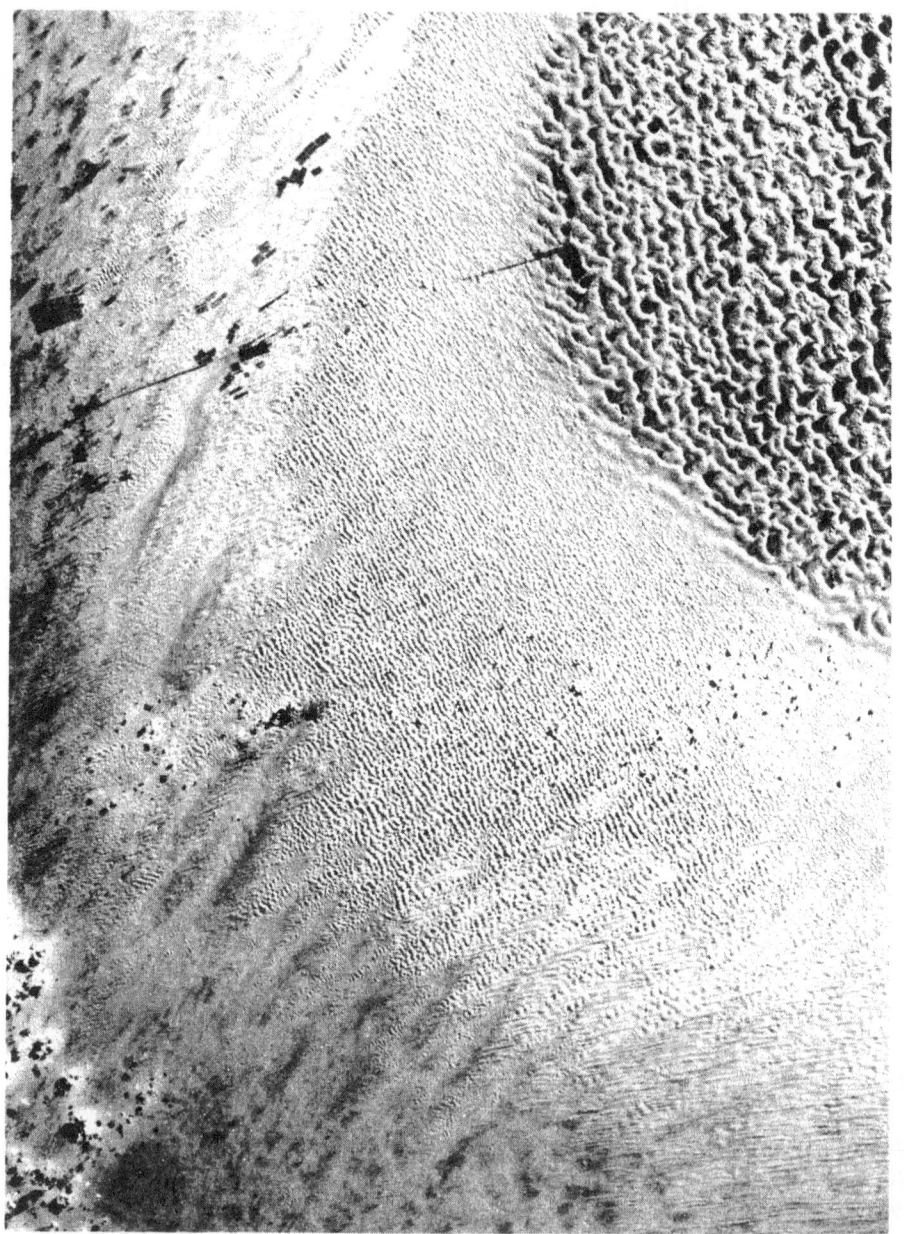

Fig. 9:   Detail of Landsat N and W of the Liwa. The NW margin of the giant barchanoid dunes of the Liwa are being encroached by small transverse dunes. To their north, the crests of NW-SE trending longitudinal megadunes have small NNW-SSE longitudinal dunes superimposed upon them. At least three phases of dune formation are involved.

Deflation of the megadunes was the natural outcome of the stopping of their supply of sand; when the wind blows at sand-transporting velocities, dunes can grow (or at least extend down wind) if there is an adequate supply of sand, but will be deflated down to the water table if that supply ceases. Along the coastal sabkhas of Abu Dhabi, the thin modern layers of algal mat and gypsum crystals are underlain by quartz-rich aeolian sand, an indication that earlier this area was dry land. Much further north, Sarnthein (1972) infers the presence of former subaerial sand-dunes beneath the waters of the Arabian Gulf. Clearly, there is considerable evidence to support the interpretation that not so long ago (geologically) the Gulf was dry, and this probably coincided with the peak of the last high-latitude glaciation some 18-20 kyrs ago (Fig. 3A).

We have already seen that "sand movement is proportional to the cube of the excess wind velocity above that at which sand begins to move". Bagnold's observation lays great stress on the ability of strong and long-lasting glacially induced winds not only to transport sand but also to build and to modify pre-existing sand-dunes. In addition, because transverse dunes are liable to be unstable in strong sand-laden winds, they may be replaced by longitudinal forms. And because wind directions may change with time, the earlier form is likely to be out of equilibrium with the new wind and thus will be modified.

Small dunes, a metre or two high and with little volume of sand, can change shape rapidly and migrate in a new direction consistent with shifts in the wind. Mega dunes (referred to by some workers as 'draas'), with a width of perhaps a kilometre and a height exceeding 100m, have a very large volume of sand to move as they change shape, and thus their morphology will retain some evidence of an earlier wind regime for perhaps thousands of years (what Cooke et al., 1993, refer to as 'dune memory'), especially if the new wind is associated with a considerable post-glacial reduction in the annual duration of strong sand-transporting winds.

Superimposed systems of dunes are not confined to Arabia, but occur also in other desert-margin locations. Examples involving both partly deflated mega longitudinal dunes, with interdune streams and narrow lakes, and smaller dunes with a 21° difference in trend between them is illustrated by Breed et al. (1979) from the southern Sahara south of Timbuktu. The smaller dunes are in approximate alignment with the modern NE trade winds. The mega dunes clearly formed under a different wind and climatic regime to that existing today.

From all the foregoing, we now have in place some important factors by which climatic change may be recognised from an association of dune types in any one area. A few descriptions follow of the evidence for changing desert conditions that have been found in dunes.

Aeolianites ('miliolite')

The Arabian Gulf dried out during the last and earlier glaciations, and had carbonate-rich dunes (now cemented into an aeolianite or, by analogy with the coastal areas of NW India, 'miliolite'; see e.g. Patel and Bhatt, 1995) migrating over its exposed floor; that supply of sand ceased with each post-Glacial flooding, causing deflation of dunes immediately down-wind of the new coast.

The miliolite underlies many of the unconsolidated sand-dunes in the near (within 50-100 km) coastal regions of SE Arabia, and also forms extensive exposures where the younger sands have been removed by deflation. In addition to grains of quartz and common ooids, the aeolianite is rich in bioclasts and a variety of foraminifera and ostracoda. Locally, the outcrops are capped by a layer of calcrete, which indicates a probable wet period following deposition of the miliolite. At any one locality, the bedding attitudes of the miliolite indicate the direction of the wind that transported the sands. For the Emirates area, some of these directions are plotted on Fig. 7. Two sets of directions show up. The first seems to match the orientations of NW-SE trending mega-longitudinal dunes (double-arrowed parallel lines is one example) and their ENE-trending counterparts in the northern Emirates. The second set are oblique to the first, and seem to match fairly closely the direction of the present Shamal (North Wind), parallel to the younger cross-cutting unconsolidated linear dunes that occur both north (160°) and west (190°) of the Liwa, for instance (Fig. 8).

The evidence provided by the miliolite indicates that the aeolian conditions in the Emirates during and after the last glaciation were preceded by similar conditions during at least one earlier glaciation. Dates provided by the Optically Stimulated Luminescence (OSL) technique from miliolite in the Emirates indicate dune activity at two periods: between about 30 and 65 kyrs ago, and greater than 125 kyrs ago (Juyal et al., 1998). The younger miliolite involved carbonate-rich sands exposed on the Gulf floor during the build-up to the last high-latitude glaciation, and the older event was one glacial cycle earlier. Gardner (1988), using ^{14}C dating, also suggested two separate periods of aeolianite deposition beneath the modern Wahiba Sands of SE Oman, where the sand-transporting wind was (and still is) mainly the summer-season Southwest Monsoon, which here blows from south to north (Fig. 3).

Following deposition, the cementation and leaching of the carbonate grains (see e.g. Glennie and Gökdag, 1998) took place under wetter conditions; the aeolianite was then peneplaned, covered by another sequence of currently uncemented dune sands (arid climate again), and then re-exposed, probably because of a later deficiency in sand supply.

Thus repeated changes in climate, from arid to humid and back to arid, can be deduced from the deposition and cementation of the miliolite found in SE Arabia.

Vegetated Dunes

With a change in climate involving a long-term increase in rainfall, dunes acquire a partial to complete cover of vegetation and become stabilised. In Central Africa (Zaire), for instance, Glacial-age longitudinal dunes, deposited in a then arid environment, are now almost completely vegetated under the influence of modern tropical rainfall.

Evidence of an older vegetation cover on sand dunes is commonly preserved as rootlet burrows (Fig. 9). If the roots had acquired a hard sheath of fine gypsum crystals, the plants probably lived (and died) in an arid environment close to a supply of saline water (eg. in an interdune area or low on the flanks of a dune). Such cemented roots (dikaka or rhizoconcretions) are clearly visible in some Miocene dune sands in the Emirates, where they are overlain by a fluvial conglomerate (Fig. 10) (e.g. Glennie and Evamy, 1968) and in Quaternary dune sands within the greater Liwa area. Deposition of the original sand dune was followed by a humid period in which the plants became established. The plants that

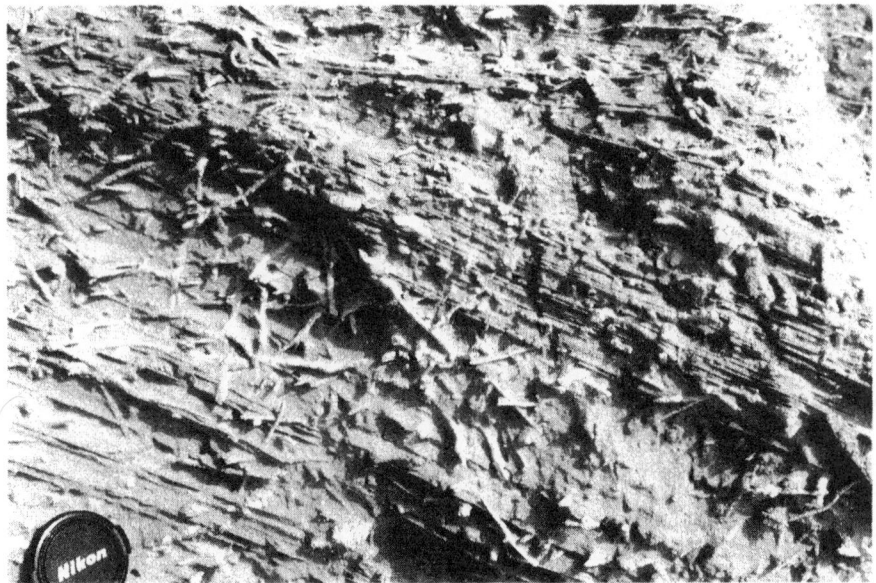

Fig. 10: Foresetted Miocene dune sand riddled with gypsum-cemented moulds of former plant roots. Jebel Barakah, SW of Jebel Dhanna, Abu Dhabi.

became *dikakah* possibly colonised pre-existing dune sand following a post-glacial rise in the water table or, as scrub vegetation, trapped aeolian sand around its roots and branches as it slowed the near-surface wind. The gypsum that encased the roots and finally killed the plants indicates proximity to a desert water table and signals a change from interglacial' humidity to a more arid climate. This interpretation also applies to the *dikaka*-riddled forsetted dune sand beneath the interdune sabkhas of the Liwa.

Permian Glaciations and Evidence of Rotliegend Wind Variability in NW Europe

The coincidence of glaciations and aeolian activity is not limited to the Quaternary, but also occurred during the Permo-Carboniferous and earlier periods. The Late Permian was a time when a major southern hemisphere ice cap over Gondwana was going into decline (the glaciers of southern Arabia had already melted: Alsharhan et al., 1993), and a part of Europe was in the same northern latitude as modern South Arabia and the Sahara Desert. Permian dune sands of the Upper Rotliegend were deposited in a subsiding basin, up to 400 km wide, which stretched eastward from the coastline of eastern England to the Polish-Russian border. The basin was flanked to the north, south and west by highlands (Fig. 11). The basin centre was occupied by a saline desert lake and fringing sabkha that eventually extended eastwards from the English coast for some 1200 km, with dune sands occupying especially the southern part of the basin.

During the Late Permian, following a long period of uplift and erosion, the Rotliegend basin began to subside rapidly, and accumulated some 2000 m of desert-lake sediments in northern Germany within about 5 or 10 million years (0.2 - 0.4 mm/year). Basin-flank dune sands accumulated with thicknesses up to 300m or more. Because the sands form reservoirs for very important accumulations of natural gas, the dune sequences have been repeatedly penetrated by the drill.

Dipmeter logs provide sequential evidence of dune bedding attitudes, which can be used statistically to deduce the palaeowind direction at each well location, and the Late Permian Rotliegend is no exception (see, eg. Glennie, 1982).

The regional palaeowind pattern over the North Sea area indicates a tight 'wheel-around' similar to that seen over Arabia today; blowing from WNW in the central North Sea and to the SW or even west in the southern North Sea (Fig. 10). In detail, however, it is often difficult to decide the reason for differences between data points. Was it the result of actual

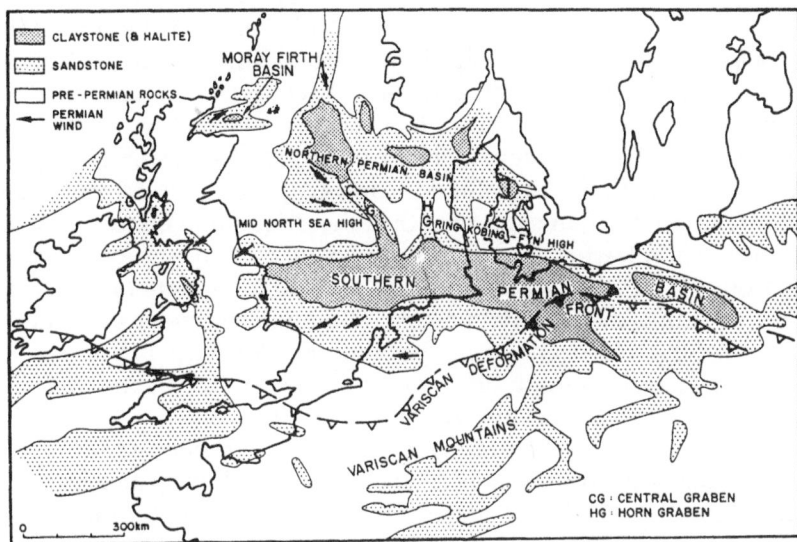

Fig. 11: Late Permian Upper Rotliegend facies in the Northern (NPB) and Southern (SPB) Permian Basins of NW Europe. Basin-centre desert lake is surrounded by a marginal sabkha facies. Dune sands, best developed along the south flanks of the basins, were deflated from fluvial sediments transported north from the Variscan Highlands. Permian wind directions indicate a tight 'wheel-around' similar to that of Quaternary Arabia. From Glennie and Provan, 1990.

changes in wind direction, or could it represent deposition with a constant wind direction but in a differentially subsiding basin where the dune's axes shifted laterally with time? This dilemma is exemplified by data from two North Sea wells, 48/19A-4 and 49/26-26 (see Table 1 below, and Glennie 1994). In 48/19A-4, the deduced wind direction varies by almost 90° with time, whereas in 49/26-26 (Fig. 12) the wind directions were much more constant from one sedimentary packet to the next.

Table 1: Palaeowind directions deduced from bedding attitudes of Late Permian Upper Rotliegend dune sands depicted on Dipmeter Logs of two wells from the southern North Sea

48/19A-4		49/26-26	
Depth Range (Ft)	Wind Direction	Depth Range (Ft)	Wind Direction
8340-8940	Average to 260 °	6380-6915	Average to 275 °
8406-8448	to 240 °	6380-6614	to 270 °
8668-8686	to 242 °	6626-6750	to 280 °
8722-8736	to 200 °	6760-6915	to 275 °
8738-8756	to 287 °		
8916-8940	to 280 °		

The highest and lowest packages in 49/19A-4 indicate a 40° change in wind direction (towards 240° and 280°); and if each of the three middle intervals is treated in isolation, they indicate winds that shift strongly with time (to about 200°, 240° and 290°). Each, however, represents only a thin interval of less than 20 feet. In a subsiding basin they could equally well represent the local development of series of migrating barchanoid dunes whose locations had shifted sideways with time, so that the drill penetrated first one flank and then the other. Although almost impossible to prove, this interpretation is supported in part by the combined dip data from all depths, which indicate a wind direction towards about 240°, thus fitting the regional pattern. Although dipmeter data indicates that most of the Rotliegend dunes beneath the North Sea are of transverse type, basin-margin longitudinal dunes are exposed in sand pits in NE England (eg. Glennie and Proven, 1990, Fig. 11).

It is also clear from the dipmeter and associated gamma ray logs that some packages of dune sand are separated by thin sequences of low-angle sands (Fig. 12). Some such sequences could represent the deposits

49/26-26

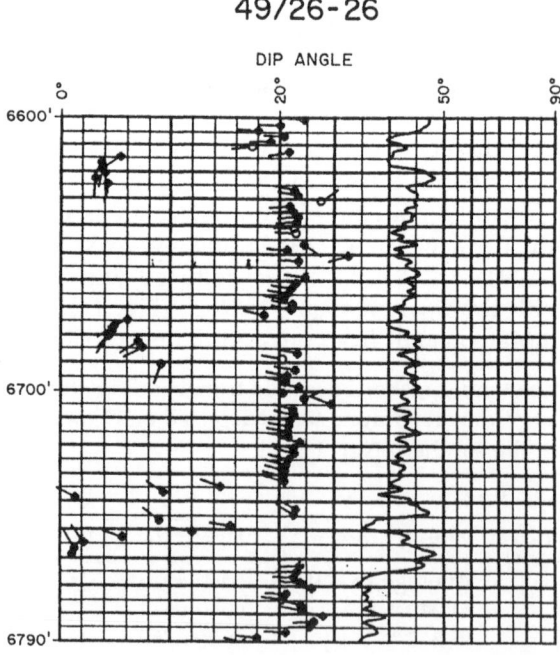

Fig. 12: Portion of a dipmeter log of well 49/26-26 showing how packages of uniformly dipping dune sands are separated by thinner sequences of low-angle sands that could represent periods of higher rainfall, higher water table, and little or no aeolian sand transport.

of a flash flood, and here, it is perhaps pertinent to point out that dune bedding is often highly oblique, even opposed, to the sense of movement of fluvial sands (Fig. 12, at a depth of about 6690 feet; also consider in Fig. 11 the direction of water flowing from the basin margin to the basin-centre lake, relative to the wind directions indicated by arrows). Alternatively, these thin non-aeolian sands may be the product of a longer period of higher rainfall, or the intersection of a bounding surface with a high water table associated with Late Permian interglacial periods, such as is believed to have been experienced during the Late Quaternary 'Climatic Optimum'.

Conclusions

In deserts, many mega-scale sand-dunes have strong affinities with one of two simple morphologies; transverse dunes whose long axes are at right angles to the prevailing sand-transporting wind, and longitudinal dunes whose axes are parallel to it. Most show some evidence of geometrical alteration, often by the superimposition of smaller, simpler dune types, caused by a later wind blowing in a different direction.

The reason for these modifications is inferred to result from changes in the Quaternary climate driven by changes in the size of high-latitude ice-caps. The presence or absence of large ice-caps controlled the locations of global centres of high and low atmospheric pressure, which shifted closer to or farther from the Equator in response to the intensity of the anticyclones located over the ice-caps. The resulting shift in location of the sub-tropical anticyclones meant that between full glacial and interglacial, the wind directions changed locally by almost 90°. With such a change, pre-existing dunes, especially if large, were no longer in equilibrium with the new wind and, if not completely destroyed, showed evidence of these changes in the form of increasingly complex morphology or new dune types and orientations. Such changes are readily recognised in many Landsat images of unconsolidated dunes, and are also recorded in the changing bedding orientations within vertical sequences of cemented dune sands of both Quaternary and Permian age.

Glacially-induced falls in sea level exposed the floor of the Arabian Gulf and other shallow continental shelves; their deflation resulted in the deposition of carbonate dune sands on what is now adjacent land areas. Interglacial increases in rainfall caused both cementation and leaching of these carbonate aeolianites.

Interglacial humidity encouraged the growth of plants in interdune areas and especially on the low flanks of dunes; this is now recognised by burrows in the dune sands. Where the ground water is sulphate-rich, as just beneath the surface of sabkhas in arid environments, the plant roots are commonly represented by moulds formed of small gypsum crystals.

Although sand-dunes are most active in areas possessing an arid climate, many contain evidence of later changes in climate, whether by changes from sand deposition to deflation, or by modifications in wind direction or caused by rainfall.

References

Alsharhan, A.S., Nairn, A.E.M. and Mohammed, A.A., (1993). Late Palaeozoic glacial sediments of the southern Arabian Peninsula: their lithofacies and hydrocarbon potential. *Marine & Petroleum Geology* 10 (1): 71-78.

Bagnold, R.A. (1941). *The Physics of Blown Sand and Desert Dunes*. Chapman and Hall. London.

Breed, C.S., Fryberger, S.G., Andrews, S., McCauley, C., Lennartz, F., Gebel, D. and Horstman, K. (1979). Regional studies of sand seas using Landsat (ERTS) imagery. In: A *Study of Global Sand Seas*. E.D. McKee, (ed.) Geological Survey Professional Paper 1052: 305-397 US Government Printing Office, Washington.

Cooke, R., Warren, A. and Goudie, A., (1993). *Desert Geomorphology*. UCL Press.

Gardner, R.A.M. (1988). Aeolianites and marine deposits of the Wahiba Sands: character and palaeoenvironments. In: The Scientific Results of the Royal Geographical Society's Oman Wahiba Sands Project 1985-1987. R.W. Dutton, (ed.) *Journal of Oman Studies Special Report No. 3*: 75-94 Ministry of National Heritage and Culture, Muscat, Sultanate of Oman.

Glennie, K.W. (1982). Early Permian (Rotliegendes) Palaeowinds of the North Sea. Sedimentary Geology 34: 245-265.

Glennie, K.W. (1994). Quaternary dunes of SE Arabia and Permian (Rotliegend) dunes of NW Europe: Some comparisons. Zentralblatt für Geologie und Paläontologie, Teil I (11/12):1199-1215. Stuttgart.

Glennie, K.W. and Evamy, B.D. (1968). Dikaka: Plants and plant-root structures associated with aeolian sand. *Palaeogeography, Palaeoclimatology, Palaeoecology* 4: 77-87.

Glennie, K.W. and Provan, D.M.J. (1990). Lower Permian Rotliegend reservoir of the Southern North Sea Gas Province. In: *Classic Petroleum Provinces*. J. Brooks, (ed.) Geol. Soc. Spe. Pub. 50: 399-416.

Glennie, K.W. and Gökdag, H. (1998). Cemented Quaternary dune sands, Ras' al Hamra housing area, Muscat. In: *Proceedings of IGCP-349 Conference on Quaternary Deserts and Climatic Change*. A.S., Alsharhan, K.W. Glennie, G.L. Whittle and C.G.St.C. Kendall (eds.) Balkema, Rotterdam, 109-116.

Juyal, N., Singhvi, A.K. and Glennie, K.W. (1998) Chronology and Paleoenvironmental significance of Quaternary desert sediments in South-eastern Arabia. In: *Quaternary Deserts and Climatic change* A.S. Alsharhan, K.W. Glennie, G.L. Whittle and C.G. St. Kendall (eds.) Balkema, Rotterdam 315-325.

Patel, M.P. and Bhatt, N. (1995). Evidence of Palaeoclimatic Fluctuations in Miliolite Rocks of Saurashtra, Western India. *J. Geol. Soc. of India*. 45: 191-200.

Sarnthein, M., (1972). Sediments and history of the postglacial transgression in the Persian Gulf and northwest Gulf of Oman. Marine Geology 12: 245-266.

United Arab Emirates University, 1993. *The National Atlas of the United Arab Emirates*.

7 Arid-Zone Rivers as Indicators of Climate Change

GERALD C. NANSON[*] AND STEPHEN TOOTH[*]

ABSTRACT

Arid-zone rivers exist under marginal fluvial conditions where a change in climate can lead to significant changes in flow regime, sediment transport and associated channel style. This chapter provides a brief review of the main characteristics of arid-zone fluvial environments (precipitation and runoff, sediment load, channel morphology, alluvial terraces and slackwater deposits) which may record changes in arid-zone climate. It also identifies a number of problems associated with using arid-zone rivers as indicators of climate change, namely: that the influence of climate on fluvial processes and landforms is poorly understood and has led to incorrect interpretations; that individual large floods and intrinsic threshold conditions can produce landforms and sedimentary sequences which are not in themselves evidence of climate change, for they are not in equilibrium with the prevailing climate; that there may be a strong element of equifinality in the nature of some channel responses; and that individual features, such as river terraces, can be formed at different stages of climate change in different environments. Attention is focused on using river floodplain and terrace stratigraphy as evidence of palaeoflow regimes and hence, by inference, climate change. As there is no general theory or extensive body of empirical research to assess how rivers respond to variations in climate, a case-study approach is adopted. Climate change is interpreted from alluvial sequences in four separate regional and temporal settings: historical changes on the Colorado Plateau of North America; Late Pleistocene and Holocene changes in the Alice Springs region of central Australia; changes over several glacial cycles of the Late to Mid Pleistocene in the Channel Country, an accretionary alluvial basin in western Queensland, Australia; and finally, the Nile River is examined briefly to provide an example of the climatic inferences that can be drawn from a well-known desert river sourced largely outside the arid-zone. These studies demonstrate that rivers in the arid-zone can provide a detailed picture of Quaternary climate change that correlates with and enhances records of regional and global climate change obtained from other sources, such as lakes, ocean and ice cores, sea-level data, and loess deposits.

[*]School of Geosciences, University of Wollongong, Wollongong, NSW, 2522, Australia

Introduction

The term 'arid-zone rivers' almost sounds like a contradiction, yet some arid regions support very large, albeit intermittently-flowing, rivers. As such rivers operate close to the limit of fluvial existence, they provide evidence of changing environmental conditions not necessarily present in other sources. Bedrock and aeolian landforms cover the greatest surface area of most arid regions (Graf, 1988a; Cooke et al., 1993), but it is ephemeral river and lake systems that are the repositories for much of the stratigraphic information on palaeoenvironmental change in the arid-zone. Indeed, a river terrace, floodplain or playa lake can be viewed as a stratigraphic 'library' of data on environmental change, but in all probability there will be certain 'volumes' missing. When collections of these landforms are systematically investigated within a drainage basin using modern techniques of palaeoenvironmental reconstruction, a detailed picture of Quaternary history can emerge. However, extracting this information is difficult. Although evidence from lakes has been used for palaeoenvironmental interpretation with considerable success (e.g. Bowler, 1981; Street-Perrott and Harrison, 1985; Sack, 1994), at this stage there are only a few examples of fluvial systems being used to provide substantial records of environmental change in arid regions. A number of these examples are reviewed here to provide direction for future research.

In order to provide evidence of Quaternary as well as older environmental change, there is a growing appreciation of the importance of understanding modern river processes and landforms on the basis of the old axiom that *'the present is the key to the past'*. However, what evidence do rivers provide of past environmental change? Sequences of river terraces have commonly been used to infer climate change but the most widespread, best preserved and detailed evidence is usually in the form of alluvial stratigraphies both within terraces and in depositional basins. The link between modern processes and older stratigraphic evidence is highlighted in the introduction to a collection of papers on floodplains where Brackenridge and Hagedorn (1992, p. iii) have stated:

"... floodplains and their underlying sediments are not merely assemblages of landforms and sedimentary facies related to process, but instead are complex records of process change through time ...
"... Perhaps the conceptual distance between the scientists who observe and measure nature as it is presently unfolding, and those who investigate the records of past processes and changes, can be bridged by an integrated 'Quaternary geomorphology' which investigates not only modern surface processes but also the preserved records left by the operation of these processes through time."

However, in a review of river landforms and sediments in deserts as evidence of climate change, Reid (1994, p. 571) suggests that *"... rivers are comparatively insensitive to changes in climate unless these changes are substantial"*, but not all would agree. Hereford (1984, p. 654) maintains subtle changes in climate are important for bringing about significant changes in river regime, but states that "... *the influence of climate on such fundamental stream processes as erosion and aggradation is not well understood."*

Confusion over the importance of climate in controlling channel changes appears to be the result of several problems. Outside of regions where climate change has been truly catastrophic, such as in Pleistocene glaciated areas, there is often little independent evidence of the nature or magnitude of climate change. If rivers are then to be used as evidence, any climatic interpretation risks becoming based on circular reasoning. Furthermore, it is now well understood that a river can exist close to a threshold condition where a small shift in flow or sediment character, possibly induced by climate, can produce a dramatic change in river style (e.g. Schumm and Kahn, 1972; Patton and Schumm, 1975; Schumm, 1977). Finally, there are numerous examples of arid-zone rivers undergoing a dramatic change in character where the cause of change, be it climate or something else, is too subtle to be identified (e.g. Cooke and Reeves, 1976).

This chapter will illustrate how knowledge both of modern fluvial processes and floodplain stratigraphies can be used to interpret Quaternary environmental change in the arid-zone. The chapter starts by providing a brief assessment of the present level of understanding of arid-zone rivers before considering the problems and potential advantages of using such rivers as indicators of palaeoenvironmental change. As there have been relatively few detailed studies of the sedimentary sequences of arid-zone rivers, the chapter will focus on four different examples to illustrate how such river landforms and sediments can be used to interpret palaeoenvironmental change. The first are relatively confined channels in canyon settings on the Colorado Plateau of North America. The second are larger, lower gradient and less confined, sand-dominated systems in the Alice Springs region of central Australia. The third are the very extensive low-gradient, unconfined systems of mud-dominated, streams within the Channel Country in western Queensland, Australia, an accretionary basin setting with relatively deep Quaternary deposits. Finally, the Nile River is briefly examined as an example of a large system only partially sourced in arid lands, but one which has been studied in detail and which illustrates major episodes of flow-regime change.

Limited Research into Arid-Zone Rivers

Although there is really a continuum of fluvial systems from hyperarid to extremely humid climatic environments (Knighton and Nanson, 1997), rivers in hyperarid, arid and semi-arid environments (arid-zone rivers) are often considered distinctly different from streams in more humid regions. However, the geomorphology, sedimentology and Quaternary history of arid-zone rivers remain poorly investigated relative to rivers in more humid regions. Useful reviews of research on arid-zone rivers are provided by Reid and Frostick (1997) and Thornes (1994a, b) but these concentrate almost entirely on modern rivers and pay little or no attention to how the rivers respond to changes in climate. Several chapters in Bull (1991) and a recent paper by Reid (1994) are the only reviews illustrating how arid-zones rivers can be used as indicators of climate change.

There are four principal reasons for the relative neglect of arid-zone rivers as a topic for study. Firstly, in a logistical sense, arid regions are often sparsely inhabited, difficult to access and can be harsh environments within which to conduct prolonged research. Secondly, arid-zone research has tended to focus on more widespread aeolian and bedrock landforms. While there is a series of books devoted to these aspects (e.g. Mabbutt, 1977; Thomas, 1997a; Cooke et al., 1993; Abrahams and Parsons, 1994), and which direct relatively little attention to rivers, only one text has so far specifically targeted arid-zone rivers (Graf, 1988a). Thirdly, arid-zone rivers are typically ephemeral and their unpredictable and infrequent flows are impediments to conducting detailed process studies. Finally, whereas studies in more humid regions have been able to establish fluvial chronologies and interpret rates of change based on radiocarbon dating of sedimentary sequences, the limited production and preservation of organic matter in arid regions has restricted the use of this technique, as has the limited preservation of other materials suitable for amino acid racemisation and uranium series dating. Only with recent advances in luminescence dating techniques (Aitken, 1998) has substantial progress been made in the dating of arid-zone fluvial sequences.

The relatively small body of research on arid-zone rivers notwithstanding, there is little doubt as to the importance of running water in shaping landforms in arid regions. Occasional but very intense rainfall and runoff events, particularly within relatively steep, poorly-vegetated headwater basins, can often accomplish a great deal of geomorphic work and result in substantial alluvial deposition. In larger, lower gradient systems, it is the magnitude of the floods and areal extent of the floodplains that makes fluvial processes so significant (e.g. Knighton and Nanson, 1994). Indeed, as a result of the poor preservation potential of organic deposits in deserts, and the often relatively limited distribution and shallow nature of lacustrine deposits, fluvial landforms and sediments may in future provide the main source of evidence for precipitation and flow regime change in the arid zone.

Parameters of Change in Arid-Zone Catchments

Of primary interest in interpreting global or regional climate change is how temperature and precipitation have varied, yet the lack of direct evidence means that proxy indicators such as sea levels, lake levels, vegetation and the isotopic composition of sea water or speleothems are commonly used to infer such changes. Outside of glaciated regions, alluvial stratigraphy cannot provide even an indirect record of temperature change, but it can reveal changes in runoff and hence precipitation. However, associated changes in climatic seasonality, vegetation, evaporation, evapotranspiration, soil permeability and water storage as snow and ice (some of which are temperature dependent) will distort any direct relationship between precipitation and runoff. As palaeo-runoff cannot be measured directly, it must be inferred from changes in channel morphology and sedimentology. The nature of the relationship between rivers and climate is clearly an indirect one, however, so an appreciation must be gained of the nature of precipitation, runoff, sediment load, channel morphology, terrace development and slackwater deposition in arid-zone environments.

Precipitation and Runoff

In arid regions, annual precipitation is not only low but also highly variable, (Bell, 1979). While many European humid-temperate stations have interannual variabilities of < 20%, those in arid regions such as the Sahara often range from 80-150% (Goudie, 1987). Pilgrim et al. (1988) give an example from the northwest coast of Australia where annual rainfall in the long-term averages about 250 mm, but in four consecutive years varied from 55 to 680 mm. There are other arid environments where the variability is even greater. Periods of drought lasting several years can greatly reduce vegetation cover and promote the erosional effectiveness of subsequent rainfall. As a consequence, arid-zone rivers are commonly adjusted to, and thus provide evidence of, less frequent flooding than is the case for rivers in humid regions. Important exceptions are where arid-zone streams respond to a strongly seasonal precipitation regime, such as those streams fed by snow-melt or a regular monsoon.

Convectional rainfall over relatively small areas (<100 km^2) can cause localised erosion and increase the unpredictability of runoff. However, while convective thunderstorms are widely considered to be a major source of desert rainfall (eg. Cooke et al., 1993), some evidence suggests otherwise. In arid western New South Wales, for instance, of all the rainfall events exceeding 5 mm, less than 10% are derived from localised convective storms (Cordery et al., 1983). Penetration of central Australia by tropical cyclones and by the northern margin of easterly-progressing frontal systems also leads to the generation of rare but important flood events (Knighton and Nanson, 1997). An understanding of this point is very important for an appreciation of the nature of climate change in those deserts that are occasionally penetrated by moist airmasses, for the changing frequency and intensity of these weather systems may be the main contributor to long-term changes in average precipitation, runoff and alluvial stratigraphy.

As arid regions are typically sparsely vegetated, Hortonian overland flow can be considerable: other things being equal, the more arid the region, the more overland flow will occur for a given rainfall event. As illustrated by Cordery et al. (1983) with data from New South Wales, runoff initiation in the more arid west required <16 mm of rain compared to >35 mm in the more humid east. This situation makes complex any interpretation of climate change, for as conditions become more arid, precipitation will decrease but due to a loss of vegetation and due to changes in infiltration, the peakedness of streamflow may increase.

Catchment scale is also important when interpreting runoff. Streamflow hydrographs in small arid-zone rivers are often flashy (Reid and Frostick, 1987; 1997), but transmission losses in very large basins can be considerable, resulting in attenuation of flood hydrographs, substantial downstream decreases in flood peaks and total runoff volumes, and very poor correlations between runoff coefficients in different parts of the same basin (Knighton and Nanson, 1994). As a consequence, environmental change may affect the upper and lower parts of large arid-zone drainage basins differently, although the complexity of such variations has never been thoroughly investigated.

In North America and Australia, mean annual runoff is about twice as variable in arid zones as in their continental areas as a whole (McMahon, 1979; Finlayson and McMahon, 1988). Relative flood magnitude curves (with floods expressed as ratios of the mean annual flood) are very steep for arid environments and, in contrast to the often dominant influence of moderate floods in humid areas, high-magnitude and low-frequency flood events appear to be more important as a control on channel development and channel form in some arid-zone rivers (Baker, 1977; Wolman and Gerson, 1978; Graf, 1988a).

Sediment Load

As interpretations of climate change from alluvial deposits almost always involve an interpretation of changed sediment characteristics, it is important to appreciate the complex nature of sediment transport in arid-zone rivers. In relatively small rivers, suspended sediment loads increase rapidly on the rising limbs of flood hydrographs and peak at very high values as a result of typically sparse vegetation cover and the availability of surface sediment. Arid-zone rivers typically have very high suspended sediment concentrations (Frostick et al., 1983; Reid and Frostick, 1987) with values in excess of 100 g/l having been reported. Furthermore, suspended sediment loads appear to be less responsive to changes in discharge than is the case for rivers in humid regions, probably indicating that sediment supply is not particularly limiting and therefore does not become exhausted during passage of a floodwave (Frostick et al., 1983). In addition, despite the fact that ephemeral arid-zone rivers only operate infrequently, Laronne and Reid (1993) have suggested that small ephemeral rivers, over time, deliver more bedload per unit channel width than their perennial counterparts, probably because the bed does not become armoured during the flash flooding that typically characterises ephemeral systems. In such small arid-zone rivers, coarse bedload may be the dominant contributor to the typically high total sediment yields. By contrast, in large, low-gradient basins, fine-grained sediment appears to be the dominant component under arid conditions. Nanson et al. (1988) have demonstrated that during the Pleistocene, Cooper Creek in western Queensland has undergone several major switches from sand-dominated, mixed-load in the more humid interglacial periods to almost entirely mud-dominated during more arid glacial phases, although importantly the mud was transported as aggregates in the form of bedload (Maroulis and Nanson, 1996). The Cooper's ability to rework its sandy substrate was greatly diminished during the arid phases compared to the more humid climatic episodes.

An oft-cited relationship between annual sediment yield and precipitation, originally proposed by Langbein and Schumm (1958), suggests that there is a maximum effective precipitation for the erosion of sediment of about 300 mm, for at such values there is sufficient runoff to induce erosion but insufficient moisture to support an effective protective cover of vegetation on hillslopes. However, even this proposal has been seriously questioned by Walling and Webb (1983) who argue that no such simple relationship exists. Clearly, sediment transport in desert streams is a complex issue that must be interpreted in relation to local conditions such as basin size, stream energy and sediment availability.

Channel Morphology

A distinctive characteristic of the morphology of many arid-zone rivers is their large width to depth ratios (Leopold et al., 1966; Frostick and Reid, 1979; Reid and Frostick, 1997). For a selection of basins less than about

10,000 km^2, Wolman and Gerson (1978) found that widths increase downstream extremely rapidly until basins reach a size of about 10-100 km^2, after which widths remain fairly constant at 100-200 m. In part, they attribute this to the possibility that drainage densities are high and large widths are sustained by rapid runoff during storm events, but that beyond a certain basin size transmission losses prevent any further increase in channel dimensions. Their suggestion that 10-100 km^2 may reflect the maximum size of convectional storm cells, thereby limiting runoff increases, is countered by evidence that convectional storms are not the major source of precipitation in some deserts (discussed above). In other areas, arid-zone rivers decrease in width downstream, presumably as a function of declining downstream discharges (Mukerji, 1976; Mabbutt, 1977; Dunkerley, 1992; Kelly and Olsen, 1993). Regardless of any primary cause, channel widths are also partly a function of the silt-clay content of the channel boundary, for in arid regions with a high mud content, channels can be very narrow and deep (Schumm, 1960a, b, 1961a, b; Mabbutt, 1977; Schumann, 1989). Given the complexity of these relationships, it is difficult to see how changes in channel cross-sectional geometry could be used to provide unambiguous evidence of climate change.

In contrast, channel planform is a readily identifiable characteristic that is known to change in response to changes in climate and associated sediment and flow regime. It is well documented that many rivers in the temperate zone switched from braided to single thread and meandering as climate ameliorated at the end of the last glacial (e.g. Fisk, 1944). Graf (1988a) maintains that many arid-zone rivers exhibit a braided pattern and that meandering in such regions is rare. A shift to braiding, in conjunction with other evidence, could indicate increasing aridity. However, in arid Australia, braided rivers are themselves rare. Many of Australia's rivers anabranch, a condition distinctly different from braiding (Nanson and Knighton, 1996). In Australia, anabranching appears to be partly a result of general aridity and in particular the continent's very low stream gradients, fine-grained sediment, stable banklines and declining downstream discharges (Nanson and Huang, 1998). The shift from meandering to anabranching in Cooper Creek in Queensland has been identified as a response to increasing aridity and a change in sediment regime from mixed-load to mud-load (Rust and Nanson, 1986; Nanson et al., 1988, 1992).

Terraces and Slackwater Deposits

In a thorough evaluation of geomorphic responses to climate change, Bull (1991) regards river terraces and alluvial fans as *alluvial geomorphic surfaces*; mappable landscape elements that are temporally discrete and that are formed during climatically-induced periods of accretion (although he acknowledges that they can also be initiated by other mechanisms such as tectonism). In this chapter we do not consider alluvial fans for these are discrete landforms which combine both fluvial and slope processes (Blair and McPherson, 1994a, b), and their role as indicators of climate change in arid environments has been reviewed by Dorn (1994). Terraces, however, are formed essentially by the abandonment of river floodplains and are an integral part of using rivers as indicators of climate change, although in different environments they appear to have responded to climate change in contrasting ways. Busche and Hagedorn (1980), Goldberg (1984) and Warren (1985) all recognise the problem of determining whether river terraces in semi-arid regions are formed during drier or wetter phases. Bull (1991) observed that in a Pleistocene glaciated river basin in New Zealand, terraces formed during full glacial conditions when hillslopes shed their sediment and valleys filled with accumulated debris, whereas in the Mojave Desert, hillslopes stored sediment during the exceptionally dry glacial times and formed valley fills when the rains returned at the end of the Pleistocene. As terraces in different areas can form at distinctly different stages within a cycle of major climate change, their use in interpreting climate change, probably even within different arid-zone regions, requires an understanding of the conditions that prevailed to induce cut-and-fill episodes. Nevertheless, river terraces are the most widely used form of fluvial evidence for climate change (e.g. Vita-Finzi, 1969; Barsch and Royse, 1972; Adamson et al., 1982), even though their mechanisms of formation are not fully understood.

In addition to the channel changes recorded by river terraces, slackwater deposits preserved in bedrock canyons provide another source of evidence for flow-regime changes in arid regions, albeit on more recent timescales. The principles behind the use of slackwater deposits are detailed in many of the chapters in the edited volume by Baker et al. (1988) and they have been widely used to unravel the history of flood events in arid-zone watersheds, in some instances extending back several thousand years (Patton et al., 1979; Kochel et al., 1982; Baker et al., 1983a, b; Ely and Baker, 1985; Partridge and Baker, 1987; Pickup et al., 1988; Smith, 1992; Ely et al., 1993; O'Connor et al., 1994; Wohl et al., 1994). In central Australia, these high-magnitude floods have been shown to have had a dramatic effect on downstream alluvial reaches (Pickup, 1991; Patton et al., 1993), a process which is likely to have affected many other arid areas, although there are few details to date.

Desert Rivers: Allogenic and Endogenic

A fundamental distinction between arid-zone rivers is that some are sourced almost entirely from outside arid regions (allogenic rivers) while others are sourced entirely from within arid regions (endogenic rivers) (Cooke et al., 1993). Allogenic streams are often very large, they usually have a relatively predictable supply of water, commonly from seasonal sources, and they typically flow considerable distances over low gradients. The Euphrates River receives almost 90% of its total runoff from the mountains of Turkey rather than from the low gradient desert it traverses in Iraq (Beaumont, 1989), and the Nile River is supplied from the east African highland and the Ethiopian highlands before flowing through a region where the mean annual rainfall is less than 50mm and gradients are about 5×10^{-5}. While these externally fed rivers often preserve evidence of fluvial change in the form of palaeochannels or terrace sequences (eg. Adamson et al., 1980, 1982; Adamson, 1982), that record must be seen as a synthesis of changes, both upstream and in the arid zone through which they flow. In contrast, the flow regime of endogenic rivers is usually less predictable, with occasional floods followed by lengthy periods without flow. In addition, endogenic rivers are commonly characterised by substantial downvalley decreases in the magnitude and frequency of flood flows.

While the two ends of the spectrum (allogenic and endogenic) may be clear enough, there are numerous examples of arid-zone rivers that exhibit both characteristics. For instance, the Channel Country rivers of western Queensland, Australia, are mainly supplied by rainfall in the headwaters originating from the northern Australian monsoon. However, these streams are not strongly allogenic, for the basins are entirely semi-arid to arid, the flow regime of the rivers is ephemeral and flood magnitude and frequency decline substantially downstream (Knighton and Nanson, 1994). The focus of the remainder of this Chapter is mainly on endogenic rivers, and those mostly so, such as the Channel Country rivers.

Problems Associated with Using Arid-Zone Rivers as Indicators of Palaeoenvironmental Change

As the preceding sections have demonstrated, studies of modern arid-zone rivers show that, for a number of reasons, interpretations of fluvial landforms and deposits in terms of palaeoclimatic change are rarely straightforward.

One problem is that the influence of climate on basic fluvial processes such as erosion and aggradation, and hence on channel morphology and sedimentary sequences, is not well understood. On the one hand, rivers may respond in a variety of ways to a given change of climate, both between different drainage basins and possibly even within the same basin. For instance, Burkham's (1972) study of the Gila River in Safford valley, documented marked changes in channel width and sinuosity over the period 1846-1970 in response to short-term variations in rainfall and flooding. In contrast, Graf's (1981) study of the Gila River 250-300 km further downstream, demonstrated that there has been little change in sinuosity over essentially the same period, despite numerous changes in channel location. On the other hand, similar channel responses are sometimes interpreted as a result of different climatic changes. For instance, widespread arroyo entrenchment in the American southwest in the late 1800s and early 1900s has commonly been attributed to secular climate change, but there is little agreement over the nature of these climate changes. Some workers have suggested that climate may have been drier at the end of the nineteenth century (eg. Antevs, 1952), others have suggested that it may have been wetter (eg. Huntington, 1914), while others still have argued that there may have been a period of relatively low frequency of light rains and a greater frequency of heavy rains, with this change in precipitation affecting both vegetation cover and discharge (eg. Leopold, 1951; Denevan, 1967; Cooke and Reeves, 1976). While some of this uncertainty probably reflects a lack of understanding of past climate change, it also suggests that there may be a strong element of equifinality in the nature of channel responses to those changes and that, given the final channel form, it may not always be possible to determine the actual cause of channel change.

A second and related problem is that certain river systems, previously chosen to represent evidence of the impact of climate change on arid-environment systems, have been misinterpreted. For instance, the rivers of the Riverine Plain of southeastern Australia have been widely cited as evidence of regional climate shifting between humid and arid phases (Schumm, 1968; Cooke et al., 1993; Reid, 1994). There has been a long and confusing history (summarised by Page et al., 1996) of interpreting the palaeochannels of this region as a response to either wetter or drier climates. Recent research (Page et al., 1996; Page and Nanson, 1996) has shown that the extensive palaeochannels, previously referred to as ancestral and prior streams, and believed to be distinct from each other, are in fact closely related and are both the product of greatly enhanced flow conditions. Distinctly arid periods during the late Quaternary (Oxygen Isotope Stage 4 as well as a limited period around the actual glacial maximum of Stage 2) have left no discernible fluvial evidence on the Plain. These rivers are not suitable indicators for variations in downstream aridity, for they are strongly allogenic systems influenced in large part by changes in the ice, snow and vegetation cover of the Great Dividing Range in which they are sourced.

A third problem is that there is considerable evidence that arid-zone channels are susceptible to the erosive effects of individual high-magnitude floods, such that longer term channel changes which occur as a result of more subtle climatic changes are easily obscured. Large-scale channel changes resulting from the impact of one or more large floods have been noted on many arid-zone rivers, with Schumm and Lichty's (1963) and Burkham's (1972) studies of the Cimarron and Gila Rivers among the most widely cited. On the Cimarron River, for instance, a high-magnitude flood in 1914 increased channel width from an average of 15m to 200m and during a further series of floods in the following 25 years, channel width increased to an average of ~360m (Schumm and Lichty, 1963). During a period of lower discharges over the next 21 years, channel width declined to ~180m as vegetation re-established, indicating that channel recovery was under way. In other arid environments, where vegetation is sparser and restorative flows less frequent, channel recovery following large floods may take much longer, if indeed it occurs at all (Wolman and Gerson, 1978). The dramatic fluctuations in the dimensions of many arid-zone channels, such as channel width, in response to short-term changes in flow regime suggests that these parameters may have only limited value as diagnostic indicators of longer term climate change (Reid, 1994).

A fourth and related problem is that there is a body of literature that suggests that the discontinuous nature of flow and sediment transport in arid regions, and the sensitivity of channel form to the impact of rare, high-magnitude floods, results in many arid-zone channels being essentially non-equilibrium systems, with transient behaviour as the norm rather than the exception (eg. Thornes, 1977, 1980; Graf, 1983). Where non-equilibrium conditions prevail, clear-cut relationships between climate and river process and form are not readily identifiable, making the interpretation of recent forms difficult and perhaps individualistic in different drainage basins (Thomas, 1997b). In arid-zone rivers, the major determinant of channel size and shape, for instance, may be the magnitude of infrequent large floods which are unrepresentative of the normal range of flows. Hence, inferences as to palaeohydrological and climatic changes on the basis of palaeochannel dimensions and sediments are fraught with difficulties. Similarly, although terrace sequences have often been interpreted as evidence of flow regime and climatic change in arid and semi-arid regions, for small catchments in extremely arid areas, Schick (1974), Schick and Magid (1978) and Bull (1991) have suggested that terraces may also result from the random occurrence of rare "superfloods". In these instances, terraces are not generally attributable to climatic changes unless these include a change in the magnitude and frequency of superfloods.

Finally, there is the problem recognised above that it is difficult to determine whether it is greater aridity or greater humidity that results in the formation of river terraces. Bull (1991) found that river terraces can aggrade under very different conditions in arid environments compared to humid environments, with a range of responses between these two extremes. This means that the existence of a river terrace in an arid environment could imply a shift to wetter conditions in some situations, and a shift to drier conditions in another. Bull also identifies the problem of diachronous terraces in regions with strongly seasonal climate and in large complex basins.

Desert River Floodplains as Indicators of Palaeoclimatic Change

Despite the sometimes equivocal relationship between climate and fluvial response, arid-zone river landforms and sediments often provide a valuable and sometimes singular guide to palaeoclimatic change. In particular, the sedimentary sequences of arid-zone rivers can harbour a record of fluvial adjustments in response to climatically-driven changes in discharge and sediment load. While within-channel deposits in arid-zone rivers can yield valuable indications of hydrodynamic and sedimentary conditions during floods (eg. McKee et al., 1967; Williams, 1971; Karcz, 1972; Pickard and High, 1973; Frostick and Reid, 1977, 1979; Langford and Bracken, 1987; Shepherd, 1987), they are easily reworked so that only the most recent events may be recorded by the stratigraphy. On the other hand, floodplain deposits record flow events occurring over longer time periods. Despite this, the floodplain deposits of arid-zone rivers are relatively poorly accounted for in the literature. Indeed, Graf (1988a, p. 217) considers that many ephemeral arid-zone rivers do not have conventional floodplains, stating that:

> "In drylands, meandering streams construct flood plains as in humid regions, but in drylands such streams are relatively rare. Braided channels, more common in drylands, are not often associated with flood plains in the geomorphologic sense. They frequently occupy the entire available space between low terraces, leaving no room for horizontal surfaces that are activated by present regime processes of the river as demanded by the definition of flood plain..."

This view is based largely on experience with arroyo-style rivers in the American southwest and highlights an important misconception associated with floodplains, for there is the implicit assumption that meandering rivers form floodplains but other river styles (particularly braided rivers) rarely do so. The implication that braided rivers do not usually form floodplains has recently been challenged by Nanson and Croke (1992) and Reinfelds and Nanson (1993), and the role of arid-zone rivers in forming floodplains has been contradicted by research into erosion and accretion of floodplains in the American southwest (Schumm and Lichty, 1963; Burkham, 1972; Hereford, 1984, 1986). To date, research into arid-zone rivers has mainly focused on relatively confined valley or piedmont settings with very few studies of channels associated with extensive floodplains, making it difficult to generalise about floodplain formation across a range of arid environments.

Another problem is that there is a wide variety of fluvial processes in arid or semi-arid regions (Olsen, 1987), whereby flood flows may be largely channelised (eg. Karcz, 1972; Picard and High, 1973), largely unchannelised (eg. Graf, 1988b), or show varying combinations of both (eg. Sneh, 1983; Tooth, 1998a, b). These processes give rise to a variety of sedimentary landforms and sequences, also rendering generalisations about floodplain forms and processes problematic.

Due to these difficulties, the remainder of this chapter adopts a case study approach in order to illustrate how floodplain sequences can be used to interpret palaeoenvironmental change over a variety of temporal and spatial scales. It examines three different arid-zone regions supporting largely endogenic drainage systems. The first of these is the Colorado River basin in the semi-arid American southwest (Fig. 1a). On the elevated Colorado Plateau, channel change has been documented for more than a century, and there are corroborating climate and stream flow records covering a substantial part of that period. In the lower Colorado River region there is a much longer regional chronology of mid to late Quaternary climate change which is being developed from alluvial evidence. The second is a study of the sand-dominated rivers draining the central Australian ranges near Alice Springs (Fig. 1b), for this region illustrates environmental change over thousands and tens of thousands of years in large single-thread and anabranching channels, some of which are confined by terraces, dunes or bedrock, and some of which disperse on to unchannelled sections of floodplain. The third is the Channel Country of western Queensland (Fig. 1b), a vast system of low-gradient, mud-transporting, anabranching channels with a detailed Quaternary record and thermoluminescence (TL) chronology extending back several hundred thousand years (more than two full glacial cycles). Finally, we review the work on the Nile River in order to illustrate the character of a large allogenic system that has undergone dramatic change during the last glacial and Holocene.

The American Southwest

Hereford's (1984, 1986) studies of the Paria River drainage basin on the Utah-Arizona border and the much larger Little Colorado River basin in

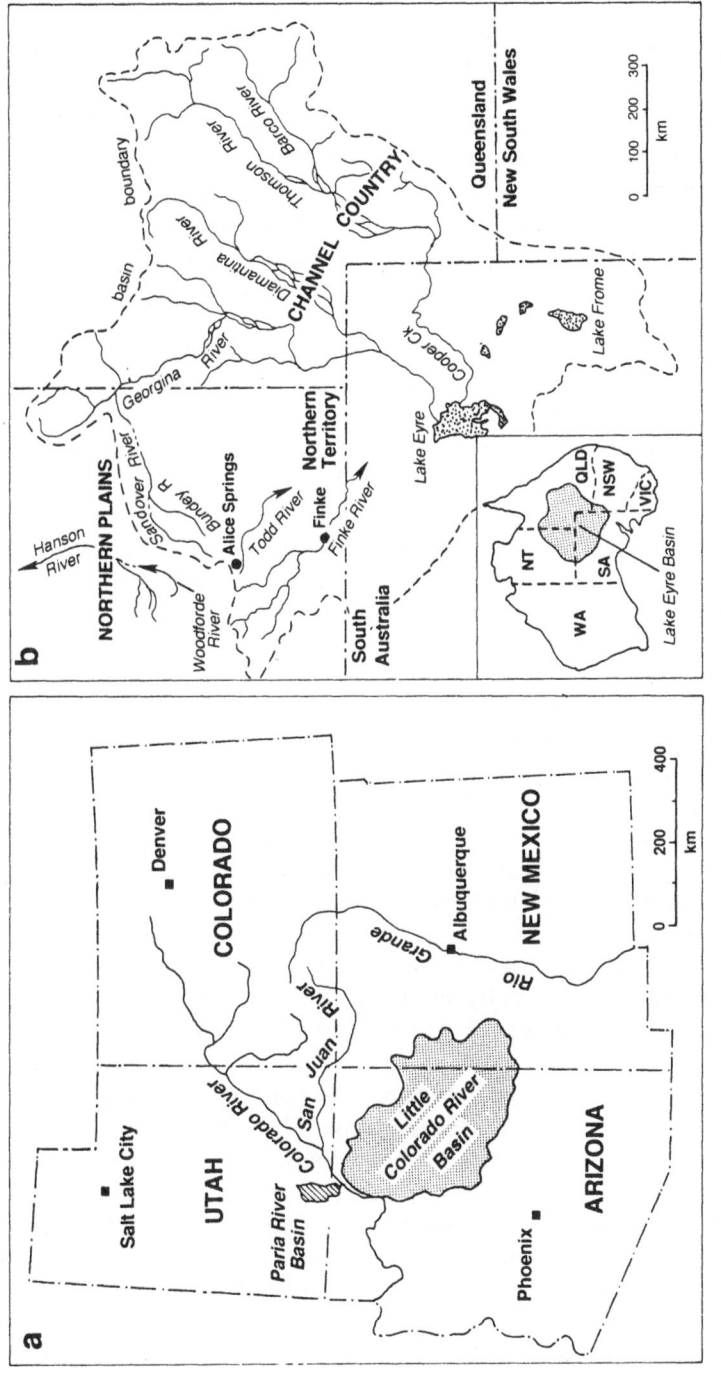

Fig. 1b: The study rivers in the Alice Springs area of central Australia, and those that constitute the Channel Country of western Queensland. All but the Woodforde River contribute to the Lake Eyre basin.

Fig. 1a: The Paria and Little Colorado River basins in the southwest United States of America.

Arizona (Fig. 1a) illustrate the important role of short-term climatic variations in channel change and floodplain formation. In both studies, instrumental records of climate and discharge are available for interpreting the changes in the alluvial record over the past century or so, and dendrochronological data supplement these. Hereford's studies are among the very few examples of documented relationships between climate and alluvial history, albeit for relatively short periods of record.

The Paria River and its tributaries drain around 3600 km^2 of the southwestern Colorado Plateau where relief in the 120 km elongated basin ranges from ~1000 m to ~3100 m in the headwaters. While gradients are not given, from these basin parameters they can be presumed to be relatively steep. The Little Colorado River drains a 400 km long basin of 44,000 km^2 that ranges in elevation from ~1830 m to ~3800 m with gradients that appear to be less than half those of the Paria River. The climate of both areas is arid to semi-arid with annual precipitation averaging 300 mm or less. These rivers have floodplains and terraces set in confining canyons. The floodplain of the Paria is particularly narrow, measurable in a few metres to a few hundred metres. Those on the Little Colorado are highly variable (<100 m to 12 km), although the sections studied were narrow floodplains in rock-walled canyons.

The Paria River is a perennial stream whereas the little Colorado is ephemeral, but both have distinctly bi-annual hydrographs peaking in February-April and August-September. Modern fluvial activity in the Paria basin consists of an erosional episode beginning in the early 1880s and a subsequent aggradational episode beginning in the early 1940s. Reports from early settlers indicate that channels throughout the basin widened and deepened in the early 1880s, attained their maximum width and depth by 1890, and remained this way until aggradation began in the early 1940s. Well-defined floodplains (Fig. 2a) occur 0.5-2 m below an older cottonwood-forested terrace. An older channel alluvium occurs beneath the floodplain and is inset within the cottonwood terrace. The geomorphic relationship between the terrace, the floodplain and the basal channel alluvium is one of cut-and-fill and one that tells a story of climate change.

The cottonwood floodplain was almost certainly the active floodplain prior to the 1880s, for its oldest cottonwood trees germinated then. The older channel alluvium consists of parallel laminated sands and gravels and is slightly coarser than the overlying floodplain alluvium. It is undated but its stratigraphic position shows it to be younger than the early 1880s terrace which it incises, and older than the overlying floodplain which from tree-ring evidence began to form about 1940. The floodplain consists of two overlying units both formed mainly by vertical accretion but under different flow regimes (Fig. 2a). Unit 1 is lowermost and is usually about

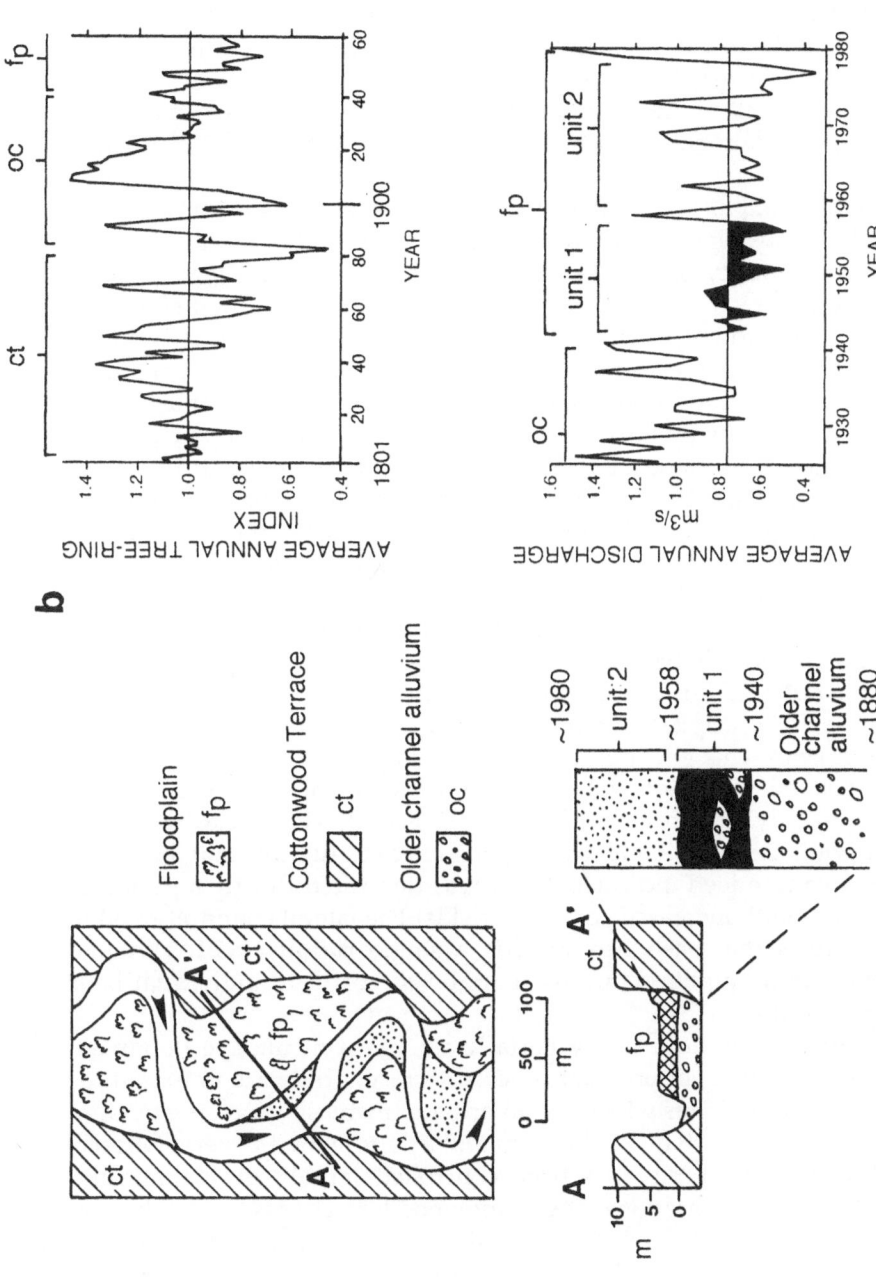

Fig. 2 (a): The planform and stratigraphy of the Paria River floodplain and terrace, giving the approximate ages of each component. (b). The average annual tree ring indices for the catchment and instrumented average annual discharges for the Paria River in relation to the formation of floodplain stratigraphic units (modified from Hereford, 1986).

1 m thick and consists of mainly thin-bedded fine sand with beds of silt but with pebble to cobble gravels occasionally present at the base. From tree-rings it dates from somewhere between 1939 and 1946, to about 1956. Unit 2 is nearly 2 m thick, consists mainly of moderately- to well-sorted fine to medium sand, and has several beds about 15 cm thick with sharp, non-gradational contacts. It was deposited after Unit 1 but before 1980.

The erosional phase from the late 1800s to the early 1940s was probably initiated and sustained by an increase in peak-flood discharge caused by increased frequency of intense rainfall. Tree-ring widths suggest a very dry phase followed by a sudden increase in precipitation that started in 1882 (Fig. 2b) and ended (from instrumental records) in the early 1940s. This appears to have resulted in deepening and widening of the channels and to have prevented floodplain formation, and it is represented in the alluvial stratigraphy by the older channel alluvium. Annual precipitation and daily and peak-flood discharges decreased between about 1942 until 1957 or 1958 (Fig. 2b), and were probably the main factors initiating floodplain aggradation. The thin beds, fine grain size, poor sorting and gradational contacts of Unit 1 suggest that it was deposited in the channel on the tops of the bars during a period of low flow and infrequent large floods, but with generally low current velocities and a lack of intense scour. After about 1958, annual precipitation and discharge increased (Fig. 2b), but without a corresponding rise in peak-flood discharges, suggesting that precipitation did not return to its earlier pattern of frequent intense storms. This change in the relationship between rainfall, runoff and associated sediment yield, appears to have caused the change in depositional style represented by Units 1 and 2. The thicker bedding, coarser grain sizes and better sorting of Unit 2 stratigraphically higher in the floodplain indeed indicates that it was deposited during a period of higher flows with increased sediment yields. The lateral continuity within beds suggests that each bed records a single runoff event, although additional stratigraphic dating is required to correlate individual beds with specific floods.

The Little Colorado, like the Paria River, has a distinctive floodplain inset 1-3 m below a cottonwood-covered terrace. There are in total three distinctive alluvial units which are, in order of increasing age, the modern floodplain, an older channel alluvium and a cottonwood-covered terrace. The alluvial stratigraphy has been controlled largely by climatically-induced discharge variations. The cottonwood terrace was the active river floodplain formed by a relatively consistent flow regime for more than 100 years prior to about 1905. Following this, a sharp increase in regional precipitation and flood discharges eroded a wide sandy channel almost free of vegetation that deposited the older channel alluvium. In the 1940s, average annual precipitation declined, reducing annual discharge by about 40% as well as reducing the frequency of large floods. The channel decreased in width and the floodplain accreted vertically in a series of depositional events until 1980 when channel behaviour changed to one of channel migration and lateral accretion. The present floodplain is now eroding and lies beyond the reach of vertical accretion, and is possibly reverting to a terrace again.

What is interesting from these two studies is that these basins, ~150 km apart, share a similar history of changing flow and sediment regime, particularly in this century. Also noteworthy is that the major episode of channel enlargement on the Little Colorado started in about 1905, some 20 years after a similar episode on the Paria River. From tree-ring evidence there is little doubt that there was an increase in precipitation in the early 1880s and a substantial one about the turn of the century. It would not be surprising if the steeper Paria River system responded first to these episodes.

Hereford concluded from these studies that floodplain development is initiated and sustained through climatic and hydrologic fluctuations. He shows that closely related historic erosional and aggradational events have occurred in a variety of basins throughout the southern Colorado Plateau (see Hereford, 1986, for references) in response to subtle changes in climate. The variable sedimentology and stratigraphy of floodplain units accumulated under different discharge regimes is a result of changes in the relationship between precipitation, runoff and sediment yield. His work shows that floodplains can, in certain situations, provide a complex but decipherable fine-resolution record of short-term climate change.

At a greatly extended temporal scale, Bull (1991) reviews terrace formation and associated climate change in the lower Colorado River region. He recognises nine alluvial geomorphic surfaces which he labels Q1 to Q4 (with subdivisions) (Fig. 3). Each surface is recognised on the basis of topographic position, surface texture, soil and rock varnish development, and various forms of dating. Q1 is a general category for all alluvial surfaces older than ~1.2 m yrs. Q2a is estimated to be 730-400 k yrs and to represent the culmination of interglacial conditions, as is Q2b which is believed to be 200-70 k yrs. Q2c consists of deposits formed during average glacial but warming conditions from 70-12 k yrs. Q3a represents the transition from full glacial to interglacial conditions from 12-8 k yrs and Q3b is the culmination of the interglacial at 8-4 k yrs. Q3c is late interglacial at 4-2 k yrs old and Q4a and b are effectively the active or immediately abandoned stream channel (0.1 - 2 k yrs). Although the age ranges for the older units are clearly very broad (Fig. 3), this reflects the less precise dating near and beyond the range of radiocarbon. Nevertheless, the scheme represents an interesting attempt at a regional

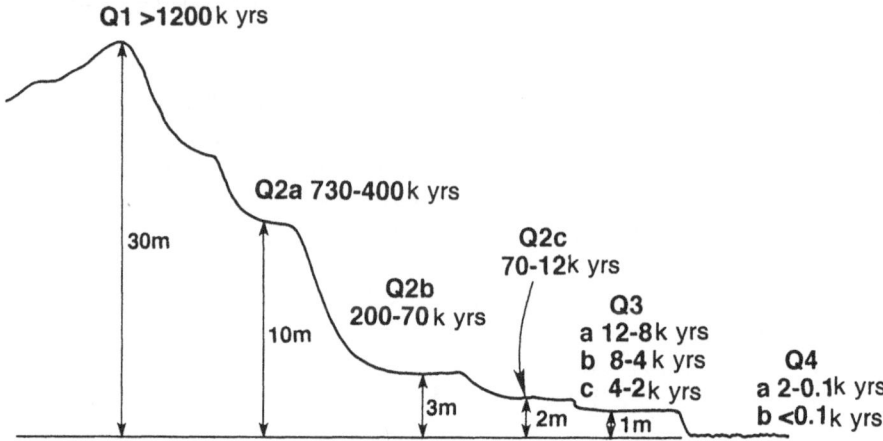

Fig. 3: Schematic diagram of the elevation and chronology of terraces south of the
Whipple Mountains, southeastern California, USA (modified from Bull, 1991,
using his Figure 2.9 and Table 2.13).

synthesis that will be revised and improved as more detail comes to
light. In confined valleys subject to intense floods typical of many
headwater or incised streams in the Colorado basin, however, the
preservation potential of an entire sequence of older deposits is likely to
be limited. While large, lower gradient rivers may be relatively insensitive
to all but major glacial-interglacial and stadial-interstadial shifts in climate,
by the same token, such rivers are likely to preserve older alluvial
successions and thus provide longer term, more continuous records of
climatic change. To illustrate the use of relatively low-gradient arid-zone
river floodplains as indicators of longer term climatic change, the following
two sections of this Chapter consider case studies from arid central
Australia.

The Alice Springs Region of Central Australia

The 'Alice Springs area' (Stewart and Perry, 1962) is an arid to semi-arid
region in the centre of the Australian continent (Fig. 1b). It is dominated
by the central ranges, a belt of sedimentary and crystalline rocks which
trend west-east across the middle of the region (Fig. 1b). The
unconsolidated alluvial sediments of the drainage systems in this
tectonically stable region are generally shallow. They provide a
fragmentary record of late Quaternary flow regime and climate change
from about 100 k yrs to the present, but with improved detail for the
Holocene.

The findings presented here are drawn from two areas. One is a
composite area of the plains to the north of the central ranges (Northern

Plains) and consists of the middle and lower reaches of the Sandover, Sandover-Bundey and Woodforde Rivers (Fig. 1b) (Tooth, 1997), three drainage systems that have catchments of approximately 10,600, 11,000 and 550 km^2, respectively. River gradients range from ~0.002 in the piedmont reaches to ~0.001 or less in the lower reaches, and planforms vary from predominantly single-thread to anabranching. The other area is a short reach of the lower Finke River southeast of the central ranges and about 200 km from Alice Springs (Fig. 1b). At that point the Finke drains an area of ~54,500 km^2 and flows at a gradient of ~0.0026.

Rainfall across the Northern Plains averages ~300 mm/yr but declines southwards to ~150 mm/yr at Finke, with pan evaporation for both areas being ~3000 mm/yr. However, there is great variability in rainfall from year to year. Particularly heavy rainfalls (>100 mm/24h) can result from incursions of monsoonal depressions or cyclones from the north and northwest of the continent, or from the easterly passage of mid-latitude frontal weather systems. The flow regime of the rivers is ephemeral, with the channels remaining dry for much of the year. There are no rated gauging stations on the rivers of the Northern Plains, or for several hundred kilometres upstream of the study area on the Finke River, but a series of floods in the last 30 years are known to have included some of the largest events in the 100-110 years of European settlement in the region (Williams, 1970; Baker et al., 1983b; Pickup et al., 1988; Pickup, 1991).

Near the township of Finke, the Finke River flows in a 0.8-2.5 km wide bedrock valley, the low but steep sides of which separate a well-defined floodplain from adjacent gently-undulating gibber plains covered with linear dunes (Nanson et al., 1995) (Fig. 4). A dense field of source-bordering transverse dunes extends about 4 km north of the river, gradually merging with the regional linear dunefield. Thermoluminescence (TL) ages indicate isolated remnants of alluvial deposits near the valley sides which date at >90 k yrs. There is little surviving evidence of fluvial activity until the Holocene, although the aeolian dunes indicate at least periodic aridity from before the Last Glacial Maximum (LGM) until the Early Holocene (~30 k yrs to ~9 k yrs) (Fig. 4) (Nanson et al., 1995). Augering and TL dating of the floodplain alluvium has revealed a period of relatively high discharges probably associated with enhanced rainfall from ~9 k yrs to ~5 k yrs, which must have reworked most of the previous valley fill and replaced it with a floodplain that has survived largely unaltered to the present. Significantly, there was a hiatus in the activity of the adjacent dunes during this period. Reduced fluvial activity, probably due to reduced rainfall, led to channel stability and reactivation of the adjacent dunefields from the Mid to the Late Holocene, and the tops of the dunes remain active today.

The rivers of the Northern Plains provide a different but largely complementary story to that obtained from the Finke River. A basic

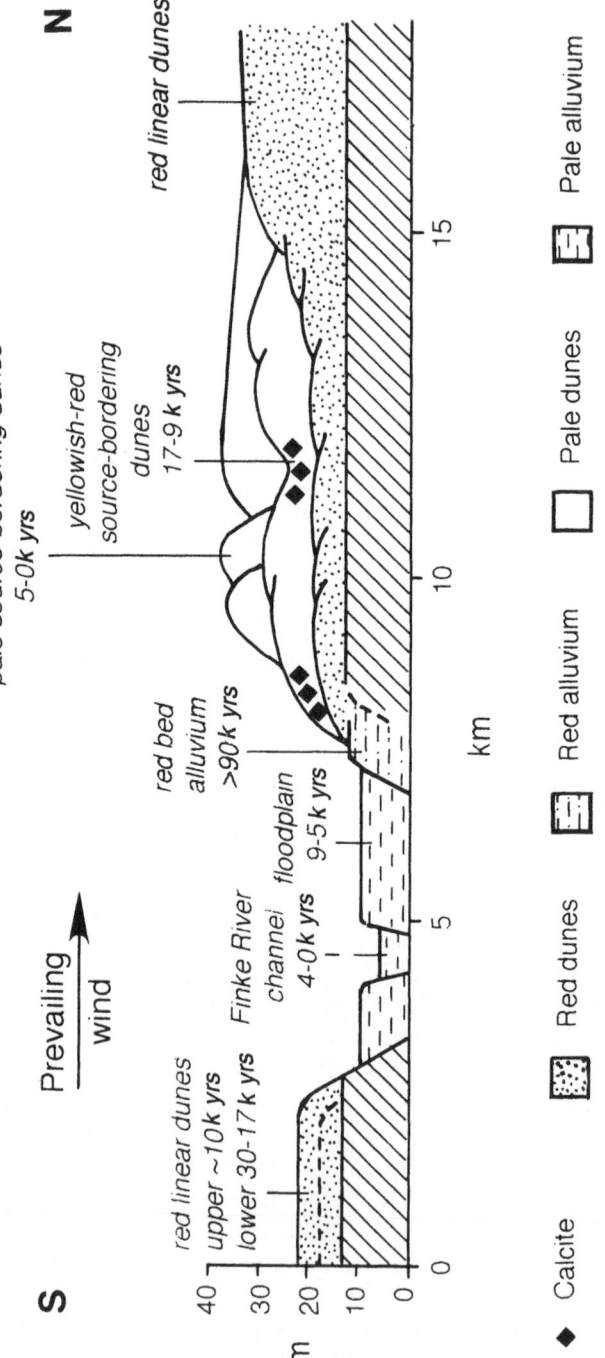

Fig. 4: A schematic section across the Finke River valley and associated dunefields, showing the stratigraphy and chronology of the depositional units (modified from Nanson et al., 1995)

distinction can be drawn between the middle reaches of the Sandover, Sandover-Bundey and Woodforde Rivers, where the channels are flanked by indurated alluvial terraces (often in association with bedrock and substantial source-bordering dunes or aeolian sandplains), and the relatively unconfined lower reaches where the channels are flanked by extensive, low-relief alluvial plains (Fig. 5) (Tooth, 1998a, b). This distinction has greatly influenced the preservation potential of alluvium in the middle compared to the lower reaches. In the former, the terraces provide a resistant boundary of indurated silts and sands and consequently restrict the width of floodplain development and limit channel migration and avulsion. In the lower reaches, the height of the terraces above the floodplain declines such that the terraces are eventually buried by younger, relatively erodible silts and sands. At this location, the channel emerges from the previously confining terraces and overbank flows spread across extensive alluvial surfaces where there are numerous splays, distributaries and abandoned channels (Fig. 5).

The geomorphology, stratigraphy and sedimentology of the terraces and floodplains in the confined and unconfined reaches of the rivers on the Northern Plains provide evidence of major flow-regime and climate change during the Late Quaternary. Meandering palaeochannels within the terrace alluvium suggest that these extensive alluvial deposits were laid down by large, sinuous, laterally migrating channels from some time prior to ~56 k yrs until ~17 k yrs. In contrast to the study site on the Finke River, there is some evidence here of fluvial activity around the Last Glacial Maximum (TL dates of ~25 k yrs and ~17 k yrs), an observation that is supported by fluvial evidence from waterfall plunge pools in tropical northern Australia (Nott and Price, 1994; Nott et al., 1996). Indeed, the present channels appear to have inherited their large meander wavelengths from these sinuous Pleistocene channels. After ~15 k yrs, decreasing flows and a reduction in channel size left the older Pleistocene alluvium as paired terraces, thus confining the middle reaches of the rivers to a narrow trench mostly less than ~300m wide. In distal reaches, shallow burial of Pleistocene alluvium by younger unconfined alluvial deposits has occurred (Fig. 5).

The situation of confined middle reaches and unconfined lower reaches has strongly influenced the preservation of the stratigraphic record in different parts of the system. In the middle reaches, enhanced Early to Mid Holocene fluvial activity must have occurred within this terrace-bounded trench in much the same way as it occurred in the considerably larger bedrock valley of the Finke River. However, the high level of confinement in the middle reaches of rivers on the Northern Plains meant that almost no floodplain alluvium from the Early to Mid Holocene was able to survive the process of fluvial reworking, a condition of dynamic

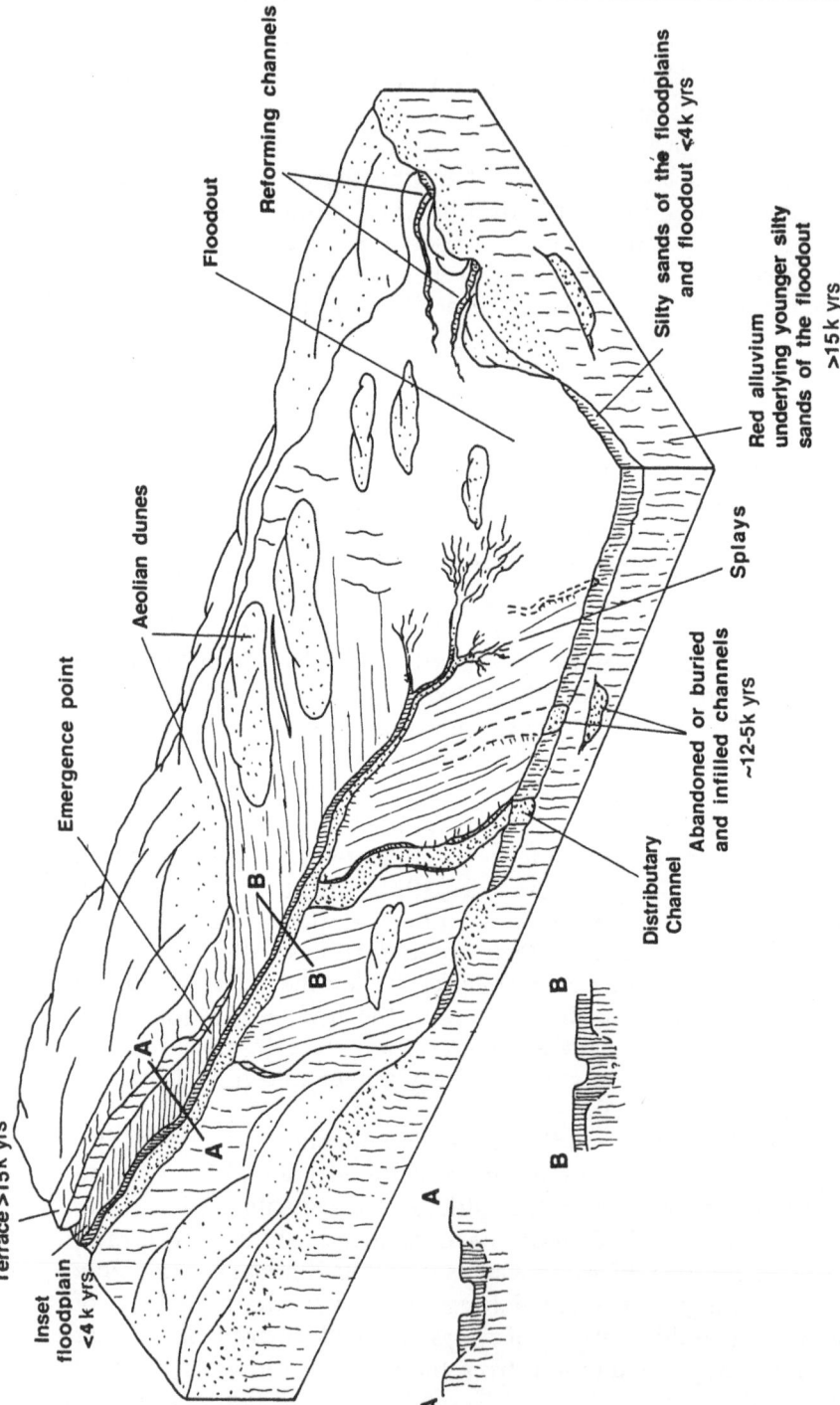

Fig. 5: A schematic illustration of channel geometry, sedimentary architecture and Quaternary chronology in the confined middle reaches and unconfined lower reaches of rivers on the Northern Plains, central Australia. In the confined lower reaches, the channels and Late Holocene floodplains are inset within Pleistocene alluvial terraces (Location A). In the lower reaches downstream of the emergence point, the Pleistocene terraces are buried by Holocene floodplains that increase in width (Location B), and channel capacities rapidly decrease downstream. In this *floodout zone* there are numerous distributaries, splays and abandoned and buried channels. Older Holocene and Late Pleistocene alluvium is partially preserved within the stratigraphy of the floodout zone where the degree of alluvial reworking has been less intense than in the confined reaches upstream.

change also observed by Bourke (1994) in some of the region's streams. Today, floodplains in the confined reaches of the Sandover, Sandover-Bundey and Woodforde Rivers consist of narrow strips of alluvium (<50 m wide) that TL date from about 4 k yrs to the present. The exponential decline in the number of samples of increasing age within these floodplains highlights the low probability of alluvium surviving any substantial length of time in this environment. However, in the unconfined distal reaches, substantial changes in the course of the channels during the Holocene resulted in partial preservation of older Holocene alluvium. Here, isolated TL dates range from ~13 to ~9.5 k yrs for deposits adjacent to the modern rivers, show that pockets of very Late Pleistocene and Early Holocene alluvium have been preserved alongside more widespread Late Holocene deposits.

Of further interest for interpreting the Late Holocene history of the rivers of this region are slackwater deposits in the headwaters of the Finke River. From the evidence for a number of high-magnitude events in the very Late Holocene (Pickup et al., 1988; Pickup, 1991; Patton et al., 1993), it appears that four of the largest events in the past 850 years have occurred since 1966. While these four latest floods have resulted in some changes in reaches close to the central ranges (Pickup, 1991; Patton et al., 1993; Bourke, 1994), they appear to have had relatively little impact in the confined middle reaches of the Sandover, Sandover-Bundey and Woodforde Rivers where aerial photographs show that the floodplains and channels have remained essentially unaltered over the past 40-50 years. The main exception has been in the unconfined lower reaches of these where floodwaters have spilled out and there has been limited activity in the form of channel widening, avulsion, and the development and enlargement of splays. Clearly, even major climatic perturbations within historical time have had a fairly minor impact along these rivers, with almost no impact in the confined reaches of channels where energy can be concentrated. This is evidence that earlier Pleistocene and Holocene flow-regime changes that had the ability to rework completely the floodplains and terraces of the Northern Plains must have been considerable indeed.

It is difficult to establish whether fluvial activity in the central ranges during the Early to Mid Holocene reflected increased summer rainfall resulting from more frequent monsoonal and cyclonic incursions, or increased winter rainfall resulting from changes in the frequency and intensity of mid-latitude frontal systems. However, the broad correspondence between the changes observed here, and the evidence for enhanced wetness and runoff in northern Australia during the same period (Nanson et al., 1993; Nott and Price, 1994; Nott et al., 1996; Wende et al., 1997), suggest that it is most likely that the rivers draining the central

Australian ranges have responded to changes in the northern Australian monsoon.

The different sequence of Late Quaternary changes recorded in the alluvium of the Northern Plains rivers compared to that on the relatively stable short study reach of the Finke River reflects very different flow-regime changes that must have been fairly similar, as both river systems originate in the central ranges. However, elsewhere on the Finke, evidence from aerial photographs flown since the 1950s indicate that some reaches have undergone widespread changes over that period. This underlines the importance of local physiography in determining the nature and distribution of the alluvial evidence that will survive. Evidence of climate change from such rivers is a function, not only of the magnitude and character of the change, but also the landscape's ability to respond to and thereby record such events. For instance, the study reach in the Finke valley is sufficiently broad to have recorded and retained alluvial evidence of major flow-regime changes from the Early Holocene, but not sufficently sensitive to have retained evidence of the lower magnitude Late Holocene changes. In contrast, the river valleys on the Northern Plains are too narrow for evidence of the Early Holocene to have survived, although they have retained at least part of the Late Holocene record. However, even the rivers of the Northern Plains have not been responsive to the series of high magnitude floods of the past 30 years. This is in contrast with the recent alluvial changes recorded on the highly responsive Paria and Little Colorado Rivers (Hereford, 1984, 1986). To illustrate the use of arid-zone rivers as an indicator of palaeoclimatic change where the level of record sensitivity is very low, but where evidence extends back to the Mid Quaternary, the following section considers the Channel Country of western Queensland.

The Channel Country of Western Queensland

The Lake Eyre basin covers an area of $1.3 \times 10^6 \, \text{km}^2$ in east-central Australia (Fig. 1c). Its main river systems on the east side of the basin, Cooper Creek and the Diamantina-Warburton and Georgina Rivers, are collectively termed the Channel Country and contain one of the most extensive and well-dated Quaternary alluvial sequences anywhere. These rivers rise to the north and northeast in bordering uplands that rarely exceed more than 500 m in elevation, and flow about 1000 km over gradients of about 0.0002 to Lake Eyre, a huge salt pan (nearly 10,000 km^2) which lies 15 m below sea level (Fig. 1b). River planforms are predominantly anabranching but with a system of braided flood-channels, and floodplains reach 70 km in width (Nanson et al., 1986; 1988).

Precipitation over the eastern basin varies from ~450 mm/yr in the headwaters to ~120 mm/yr in the Simpson Desert, with most of this area receiving less than 250 mm/yr. Annual pan evaporation varies from ~2400 to over 3500 mm/yr. A common pattern of precipitation in the basin occurs when moist, tropical air spills over the northeastern Great Dividing Range or the northern tablelands during the monsoon (November to March). This leads to intense but erratic rainfall that brings floods to the rivers in most years (Kotwicki, 1986), but major flooding episodes occur only once every twenty years or so and are often associated with La Niña phases of the El Niño Southern Oscillation (ENSO) (Allan, 1988). Less intense than the present monsoon influences are frontal rainfall events that occur mostly during the winter months in the central and southern parts of the basin. Latitudinal shifts in the pressure belts have caused variations in the intensity of the northern monsoon and westerly frontal system within the basin during the Quaternary.

The Channel Country contains a detailed alluvial and aeolian sedimentary record of Quaternary environmental change, for the region has responded to global changes in climate and the associated regional changes in water discharge, sediment load and vegetation (Nanson et al., 1992; Callen and Nanson, 1992). Cooper Creek and its upstream tributaries is the system which has been studied in most detail (Nanson et al., 1988). This gradually accreting alluvial system contains over 100 m of Late Tertiary and Quaternary alluvium (Senior et al., 1978). Although the older and deeper Quaternary record is beyond the range of TL dating, deposits laid down in the upper 8-12 m over the past ~300 k yrs provide a picture of markedly fluctuating environmental conditions. At present, Cooper Creek transports very little sand. The channels are mud-lined and are laterally very stable, and extensive floodplains are formed of mud to depths of 2-8 m (Fig. 6) (Nanson, et al., 1988). The base of these surficial muds TL date at about 80 k yrs, the end of the last interglacial (Stage 5a or 5b), and they appear to have been deposited throughout the subsequent period, including the last glacial through to the present. However, along the middle and lower reaches of Cooper Creek, aerial photographs and shallow augering have revealed remnant scroll-bars and palaeochannels beneath this mud unit. These features are scaled to river meanders far larger than any present in the system today (Fig. 7) (Rust and Nanson, 1986, Nanson et al., 1988), and were formed by mixed-load, laterally-migrating rivers that deposited the Katipiri Formation, extensive sandy units with abundant flow structures. It has been possible to TL date these alluvial deposits (Fig. 8b) and also to obtain U-series dates for pedogenic minerals (ferricrete, calcrete and gypcrete) that have formed in the same units at or soon after alluvial deposition (Fig. 8a).

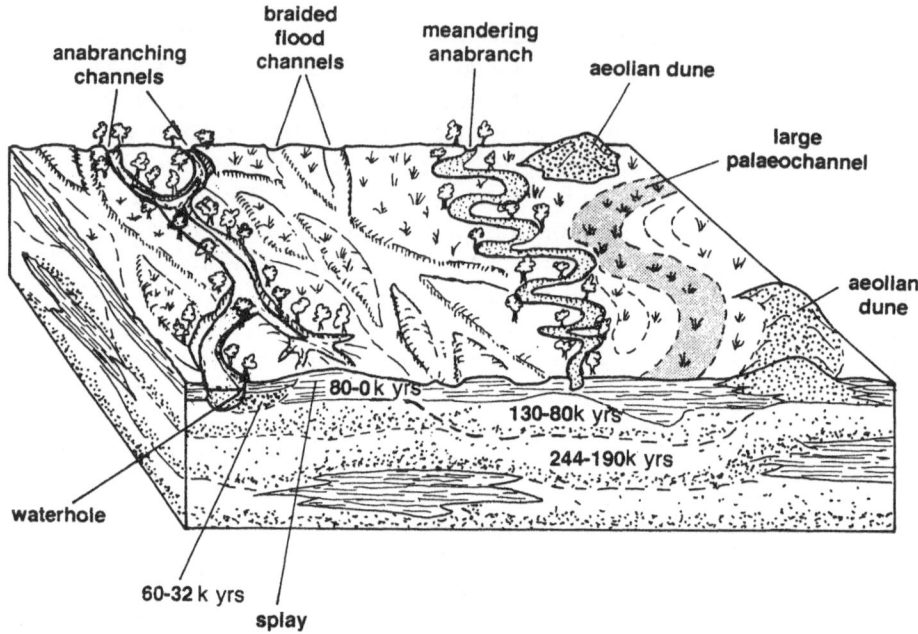

Fig. 6: A schematic illustration of the landforms, palaeochannels, stratigraphic architecture and Quaternary chronology of the middle reaches of the Cooper Creek floodplain. Anabranching channels, waterholes, splays, braided flood channels, palaeochannels and aeolian dunes characterise the mud-covered floodplain surface. Katipiri Formation sand-bodies are evidence of lateral accretion during the interglacials with surface and remnant subsurface mud units linked to the more arid glacials.

Preliminary TL dating at depth reveals that the Katipiri Formation is at least 0.5-0.75 m yrs (Maroulis, in prep.), although a reasonably detailed chronology exists for only the last two glacial cycles (250-0 k yrs) (Fig. 8b). The last two interglacials (Oxygen Isotope Stage 7 and Stage 5) in the eastern half of the Lake Eyre basin were characterised, at least in part, by pronounced pluvial episodes during which the majority of these uppermost sands were deposited (Nanson et al., 1988, 1992) (Fig. 6 and 8b). It is noteworthy that these pluvials were associated with low dust flux in the North Pacific Ocean and in the Vostok ice core of Antarctica, and in China (Fig. 8c, d and f). The intervening glacials are very broadly interpreted as arid events and were probably associated with mostly overbank deposition along the rivers, for mud units certainly characterised the last glacial (the period after about 80 k yrs) and remnants of similar muds are found overlain by sand at depth (Nanson et al., 1988, 1992). Not surprisingly, these fluvially relatively-inactive dry periods are recorded as being periods of high dust flux in the ocean and ice cores and on the loess plateau in China (Fig. 8c, d and f).

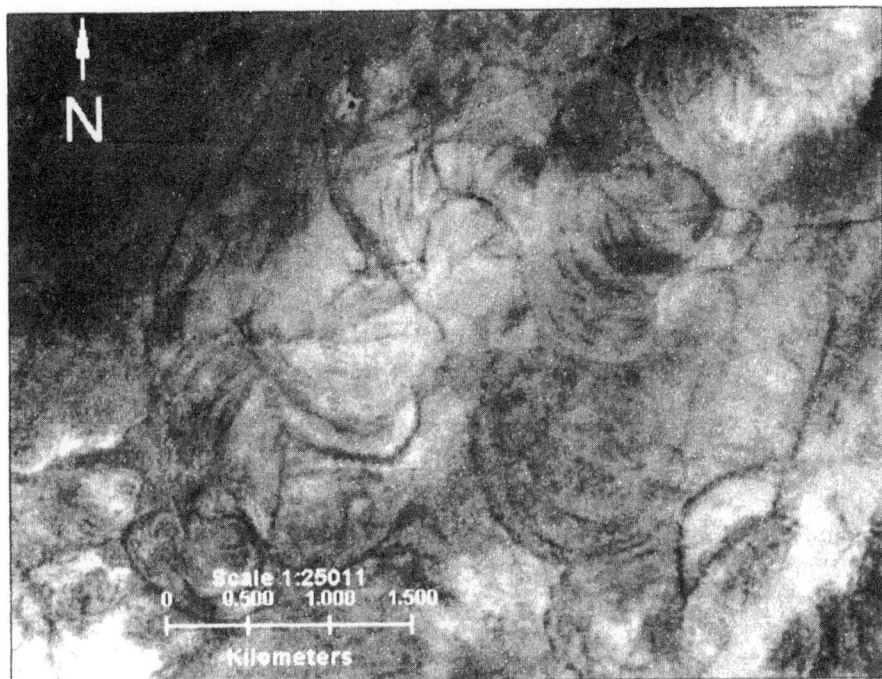

Fig. 7: A SPOT satellite image of very large palaeochannels on the Cooper floodplain near Mt. Howitt, Queensland, enhanced using NW, NE & EW directional filters (CNES 1988 Distribution SPOT image) The meander wavelength of the palaeochannels is about 1800 m, many times larger than the modern meander wavelengths in the same region of Cooper Creek (acknowledgements to Dr Toni O'Neill and Ingrid Wootten for processing and provision of this enhanced satellite image).

The TL dating record for Stage 7 (the penultimate interglacial: 244-190 k yrs) is fairly sparse, but river terraces in the headwater areas and extensive buried alluvium along the middle and lower reaches can be ascribed to a Stage 7 phase of the Katipiri Formation, during which pedogenic mineralisation also occurred (Fig. 6 and 8a and b) (Nanson et al., 1988, 1992). This period was followed by a period of relative fluvial inactivity, the Stage 6 glacial (190-130 k yrs), although recent dating suggests there may have been some interstadial fluvial activity at ~160-170 k yrs (Maroulis, in prep.). During the last interglacial (Stage 5, 130-74 k yrs), the Channel Country rivers again reworked coarse sands and overbank muds, laying down a Stage 5 phase of the Katipiri sands, with the peak of fluvial activity occurring probably at ~110 k yrs or slightly younger (Fig. 8b). This too was a period of pedogenic mineralisation (Fig. 8a). Deposition near Lake Eyre continued to late Stage 5 (Magee, 1995), following which Stage 4 was largely arid. Early to middle Stage 3 (60-32 k yrs) has been recognised across Australia as a fluvially enhanced

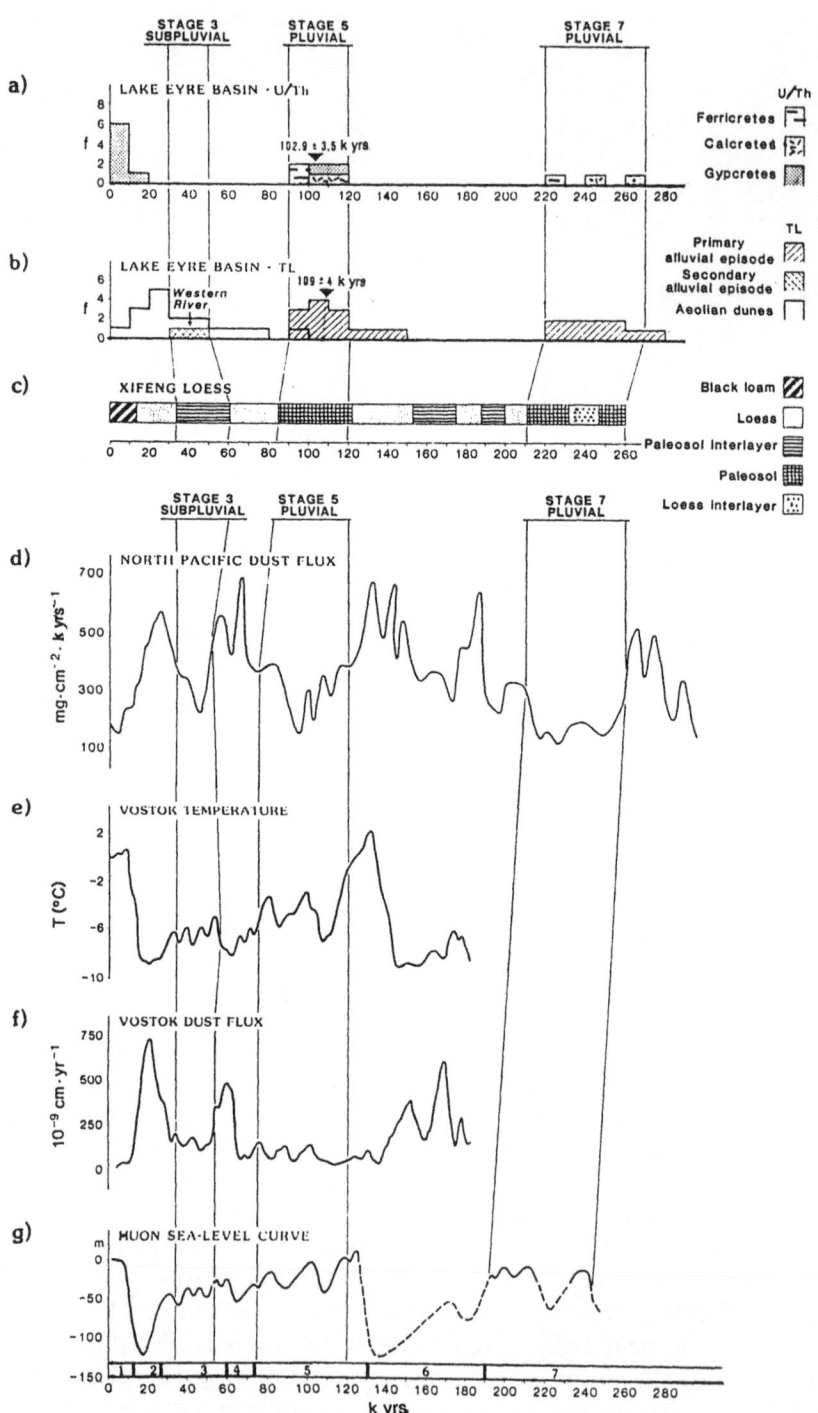

period (Nanson et al., 1992). However, the stratigraphy of the Channel Country suggests that it was most active in the steeper headwater streams (Nanson et al., 1988, 1992) and less so near the arid centre of the basin. Following Stage 3 it appears that most of the Cooper became a relatively low-energy, mud-dominated system again.

While there is no evidence for widespread fluvial activity on the Cooper after about 32 k yrs, Wasson (1983) recognised a subsurface sand body at Moomba near Innamincka which radiocarbon dating suggests could be about Last Glacial Maximum (LGM) in age. If the Cooper was fluvially active at that time then clearly it was a not a major event compared to the activity that took place during the interglacials, for the LGM was not a period of extensive sand deposition here. Similarly, so far there is almost no record of Holocene climate change in the alluvium of the Channel Country. The cohesive, very low-gradient, mud-bound channels must have been largely resistant to any Holocene flow-regime changes.

If the Channel Country had been a rapidly subsiding basin, it would probably have preserved a stratigraphic sequence of alternating sand and mud units. However, despite earlier reports based on industry bore logs of an alternating sequence of sand and mud units (Rust, 1981; Rust and Nanson, 1986; Nanson et al., 1988), a detailed drilling programme to depths of over 40 m (Maroulis, in prep.) has revealed a vertically and laterally extensive sand-body beneath the muddy floodplain surface, with intervening mud units only locally encountered (Fig. 6). It appears that very slow vertical accumulation rates and prodigious lateral activity of the channels during the sand-dominated phases largely reworked the overbank deposits, with only the present floodplain surviving as an extensive mud unit.

Despite problems of slow vertical accretion rates, and of cohesive muds preventing the recording of detailed flow-regime changes during the mud-dominated periods, a well defined alluvial stratigraphy combined with intensive TL and U-series dating has revealed a remarkable long-term record of major Quaternary climate change extending over nearly 300 k yrs in the Channel Country (Fig. 8). Relatively inactive, mud-dominated systems operative during the arid to semi-arid glacial periods

Fig. 8: Frequency histograms (a) of pedogenic U/Th dates and (b) sedimentary TL dates from aeolian and alluvial deposits along Cooper Creek and the Diamantina River (including Warburton Creek) compared to evidence of climate change from (c) Chinese Xifeng loess (Kukla, 1987), (d) the north Pacific marine core V21-146 (Hovan et al., 1989), (e&f) an Antarctic ice core (Petit et al., 1990) and (g) New Guinea sea-level data (Chappell and Shackelton, 1986). TL dates are plotted using a class interval of 10 k yrs for dates <200 k yrs, and 20 k yrs for dates >200 k yrs; U/Th dates are are plotted with a class interval of 10 k yrs; pooled mean ages are given for TL dates in Stage 5 pluvial; **Numbers 1-7** at the base are marine oxygen isotope Stages. Australian pluvial stages associated with the marine Stages 3, 5, and 7 are interpolated with vertical lines (modified from Nanson et al., 1992).

have given way to more fluvially active, sand-dominated systems during what are broadly considered to be the wetter interglacial periods. The drier periods broadly correlate with periods of enhanced dust levels in the Antarctic ice cap, the loess plateau of China and ocean basins, and the wetter periods with reduced dust levels.

In detail, however, the picture is more complex. Of particular note is that the fluvially active episodes did not correspond with Quaternary temperature maxima known from sea level and ice-core data. Fluvial activity was most pronounced following the temperature peak of the last interglacial (i.e. at about 110 k yrs compared to a temperature peak at about 125 k yrs), and less pronounced fluvial episodes have also been associated with the interstadials, such as that during Stage 3 (Fig. 8) (Nanson et al., 1992; Kershaw and Nanson, 1993). This lack of correspondence supports the observation that, while we are presently in an interglacial, there is as yet no sign of enhanced fluvial activity usually signified by pronounced channel migration and sand deposition. Such fluvial activity should follow the present interglacial maximum, although at this time there is no physical explanation at to why the temperature and moisture regimes should be so significantly offset. In the Channel Country of Queensland, the past is the key to understanding the context of present conditions, but it may also provide an important basis for predicting future changes.

The Nile River

The Nile drains a 3.35×10^6 km^2 basin, which responds to climate and associated moisture regime changes over 35° of latitude (more than 6500 km) from just south of the Equator to the Mediterranean Sea. The river and its tributaries pass through glaciated highlands, montane forests, alluvial plains, swamps, and desert. Its hydrology is dominated by two separate systems; the Blue Nile and Atbara River drain the Ethiopian highlands, while the White Nile drains the east African highlands. Research over the past several decades has yielded one of the more detailed alluvial records of Late Quaternary environmental change. However, what makes this alluvial record for the Nile River particularly interesting and useful is that it is supported by pollen and other micro-fossil records from swamps and lakes for the region as a whole (Fig. 9). While the hydrology of the Nile is allogenic to the deserts of Sudan and Egypt, it clearly reflects, in general terms, a record of major climate change over much of northeast Africa including these desert regions (Williams and Faure, 1980; Adamson et al., 1980).

In combination, these data reveal a period of colder drier climate between ~26 to 12.5 k yrs when the Blue Nile and Atbara were highly

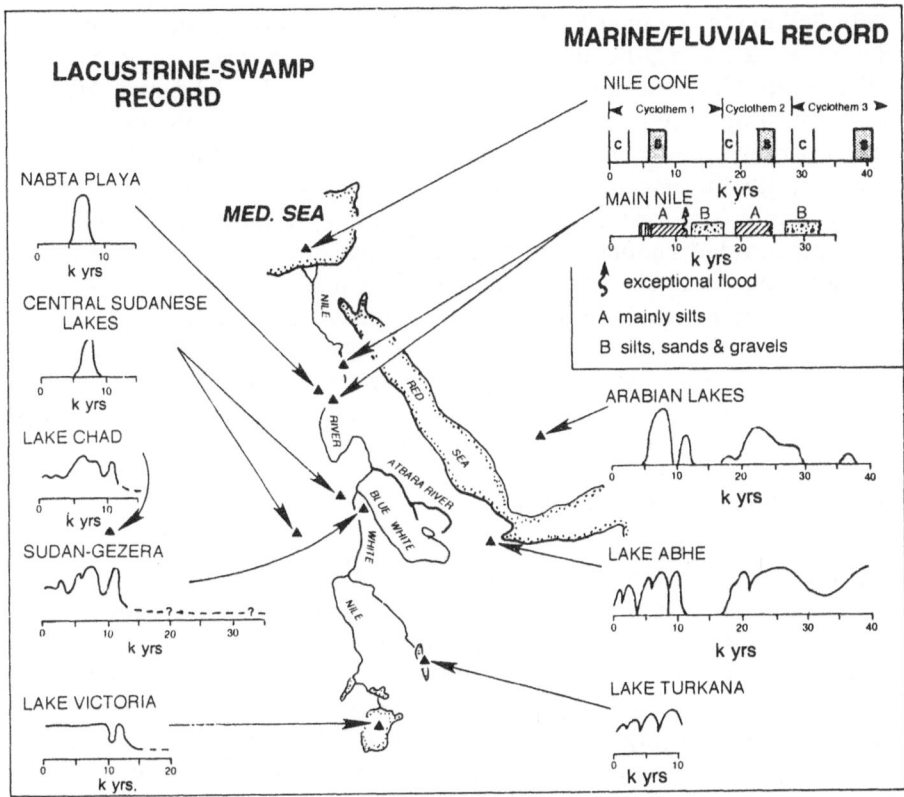

Fig. 9: Late Quaternary alluvial and lacustrine records in and around the Nile basin. Relative water levels are indicated in the lakes with higher levels during wetter periods. On the Nile cone, each cyclothem comprises a sapropel unit with muds and turbidites (s), which correspond to high lake levels in the basin, and a calcareous unit (c) that corresponds with times of higher aridity in the basin. On the main Nile River, the wetter periods correspond with a high silt content, and the more arid periods with a higher sand and gravel content (modified from Adamson et al., 1980).

seasonal, braided rivers but with lower average discharges, transporting high sediment yields from climatically deforested and periglaciated headwater catchments. Dolomite and calcite evaporites accumulated in places along their length and their braid patterns are still visible in the form of palaeochannels. Braiding also characterised the main Nile in Sudan and Egypt, and in many places bedload sands were reworked from the channels into aeolian dunes at times of low flow. Periodically, outflow from the east African lakes in the headwaters of the White Nile ceased and it became a seasonally intermittent river.

At the end of the Pleistocene, major climate and other environmental changes started to take place (Williams and Faure, 1980; Adamson et al.,

1980). Following the overflow of the headwater lakes on the White Nile after about 12.5 k yrs, the main Nile became perennial again. Reactivation of the White Nile, and larger discharges from the Blue Nile and Atbara Rivers resulting from increased rainfall in the Ethiopian highlands, caused major flooding on the main Nile in the Late Pleistocene and Early Holocene (Fig. 9). With the re-establishment of upland vegetation resulting from increased moisture through to the Mid Holocene, sediment loads became finer and braided channels were replaced by sinuous, single-thread or anabranching channels. After about 5 k yrs, the climate in Egypt and Sudan became increasingly arid and the present conditions on the Nile started to prevail. While these results are not specific to the desert regions along the Nile, they do provide a guide as to the broad-scale climatic changes of northeast Africa.

Discussion and Conclusion

Rivers are the product of their environment and, as regional climates change, so rivers respond with variations in planform, channel geometry, and alluvial sedimentology and stratigraphy. Arid-zone rivers are part of a continuum of river types, but they show a number of distinctive characteristics that make them useful for interpreting environmental change. The relationships between rainfall, vegetation, runoff and sediment loads being sensitive, arid-zone rivers respond readily to changes in the often seasonal rainfall patterns that characterise such regions. However, as a group, arid-zone rivers are inherently diverse and difficult to typify (Knighton and Nanson, 1997), and as shown above, geomorphic setting and basin scale are important determinants of how different basins will respond to changing climate.

In predominantly erosional areas, such as in headwater or piedmont zones, climate change can cause channel incision and the abandonment of floodplains to form terraces that can be inspected along eroded exposures. By contrast, in accretionary sinks such as the Channel Country of Queensland, Quaternary deposits accumulate as extensive subsurface deposits that must be excavated or drilled to obtain evidence of climate and flow-regime changes.

Generally speaking, smaller, steeper headwater basins are sensitive to relatively minor changes in climate and, as a consequence, their terraces and floodplains have a short-term alluvial 'memory', sometimes measurable in individual years but more commonly in decades or centuries (eg. Hereford, 1986). Sediments are coarse or mixed load, and channels, floodplains and terraces tend to be clearly defined and readily recognisable, both topographically and stratigraphically. However, such systems, being sensitive to change, respond to individual disturbances such as landslides, debris jams, isolated but extreme floods, or runoff after fires, and they can also be affected by complex responses triggered by these localised events. This type of evidence can make the alluvial record confusing if it is to be used to identify larger-scale and longer-term climate changes. In these small and dynamic basins, the relatively high magnitude and frequency of erosional events can prevent the accumulation of any long-term record.

Intermediate-sized and moderate-gradient arid-zone rivers tend to be dominated by sand loads. Where these are unconsolidated, then such rivers can be highly responsive to even minor changes in flow regime (eg. Schumm and Lichty, 1963; Burkham, 1972; Hereford, 1984), behaving rather like their headwater counterparts. In contrast, similar intermediate-sized systems in central Australia are relatively unresponsive, probably because of the typically lower gradients and the higher resistance to erosion resulting from higher mud contents in the banks and an abundance of riparian vegetation (Tooth, 1997). Very important is the degree of confinement of the channel and floodplain. Entrenched and narrow floodplains concentrate flow energy, leading to reworking of floodplains and possibly removing evidence of earlier Quaternary events, while larger floodplains often retain such evidence, even if only in residual pockets. By way of contrast, in confined bedrock reaches, usually found in small and intermediate-sized arid-zone basins, slackwater deposits can provide a valuable record of extreme floods in Late Holocene and historical times (eg. Patton et al., 1979), as can alluvial deposits on broad piedmont zones (Pickup, 1991; Patton et al., 1993).

Large, low-gradient, slowly accreting arid-zone basins, such as those that constitute much of the Channel Country of Queensland, are fine grained and unresponsive to short-term changes in climate and flow regime, but they retain a valuable record of major long-term trends (Nanson et al., 1988, 1992). In that sense, records from such basins are comparable to records from the ocean basins (eg. Hovan et al., 1989), ice cores (Petit et al., 1990), or the loess deposits of China (Kukla, 1987), in that they contain evidence extending back over several glacial cycles (Fig. 8). Furthermore, like the loess record, the evidence is for terrestrial change in arid-zone regions. However, alluvial deposits are more fragmentary, both spatially and temporally, than are these other long-term records, and as a consequence, observations usually need to be patched together from basin-wide or interbasin locations in order to obtain a detailed and extended record.

The use of rivers and alluvial deposits as palaeoclimate indicators in arid-zone basins is still in the early stages of development. Extensive stratigraphic studies relying on detailed chronologies from the Nile in

Africa and the Channel Country in Australia, and an expanding record of chronological and stratigraphic data from the American southwest (eg. Bull, 1991), are evidence of how effective such data can be for reconstructing regional, and eventually global, long-term palaeoclimates. Recent developments in luminescence dating will ensure the rapid expansion of this field of research and will greatly extend the palaeoclimatic information obtained from palaeolakes and dunefields in the arid-zone.

References

Abrahams, A.D. and Parsons, A.J. (eds.) (1994). *Geomorphology of Desert Environments*. Chapman and Hall, London.

Adamson, D.A. (1982). The integrated Nile, In: *A Land Between Two Niles*, M.A.J. Williams and D.A. Adamson, (eds.) pp.221-34., Balkema, Rotterdam.

Adamson, D.A., Gasse, F., Street, F.A. and Williams, M.A.J. (1980). Late Quaternary history of the Nile, Nature, 288: 50-5.

Adamson, D.A., Gillespie, R. and Williams, M.A.J. (1982). Palaeogeography of the Gezira and of the lower Blue and White Nile valleys, In: *A Land Between Two Niles*, M.A.J. Williams and D.A. Adamson (eds.) pp. 165-219, Balkema, Rotterdam.

Aitken, M.J. 1998. *An Introduction to Optical dating*. Oxford University Press, Oxford, UK.

Allan, R.J. (1988). El Nino Southern Oscillation influences in Australasia, *Progress in Physical Geography* 12: 4-40.

Antevs, E. (1952). Arroyo cutting and filling, *J. Geol.*, 60: 375-85.

Baker, V.R. (1977). Stream-channel response to floods, with examples from central Texas, *Geol. Soc. America Bull.*, 88: pp.1057-71.

Baker, V.R., Kochel, R.C. and Patton, P.C. (eds.) (1988). *Flood Geomorphology*. John Wiley and Sons. New York.

Baker, V.R., Kochel, R.C., Patton, P.C. and Pickup, G. (1983a). Palaeohydrologic analysis of Holocene flood slack-water sediments, In: *Modern and Ancient Fluvial Systems*, International Association of Sedimentologists, Special Publication 6, In: J.D. Collinson, and J. Lewin (eds.), pp. 229-39. Blackwell Scientific Publications, Oxford.

Baker, V.R., Pickup, G. and Polach, H.A. (1983b). Desert palaeofloods in central Australia, *Nature*, 301: 502-4.

Barsch, D. and Royse, C.F. (1972). A model for the development of Quaternary terraces and piedmont-terraces in the southwestern United States, *Zeitscrift für Geomorphologie*, 16: 54-75.

Beaumont, P. (1989). *Drylands. Environmental Management and Development*, Routledge, London.

Bell, F.C. (1979). Precipitation, In: *Arid land ecosystems*, D.W. Goodall and R.A. Perry (eds.) pp. 373-93, Cambridge University Press, Cambridge.

Blair, T.C. and McPherson, J.G. (1994a). Alluvial fans and their natural distinction from rivers based on morphology, hydraulic processes, sedimentary processes and facies assemblages, *J. Sedimentary Res.*, A64: 450-89.

Blair, T.C. and McPherson, J.G. (1994a). Alluvial fan processes and forms, In: *Geomorphology of Desert Environments*: A.D. Abrahams and A.J. Parsons (eds.), pp. 354-402, Chapman and Hall, London.

Bourke, M.C. (1994). Cyclical construction and destruction of flood dominated flood plains in semi arid Australia, In: *Variability in Stream Erosion and Sediment Transport*, IAHS Publication No. 224: p. 113-23.

Bowler, J.M. (1981). Australian salt lakes: a palaeohydrological approach. In: *Salt Lakes. Developments in Hydrobiologia.* Vol. 5, W.D. Williams, (ed.) pp. 431-44, The Junk, Hague.

Brackenridge, G.C. and Hagedorn, J. (1992). Introduction, In: *Floodplain Evolution* Brackenridge, G.C. and Hagedorn, J. (eds.), *Geomorphology*, 4, p. iii.

Bull, W.B. (1991). *Geomorphic Responses to Climatic Change.* Oxford University Press, New York.

Burkham, D.E. (1972). Channel changes of the Gila River in Safford Valley, Arizona 1846-1970, *US Geo. Sur. Professional Paper*, 655-G.

Busche, D. and Hagedorn, H. (1980). Landform development in warm deserts - the central Saharan example, *Zeitschrift für Geomorphologie*, Supplementband 36: 123-39.

Callen, R.A. and Nanson, G.C. (1992). Discussion-Formation and age of dunes in the Lake Eyre depocentres, *Geologische Rundschau*, 81(2): 589-93.

Chappell, J. and Shackleton, N.J. (1986). Oxygen isotopes and sea level, *Nature*, 324: pp. 137-140.

Cooke, R.U. and Reeves, R.W. (1976). *Arroyos and Environmental Change in the American South-West.* Clarendon Oxford.

Cooke, R.U., Warren, A. and Goudie, A.S. (1993). *Desert geomorphology.* University College London Press, London.

Cordery, I., Pilgrim, D.H. and Doran, D.G. (1983). Some hydrological characteristics of arid western New South Wales, *Hydrology and Water Resources Symposium 1983*, Institution of Engineers, Australia, pp. 287-92.

Denevan, W.M. (1967). Livestock numbers in Nineteenth-Century New Mexico, and the problem of gullying in the Southwest, *Annals of the Asso. of American Geographers*, 57: 691-703.

Dorn, R.I. (1994). The role of climatic change in alluvial fan development, In: *Geomorphology of Desert Environments*, A.D. Abrahams, and A.J. Parsons (eds.), pp. 593-615 Chapman and Hall, London.

Dubief, J. (1953). *Essai sur L'Hydrologie Superficielle Au Sahara.* Birmandreis: Gouvernement General de L'Algerie, Service des Etudes Scientifiques.

Dunkerley, D.L. (1992). Channel geometry, bed material, and inferred flow conditions in ephemeral stream systems, Barrier Range, western N.S.W., Australia, *Hydrological Processes*, 6: 417-33.

Ely, L.L. and Baker, V.R. (1985). Reconstructing palaeoflood hydrology with slackwater deposits: Verde River, Arizona, *Physical Geography*, 5: 103-26.

Ely, L.L., Enzel, Y., Baker, V.R. and Cayan, D.R. (1993). A 5000-year record of extreme floods and climate change in the southwestern United States, *Science*, 262: 410-12.

Finlayson, B.L. and McMahon, T.A. (1988). Australia v. the world: A comparative analysis of streamflow characteristics, In: *Fluvial Geomorphology of Australia*, R.F. Warner (ed.), pp.17-40: Academic Press Sydney.

Fisk, H.N. (1944). *Geological investigation of the alluvial valley of the Mississippi River.* Vicksburg: Mississippi River Commission.

Frostick, L.E. and Reid, I. (1977). The origin of horizontal laminae in ephemeral stream channel-fill, *Sedimentology*, 24: 1-9.

Frostick, L.E. and Reid, I. (1979). Drainage-net control of sedimentary parameters in sand-bed ephemeral streams, In: *Geographical Approaches to Fluvial Processes*, A.F. Pitty (ed.) pp. 173-201: Geo Abstracts, Norwich.

Frostick, L.E., Reid, I. and Layman, J.T. (1983). Changing size distribution of suspended sediment in arid-zone flash floods, In: *Modern and Ancient Fluvial Systems*, J.D. Collinson and J. Lewin (eds.), pp. 97-106, International Association of Sedimentologists, Special Publication 6: Blackwell Scientific Publications, Oxford.

Goldberg, P. (1984). Late Quaternary history of Qadesh Barnea, Northeastern Sinai, *Zeitschrift für Geomorphologie*, 28: 193-217.

Goudie, A.S. (1987). Change and instability in the desert environment, In: *Horizons in Physical Geography*, M.J. Clark, K.J. Gregory and A.M. Gurnell (eds.), pp. 250-67: Macmillan Education, London.

Graf, W.L. (1981). Channel instability in a braided, sand-bed river, *Water Resources Research*, 17: 1087-94.

Graf, W.L. (1983). Flood-related channel change in an arid region river, *Earth Surface Processes and Landforms*, 8: 125-39.

Graf, W.L. (1988a). *Fluvial Processes in Dryland Rivers*: Springer-Verlag, Berlin.

Graf, W.L. (1988b). Definition of floodplains along arid-region rivers, In: *Flood geomorphology*, V.R. Baker, R.C. Kochel and P.C. Patton (eds.), pp. 231-242: John Wiley and Sons, New York.

Hereford, R. (1984). Climate and ephemeral-stream processes: twentieth-century geomorphology and alluvial stratigraphy of the Little Colorado River, Arizona, *Geol. Soc. of America Bull.* 95: 654-68.

Hereford, R. (1986). Modern alluvial history of the Paria River drainage basin, southern Utah, *Quat. Res.*, 25: 293-311.

Hovan, S.A., Rea, D.K., Pisias, N.G. and Shackleton, N.G. (1989). A direct link between China loess and marine ^{18}O records: aeolian flux to the north Pacific, *Nature*, 340: 296-8.

Huntington, E. (1914). *The climatic factor as illustrated in arid America.* Carnegie Institution of Washington, Publication 162.

Karcz, I. (1972). Sedimentary structures formed by flash floods in southern Israel, *Sedimentary Geol.*, 7: 161-82.

Kelly, S.B. and Olsen, H. (1993). Terminal fans — a review with reference to Devonian examples, In: *Current Research in Fluvial Sedimentology*, C.R. Fielding (ed.), *Sedimentary Geology* 85: 339-74.

Kershaw, A.P. and Nanson, G.C. (1993). The last full glacial cycle in the Australian region, *Global and Planetry Change*, 7: 1-9.

Knighton, A.D. and Nanson, G.C. (1994). Flow transmission along an arid zone anastomosing river, Cooper Creek, Australia, *Hydrological Processes*, 8: 137-54.

Knighton, D. and Nanson, G.C. (in press), Distinctiveness, diversity and uniqueness in arid zone river systems. In: *Arid Zone Geomorphology (2nd. Edition)*, D.S.G. Thomas (ed.): John Wiley and Sons, London.

Kochel, R.C., Baker, V.R. and Patton, P.C. (1982). Palaeohydrology of southwestern Texas, *Water Resources Research*, 18: 1165-83.

Kotwicki, V. (1986). *Floods of Lake Eyre*. Adelaide: Engineeering and Water Supply Department.

Kukla, G. (1987). Loess stratigraphy in central China, *Quat. Sci. Rev.*, 6: 191-219.

Langbein, W.B. and Schumm, S.A. (1958). Yield of sediment in relation to mean annual precipitation, *Trans. American Geophy. Union*, 39: 1076-1084.

Langford, R. and Bracken, B. (1987). Medano Creek, Colorado, a model for upper-flow-regime fluvial deposition, *J. Sedimentary Petrology*, 57: 863-70.

Laronne, J.B. and Reid, I. (1993). Very high rates of bedload sediment transport by ephemeral desert rivers, *Nature*, 366: 148-150.

Lees, B.G. (1992). Geomorphological evidence for Late Holocene climate change in northern Australia, *Australian Geographer*, 23(1): 1-10.

Leopold, L.B. (1951). Rainfall frequency: an aspect of climatic variation, *American Geoph. Union Transactions*, 32: pp. 347-57.

Leopold, L.B., Emmett, W.W. and Myrick, R.M. (1966). Channel and hillslope processes in a semi-arid area, New Mexico, *US Geol. S. Prof. Paper*, 352-G.

Mabbutt, J.A. (1977). *Desert Landforms*. Australian National University Press, Canberra.

Magee, J.W., Bowler, J.M., Miller, G.H. and Williams, D.G.L. (1995). Stratigraphy, sedimentology, chronology and palaeohydrology of Quaternary lacustrine deposits at Madigan Gulf, Lake Eyre, South Australia, *Palaeogeography, Palaeoclimatology, Palaeoecology* , 113: 3-42.

Maroulis, J.C. (in prep.). *Stratigraphy and Late Quaternary Chronology of the Cooper Creek Floodplain, Southwestern Queensland*. PhD. Thesis, University of Wollongong.

Maroulis, J.C. and Nanson, G.C. (1996). Bedload transport of aggregated muddy alluvium from Cooper Creek, central Australia: a flume study, *Sedimentology*, 43: 771-90.

McKee, E.D., Crosby, E.J. and Berryhill, H.L.J. (1967). Flood deposits, Bijou Creek, Colorado, June 1965, *J. Sed. Petrology*, 37: 829-51.

McMahon, T.A. (1979). Hydrological characteristics of arid zones, *The Hydrology of Areas of Low Precipitation*, Proceedings of the Canberra Symposium, Canberra, IAHS-AISH Publication No.128: 105-23.

Mukerji, A.B. (1976). Terminal fans of inland streams in Sutlej-Yamuna Plain, India, *Zeitschrift für Geomorphologie NF*, 20: 190-204.

Nanson, G.C., Rust, B.R. and Taylor, G. (1986). Coexistent mud braids and anastomosing channels in an arid-zone river: Cooper Creek, central Australia, *Geology*, 14, pp. 175-8.

Nanson, G.C., Young, R.W., Price, D.M. and Rust, B.R. (1988). Stratigraphy, sedimentology and late-Quaternary chronology of the Channel Country of western Queensland, In: *Fluvial Geomorphology of Australia*, R.F. Warner (ed.), pp. 151-75: Academic Press, Sydney.

Nanson, G.C. and Croke, J.C. (1992). A genetic classification of floodplains, In: Floodplain Evolution, G.R. Brackenbridge and J. Hagedorn (eds.), *Geomorphology*, 4: 459-86.

Nanson, G.C., Price, D.M. and Short, S.A. (1992). Wetting and drying of Australia over the past 300 k yrs, *Geology*, 20: 791-4.

Nanson, G.C., East, T.J. and Roberts, R.G. (1993). Quaternary stratigraphy, geochronology and evolution of the Magela Creek catchment in the monsoon tropics of northern Australia, *Sedimentary Geology*, 83: 277-302.

Nanson, G.C., Chen, X.Y. and Price, D.M. (1995). Aeolian and fluvial evidence of changing climate and wind patterns during the past 100 k yrs in the western Simpson Desert, Australia, *Palaeogeography, Palaeoclimatology, Palaeoecology*, 113, pp. 87-102.

Nanson, G.C. and Knighton, A.D. (1996). Anabranching rivers: their cause, character and classification, *Earth Surface Processes and Landforms*, 21: 217-39.

Nanson, G.C. and Huang, H.Q. (in press). Anabranching rivers: divided efficiency leading to fluvial diversity, In: *Varieties of Fluvial Form*, A. Miller, and A. Gupta, (eds.), John Wiley and Sons.

Nott, J.F. and Price, D.M. (1994). Plunge pools and palaeoprecipitation, *Geology*, 22: 1047-50.

Nott, J.F., Price, D.M. and Bryant, E.A. (1996). A 30,000 year record of extreme floods in tropical Australia from relict plunge-pool deposits: implications for future climate change, *Geophysical Research Letters*, 23(4): 379-82.

O'Connor, J.E., Ely, L.L., Wohl, E.E., Stevens, L.E., Melis, T.S., Kale, V.S. and Baker, V.R. (1994). A 4500-year record of large floods on the Colorado River in the Grand Canyon, *Arizona, J. Geol.*, 102: 1-9.

Olsen, H. (1987). Ancient ephemeral stream deposits: a local terminal fan model from the Bunter Sandstone Formation (L. Triassic). in the Tonder-3, -4 and -5 wells, Denmark, In: *Desert Sediments: Ancient and Modern*, L. Frostick, and I. Reid (eds.), pp. 69-86: Geological Society, Special Publication No. 35.

Page, K.J. and Nanson, G.C. (in press). Stratigraphic architecture resulting from late Quaternary evolution of the Riverine Plain, southeastern Australia, *Sedimentology*.

Page, K.J., Nanson, G.C. and Price, D.M. (1996). Chronology of Murrumbidgee River palaeochannels on the Riverine Plain, southeastern Australia, *J. Quat. Sci.*, 11(4): 311-26.

Partridge, J. and Baker, V.R. (1987). Palaeoflood hydrology of the Salt River, Arizona, Earth *Surface Processes and Landforms*, 12: 109-25.

Patton, P.C., Baker, V.R. and Kochel, R.C. (1979). Slackwater deposits: a geomorphic technique for the interpretation of fluvial paleohydrology, In: *Adjustments of the Fluvial System*, D.D. Rhodes and G.P. Williams (eds.), pp. 225-53: Kendall/Hunt, Dubuque, Iowa.

Patton, P.C., Pickup, G. and Price, D.M. (1993). Holocene palaeofloods of the Ross River, central Australia, *Quaternary Res.*, 40: 201-212.

Patton, P.C. and Schumm, S.A. (1975). Gully erosion, northwestern Colorado, *Geology*, 3: 88-90.

Petit, J.R., Mounier, L., Jouzel, J., Korotkevich, Y.S., Kotlyakov, V.I. and Lourius, C. (1990). Palaeoclimatological and chronological implications of the Vostok core dust record, *Nature*, 343: 56-8.

Picard, M.D. and High, L.R.J. (1973). *Sedimentary Structures of Ephemeral Streams.* Elsevier Scientific Publishing, Amsterdam Company, Amsterdam.

Pickup, G. (1991). Event frequency and landscape stability on the floodplain systems of arid central Australia, *Quaternary Sci. Rev.*, 10: 463-73.

Pickup, G., Allan, G. and Baker, V.R. (1988). History, palaeochannels and palaeofloods of the Finke River, central Australia, In: *Fluvial Geomorphology of Australia*, R.F. Warner, (ed.), pp. 105-27: Academic Press, Sydney.

Pilgrim, D.H., Chapman, T.G. and Doran, D.G. (1988). Problems of rainfall-runoff modelling in arid and semi-arid regions, *Hydrological Sci. J.*, 33, 379-400.

Reid, I. (1994). River landforms and sediments: evidence of climatic change, In: *Geomorphology of Desert Environments*, A.D. Abrahams and A.J. Parsons (eds.), pp. 571-92: Chapman and Hall, London.

Reid, I. and Frostick, L.E. (1987). Flow dynamics and suspended sediment properties in arid zone flash floods, *Hydrological processes*, 1: 239-53.

Reid, I. and Frostick, L.E. (1989). Channel form, flows and sediments in deserts, In: *Arid Zone Geomorphology*, D.S.G. Thomas (ed.), pp. 117-35: Belhaven Press and Halsted Press, London and New York.

Reinfelds, I. and Nanson, G.C. (1993). Formation of braided river floodplains, Waimakariri River, New Zealand, *Sedimentology*, 40: 1113-27.

Rust, B.R. (1981). Sedimentation in an arid-zone anastomosing fluvial system: Cooper Creek, central Australia, *Journal of Sedimentary Petrology*, 51: 745-55.

Rust, B.R. and Nanson, G.C. (1986). Contemporary and palaeochannel patterns and the late Quaternary stratigraphy of Cooper Creek, SW Queensland, Australia, *Earth Surface Processes and Landforms*, 11: 581-90.

Sack, D. (1994). Geomorphic evidence of climate change from desert-basin palaeolakes, In: *Geomorphology of Desert Environments*, A.D. Abrahams and A.J. Partons (eds.) pp. 616-30: Chapman and Hall, London.

Schick, A.P. (1974). Formation and obliteration of desert stream terraces — a conceptual analysis, *Zeitschrift für Geomorphologie*, Supplementband 21: 88-105.

Schick, A.P. and Magid, D. (1978). Terraces in arid stream valleys: a probability model, *Catena*, 5: 237-50.

Schumann, R.R. (1989). Morphology of Red Creek, Wyoming, an arid-region anastomosing channel system, *Earth Surface Processes and Landforms* 14: 277-88.

Schumm, S.A. (1960a). The effect of sediment type on the shape and stratification of some modern fluvial deposits, *Am. J. Sci.*, 258: 177-84.

Schumm, S.A. (1960b). The shape of alluvial channels in relation to sediment type: erosion and sedimentation in a semiarid environment, *US Geol. Sur. Professional Paper*, 352-B: 17-30.

Schumm, S.A. (1961a). A reply: the effect of sediment type on the shape and stratification of some modern fluvial deposits, *Am. J. Sci.*, 259: 234-39.

Schumm, S.A. (1961b). The effect of sediment characteristics on erosion and deposition in ephemeral stream channels, *US Geol. Sur. Profess. Paper*, 352-C: 31-70.

Schumm, S.A. (1968). River adjustment to altered hydrologic regimen, Murrumbidgee River and palaeochannels, Australia, *US Geolo. Sur. Profess. Paper*, 598.

Schumm, S.A. (1977). *The Fluvial System*. Wiley-Interscience, New York.

Schumm, S.A. and Kahn, H.R. (1972). Experimental study of channel patterns, *Geol. Soc. America Bull.*, 83: 1755-1770.

Schumm, S.A. and Lichty, R.W. (1963). Channel widening and floodplain construction along Cimarron River in southwestern Kansas, *US Geol. Sur. Profess. Paper*, 352-D: 71-88.

Senior, B.R., Mond, A. and Harrison, P.H. (1978). Geology of the Eromanga Basin, *Bureau of Mineral Resources Bulletin*, 167, 102pp.

Shepherd, R.G. (1987). Lateral accretion surfaces in ephemeral-stream point bars, Rio Puerco, New Mexico, In: *Recent Developments in Fluvial Sedimentology*, F.G. Ethridge, R.M. Flores, and M. Harvey (eds.) pp.93-8, Special Publication No. 39, Society of Economic Palaeontologists and Mineralogists.

Smith, A.M. (1992). Holocene palaeoclimatic trends from palaeoflood analysis, *Palaeogeography, Palaeoclimatology, Palaeoecology*, 97: 235-40.

Sneh, A. (1983). Desert stream sequences in the Sinai Peninsula, *J. Sedimentary Petrology*, 53: 1271-9.

Stewart, G.A. and Perry, R.A. (1962). Part I. Introduction and summary description of the Alice Springs area, In: *General Report on Lands of the Alice Springs Area, Northern Territory, 1956-57* R.A. Perry (ed.), CSIRO Land Research Series, No. 6: 9-19.

Street-Perrott, F.A. and Harrison, S.P. (1985). Lake levels and climate reconstruction. In: *Palaeoclimate Analysis and Modeling*, A.D. Hecht (ed.), John Wiley and Sons, New York pp.

Thomas, D.S.G. (ed.) (1997a). *Arid Zone Geomorphology*. Belhaven Press and Halsted Press, London and New York.

Thomas, D.S.G. (1997b). Reconstructing ancient arid environments, In: *Arid Zone Geomorphology*, D.S.G. Thomas (ed.), pp. 311-34: Belhaven Press and Halsted Press, London and New York.

Thornes, J.B. (1977). Channel changes in ephemeral streams: observations, problems and models, In: *River Channel Changes*, K.G. Gregory (ed.), pp. 317-35, John Wiley and Sons, Chichester.

Thornes, J.B. (1980). Structural instability and ephemeral channel behaviour, *Zeitschrift für Geomorphologie*, Supplementband 36: 233-44.

Thornes, J.B. (1994a). Catchment and channel hydrology, In: *Geomorphology of Desert Environments*, A.D. Abrahams, and A.J. Parsons (eds.), pp. 257-87: Chapman and Hall, London.

Thornes, J.B. (1994b). Channel processes, evolution and history, In: *Geomorphology of Desert Environments*, A.D. Abrahams and A.J. Parsons (eds.), pp. 288-317, Chapman and Hall, London.

Tooth, S. (in press). Floodouts in central Australia, In: *Varieties of Fluvial Form*, A. Miller, and A. Gupta (eds.), John Wiley and Sons.

Tooth, S. (in prep.) *The Morphology, Dynamics and Late Quaternary Sedimentary History of some Ephemeral Drainage Systems in Arid Central Australia*. Ph.D. Thesis, University of Wollongong.

Vita-Finzi, C. (1969). *The Mediterranean Valleys*. Cambridge University Press, Cambridge.

Walling, D.E. and Webb, B.W. (1983). Patterns of sediment yield, In: *Background to Palaeohydrology: A Perspective*, K.G. Gregory (ed.), pp. 69-100, John Wiley and Sons, Chichester.

Warren, A. (1985). Arid geomorphology, *Progress in Physical Geography*, 9: p. 434-41.

Wasson, R.J. (1983). The Cainozoic history of the Strzelecki and Simpson dunefields (Australia), and the origin of the desert dunes, *Zeitschrift für Geomorphologie*, Supplementband 45: 85-115.

Wende, R., Nanson, G.C. and Price, D.M. (in press). Aeolian and fluvial evidence for late Quaternary environmental change in the east Kimberley of Western Australia, *Aust. J. Earth Sci.*

Williams, G.E. (1970). The central Australian stream floods of February-March 1967, *J. Hydrol.*, 11: pp. 185-200.

Williams, G.E. (1971). Flood deposits of the sand-bed ephemeral streams of central Australia, *Sedimentology*, 17: 1-40.

Williams, M.A.J. and Faure, H. (Eds.) (1980). *The Sahara and the Nile*. Balkema, Rotterdam.

Wohl, E.E., Greenbaum, N., Schick, A.P. and Baker, V.R. (1994). Controls on bedrock channel incision along Nahal Paran, Israel, *Earth Surface Processes and Landforms*, 19: 1-13.

Wolman, M.G. and Gerson, R. (1978). Relative scales and time and effectiveness of climate in watershed geomorphology, *Earth Surface Processes and Landforms*, 3: 189-208.

8 Arid Zone Palaeoenvironmental Records from Cave Speleothems

GEORGE A. BROOK[*]

ABSTRACT

Solution caves and speleothems are common in many of the world's arid areas. Speleothems can be accurately dated to at least 500 k yrs B.P. and can provide a wealth of palaeoenvironmental data. In areas almost too dry for speleothem growth today, the frequency of ages for relict formations provides evidence of past periods of increased moisture. Speleothems now under water because of a rise in the groundwater level, provide information on past drier periods of climate when water levels were much lower than they are today. Speleothems may contain pollen and other plant microfossils even when the clastic cave sediments and sediments outside the cave do not. In deserts, where pollen grains are rarely preserved in the highly oxidizing environment, cave speleothems are potentially a major source of palaeovegetation data. Many arid-land caves have poorly ventilated passages with close to 100% humidity. These passages may contain stalagmites laid down in isotopic equilibrium with the precipitating waters. The oxygen, carbon and hydrogen isotope characteristics of these formations can provide relative or absolute palaeotemperature data, information about the air masses that brought rainfall to the cave in the past, and information on the relative importance of C_3 and C_4 plants in the vegetation cover above the cave over time. Finally, it is now clear that caves in semi-arid areas may contain stalagmites with annually deposited layers, the thickness of which may be a proxy of annual precipitation. Such deposits may eventually provide high-resolution precipitation records that can be used to predict climate into the 21st century.

Introduction

There is now convincing evidence that the arid and semi-arid regions of the world have not always been as they are today. Rock paintings, fossils of former animals, ancient lake sediments, and satellite and Space Shuttle images showing buried river systems, have provided convincing evidence that at times in the past, many areas were much wetter compared to the

*Department of Geography, University of Georgia, Athens, GA 30602, USA.

present. It is also clear that some present-day drylands were even more arid in the past. A major problem in reconstructing the past environments of the world's deserts has been the poor preservation of sediments recording the former presence of rivers and lakes. Many such sediments have been deflated under the present arid conditions. Where such sediments have been preserved they provide irrefutable evidence of a well-watered, well-vegetated land often eminently suitable for human habitation. In other areas relict dunes are testament to more arid conditions in the past. Until very recently, however, the ages of these widespread and spectacular landforms were difficult to determine. One of the most exciting directions in desert research in recent years has been the development of thermoluminescence (TL) and optically stimulated luminescence (OSL) dating methods that can be used to date sand. As a result, ancient dunes are slowly revealing their past histories. The poor preservation of pollen grains in the generally highly oxidizing environments of the world's deserts has also hindered the modeling of past vegetation distributions. Despite these problems, our understanding of the effects of Quaternary climatic change on the world's drylands increases rapidly.

It is probably true to say that the focus of desert research in the past has been upon the spectacular dune systems or on ancient river and lake sediments. It is true that these have provided a wealth of data but these are often limited by dating potential or by incomplete records. Solution caves are quite common in deserts and probably constitute one of the most neglected sources of palaeoenvironemental information for these regions. Solution cavities are usually protected environments which frequently preserve evidence of past climatic/environmental events is no longer preserved at the surface. This is why caves have provided some of the earliest and best-preserved examples of early humans and the remains of many extinct animals. A wealth of palaeonvironmental data can be obtained from cave clastic sediments and cave speleothems (stalagmites, stalactites, flowstones). This paper will deal only with the data that cave speleothems can provide.

Palaeoenvironmental information from cave speleothems would be of limited use if the data could not be fitted to a reliable chronology. Fortunately, speleothems can be dated by a variety of proven methods including ^{14}C to about 35 k yrs B.P., $^{230}Th/^{234}U$ to about 500 k yrs B.P. and TL, OSL and electron spin resonance (ESR) also to about 500 k yrs B.P. (eg. Hennig et al., 1980; Ivanovic and Harmon, 1982; Gascoyne, 1984; Bluszcz et al., 1988; Grün, 1989). A relatively recent approach is the application of thermal ionization mass spectrometry (TIMS) in U-series dating, with the advantages of reduced sample size and much greater analytical precision (Li et al., 1989). Furthermore, Lauritzen et al. (1994)

have recently shown that isoleucine epimerization of amino acids in speleothems, calibrated against U-series ages, can be used to extend the U-series timescale in some formations. The timing of periods of substantial sediment input to a cave from outside can be determined by dating these sediments by the TL or OSL methods. Collagen in bones preserved in caves can provide reliable ^{14}C ages for the bone and possibly also the sediments in which they occur. Dating of animal or human tooth enamel by ESR can provide ages to one million yrs (eg. Kai et al., 1988; Rink et al., 1994). As speleothems may also record the geomagnetic field at the time of deposition, correlation with known master curves provides a means of dating (eg., Latham et al., 1979; 1982; 1986). Speleothems and cave clastic sediments also record magnetic reversals (the most recent being the Matuyama reversed to Brunhes normal at 780 k yrs B.P.), so that they can, in some cases, be used to approximately date sediments well beyond the range of the other dating methods (eg. Schmidt, 1982; Schmidt et al., 1984; Jones et al., 1986).

Caves and cave speleothems have the potential to provide a wide variety of palaeoecological information for an area and this information can frequently be placed into a long, accurate chronological framework. This paper will attempt to describe in detail the methods that can be employed to obtain palaeoenvironmental data from cave speleothems in arid and semi-arid environments.

Sampling Speleothems in a Cave

Cave speleothems in caves are one of the attractions of these underground landforms and unfortunately they are often damaged either accidentally or deliberately by visitors. In recent years, the value of speleothems in scientific research has been realized and so it has been necessary for scientists to remove whole formations or parts of them for study. Generally this is done carefully so as not to damage the cave unnecessarily. Large speleothems up to several metres in diameter and height are ideal for study as they could contain lengthy records of environmental change. To utilize the potential of large speleothems without damaging them significantly, a drilling rig can be used to extract horizontal or vertical cores. An electrical drilling rig developed at the University of Georgia can drill cores of 5 cm diameter up to about 3 m in length (Fig. 1). The device is light-weight and can be broken down into pieces that can be carried into and out of caves easily. The modified industrial drill that is used is mounted on a collapsible aluminium frame with four telescoping legs. The drill stem consists of a series of threaded extension pipes each about 30 cm long. A diamond-impregnated cutting edge has been added to one of these and this serves as the drill bit. The drill can be rolled back

Fig. 1: Portable electric drilling rig used to recover 5 cm-diameter cores from cave speleothems. The drill is being used to recover a 80 cm long core from a speleothem in Russell Cave National Monument, Alabama, USA. The legs of the aluminium frame have been extended to make drilling possible even 3 m above the floor of the cave.

and forth along a platform. The drill bit slowly abrades the speleothem while the drill stem and cutting surface are cooled and lubricated by water flushed down the center of the drill barrel. The drill can be powered by a portable generator linked to a transformer when the drilling site is relatively close to the entrance or by two 12 volt batteries connected in series when it is deep in the cave. Use of a generator is only possible within about 50-100 metres of the cave entrance as the generator must be located outside the cave because of the danger posed by toxic exhaust fumes in passages with little aeration. The system is simple to use and has been employed for the recovery of cores in a variety of caves in the USA, Madagascar and six countries in Africa. After drilling, the hole can be sealed with plaster or concrete so that there is very little evidence of the sampling.

Unfortunately, for stable isotope studies of speleothems it is preferable to collect whole formations, particularly whole stalagmites. This is because tests for isotopic equilibrium require sampling for considerable distances along individual growth layers. If a vertical core is drilled down the axis of growth, this provides a means of sampling for variations in stable oxygen and carbon isotopes during the period of stalagmite growth. Talma and Vogel (1992) argue that the carbonate precipitated on the upper, flat part of a stalagmite is likely to be precipitated in isotopic equilibrium while that deposited down the steep sides is not. If this is the case, axial cores are all that are needed for isotopic studies. If there is uncertainty about isotopic equilibrium, then whole formations are needed. The best way to obtain these is to use an electric masonry saw powered by a portable generator or by batteries connected in series. The blade should have a diamond-impregnated cutting edge and be lubricated and cooled by water. By cutting carefully and covering the stump of the formation afterwards, the removal of a single stalagmite from a cave can go almost unnoticed.

Age Data

Age Frequency Data of Speleothems in the Present-day Vadose Zone

Speleothems can be dated by ^{14}C, U-series, TL/OSL and ESR. Following Franke and Geyh (1971), Cooke and Verhagen (1977), Hennig et al. (1980) and Bastin and Gewelt (1986), speleothems are generally assumed to accumulate with 85% rather than 100% modern carbon and their ^{14}C ages are corrected accordingly. Unfortunately, ^{14}C ages cannot be compared directly with U-series, TL/OSL or ESR ages. This is because there have been variations in atmospheric ^{14}C content over time to the extent that radiocarbon ages are consistently and significantly younger than the other calender ages through the effective time range of the radiocarbon method (Vogel, 1983; Bard et al., 1990; Holmgren et al., 1994). The relationship between U-series ages and ^{14}C ages has been thoroughly documented up to about 30 k yrs B.P. by Bard et al. (1990). Using their data of 19 data pairs, the following regression equation was calculated: ^{14}C age (yr B.P.) = 54 + 0.85 U-series age (yr B.P.). The correlation coefficient for the relationship is 0.99. ^{14}C and U-series ages for speleothems from Cango Caves in South Africa (Vogel, 1983) suggest that the Bard et al. (1990) relationship, developed from the study of corals in Barbados, can be extended to 35 k yrs B.P. The equation suggests that ^{14}C ages of 10, 20, 30, and 35 k yrs B.P. should be 11.7, 23.5, 35.2, and 41.1 k yrs B.P., respectively, on the U-series timescale. Vogel (1983) suggests that a plateau is reached in the difference between ^{14}C and U-series ages at about 30-35 k yrs B.P. In their studies of speleothem and tufa ages in Africa, Brook et al. (1996, 1997) corrected ^{14}C ages to 35 k yrs B.P. using the above regression equation. ^{14}C ages of 35-45 k yrs B.P. were corrected by adding 6 kyrs to the age. Ages above 45 k yrs B.P. were not used. Whenever radiocarbon and U-series analyses are conducted on the same sample, it is preferable to use only the U-series age because of the uncertainty in the exact percentage of modern carbon in the sample under study. Unless otherwise stated, ages mentioned in this paper are according to the U-series timescale.

In northern high latitudes speleothem age frequencies provide information on past temperature conditions (Franke and Geyh, 1971; Hennig et al., 1983; Gascoyne et al., 1981, 1983a, 1983b; Harmon et al., 1977; Kashiwaya et al., 1991). During glacial times, the intense cold and frozen soils prevented water from penetrating to subsurface cavities. Therefore, speleothems were not formed at these times. It is also likely that speleothem deposition was suppressed by low temperatures in the cold desert regions of the world even if moisture was more available at these times than it is today. Speleothem deposition at high latitudes indicates warmer conditions in the past. In the warm, dry areas of the world it is a lack of moisture, not intense cold that limits the growth of speleothems in caves. In many arid and semi-arid areas there are few or no active speleothems today, but many caves contain large formations that are imposing evidence of former periods when moisture was more abundant than now.

Speleothem age data have been used by Brook et al. (1990b, 1996, 1997) to develop records of wet and dry periods for the drylands of eastern and southern Africa and the southwestern U.S.A. to 300 k yrs B.P. (Fig. 2). There are a limited number of ages in the interval 300-50 k yrs B.P. for the summer rainfall region of southern Africa. However, speleothem growth in the Transvaal region of South Africa, and in

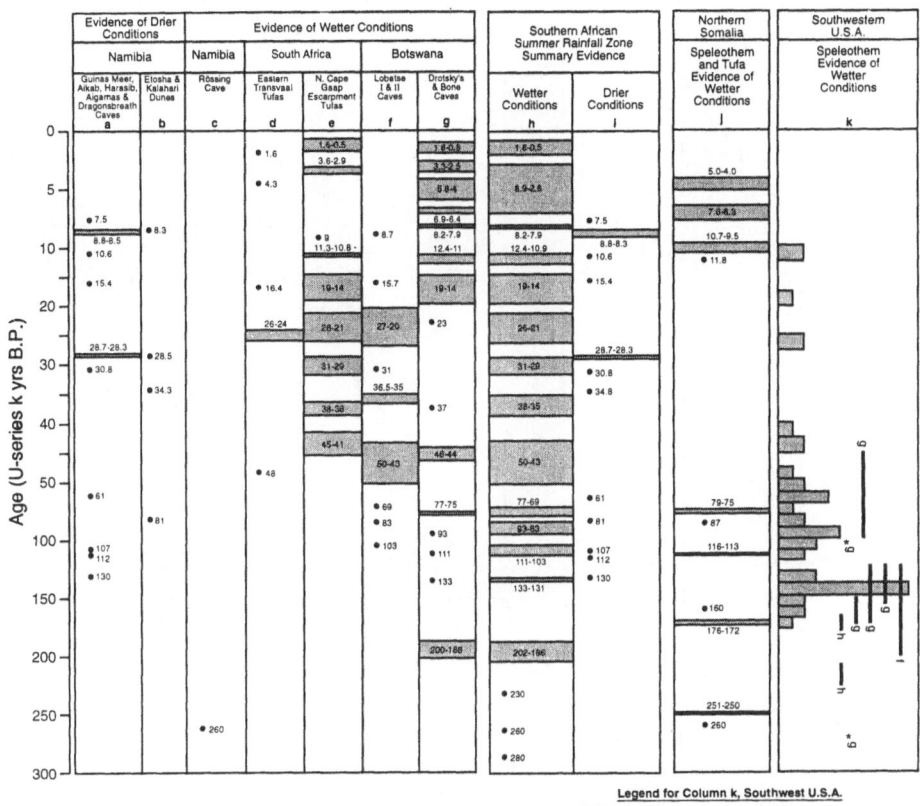

Fig. 2: Wet and dry phases of climate over the last 300 k yrs B.P. in the summer rainfall zone of southern Africa, in northern Somalia, and in the southwestern U.S.A. (modified after Brook et al., 1990b, 1996, 1997).

Botswana, points strongly to wet phases of climate in both areas from 200-186, 133-131, 111-103, 93-83, and 77-69 k yrs B.P. A carbonate raft deposit in Rössing Cave, Namibia indicates wetter conditions at about 260 k yrs B.P. The time interval 50-13 k yrs B.P. was apparently characterized by five wet intervals at 50-43, 38-35, 31-29, 26-21, and 19-14 k yrs B.P. The interval 13 k yrs B.P. to the present was punctuated by significant Mid and Late Holocene wet periods lasting from 6.9-2.6 and 1.6-0.5 k yrs B.P., respectively (Fig. 3). Conditions from 13-10 k yrs B.P. were apparently much drier than the later part of the Holocene. U-series ages for speleothems from two caves in northern Somalia, and ages of relict spring tufa deposits at several locations in the Golis Mountains, suggest significantly wetter conditions in eastern Africa during 260-250,

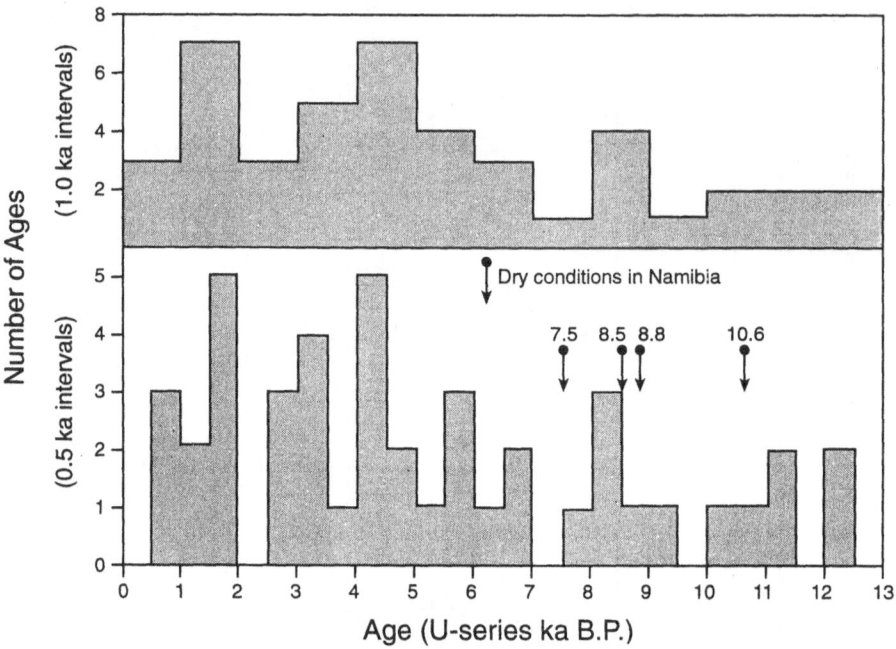

Fig. 3: Speleothem and tufa age frequencies for the summer rainfall region of southern Africa 13 k yrs B.P. Present (after Brook et al., 1996).

176-172, 160, 116-113, 87-75, 12, 10.7-9.5, 7.6-6.3, and 5-4 k yrs B.P. The ages of speleothems from Carlsbad Cavern and Ogle Cave in the southwestern U.S.A. suggest periods of enhanced growth at ca. 270, 230-210, 174-134, 119-83, 71-57, 50-42, ca.25, ca.18, and ca.11.5 k yrs B.P.

Comparison of the records for the three regions reveals some interesting points. In the southwestern U.S.A. there was apparently extensive speleothem growth in Carlsbad Cavern and Ogle Cave during the glacial periods represented by deep-sea isotope Stages 6, 4 and 2 and to a lesser extent during interstadial periods such as isotope Stage 3. Isotope sub Stages 5e and 5a (and possibly also 5c) were dry while substages 5d and 5b appear to have been relatively wet compared to today. This evidence contrasts markedly with that for wet periods in eastern Africa. Inspection of Fig. 2 reveals that speleothem and tufa deposition in northern Somalia at 260-250, 176-160, 116-113, 87-75, and 11.8-4.0 k yrs B.P. occurred at times when there was significantly reduced deposition in the southwestern U.S.A. The absence of speleothems and tufas in northern Somalia dating to isotope Stages 4 and 2 may indicate that eastern Africa was dry during northern hemisphere glacial maxima, while at the same time in the southwestern U.S.A. water was much more abundant than now. In southern Africa, the early Holocene was relatively dry with marked wet intervals during the Mid and Late Holocene. The five wet intervals in the period 50-14 k yrs B.P., with strong evidence of dry intervals at 35-30 and 29-26 k yrs B.P., suggest that warmer intervals during interstadials were moist, and cooler intervals were relatively dry. Evidence of dryness at about 61 and 20 k yrs B.P. suggests that glacial maxima in the northern hemisphere (during isotope Stages 4 and 2 for example) brought reduced levels of moisture to southern Africa.

Submerged Speleothems

As the vast majority of speleothem deposits are subaerial in origin, submerged formations provide evidence of a rise in the water table in the vicinity of the cave after speleothem deposition. In many coastal environments submerged speleothems can provide evidence of lower glacial or interstadial sea levels. Spalding and Mathews (1972) obtained a U-series date of 22 k yrs B.P. for a submerged stalagmite from 12 m below sea level in the Bahamas. This suggested a sea level at least 12 m lower during the Late Wisconsin glacial maximum. Hiatuses in the growth of submerged speleothems can indicate times when sea level was higher than the elevation of the speleothem. Other studies of submerged speleothems that have provided information on past sea levels include those of Gascoyne et al. (1979), Harmon et al. (1978, 1981, 1983), Li et al. (1989) and Richards et al. (1994). As many of the world's deserts, such as the Atacama and Namib, stretch for long distances along coasts, submerged speleothems could provide palaeosealevel data for these regions following the methods discussed in the aformentioned papers. As Gascoyne (1992) points out, the search for datable submerged speleothems is worthwhile because they provide firm data on the extent of sea level lowering in the past.

Submerged speleothems may also be found far from the sea, as in the Florida peninsula of the U.S.A. Here, land is not much above sea level and submergence resulted from a rise in the ground water table triggered by the post-glacial rise in sea level. In some cases, however, submergence of caves and speleothems is not controlled by sea level and in these cases speleothem ages provide reliable information about drier periods of the past. This is the case in the Otavi Mountain Land of Namibia, southeast of Etosha Pan, where the annual rainfall is 600-700 mm. Here, many of the caves are water-filled and contain numerous submerged stalactites and stalagmites in growth position down to depths of 40 m. The roofs of some of these caves have collapsed to form cenotes or hemi-cenotes (Fig. 4). Brook et al. (1996, 1997) report 18 U-series ages for speleothems recovered by divers from five of these caves (Figs. 2 and 3). They range

Fig. 4: Guinas Meer Cenote, Otavi Mountain Land, Namibia. Submerged speleothems ranging in age from 13.5 ± .3 to 61.4 ± 3.6 k yrs B.P. have been recovered from depths of 5-11 m.

from 7.5 to 129.9 k yrs B.P. for speleothems recovered from depths of 3-40 m. There are two distinct age groupings at 8.8-7.5 k yrs B.P. (3 ages) and 30.8-28.3 k yrs B.P. (4 ages). If dryness of climate can be interpreted from the minimum amount of groundwater lowering necessary for speleothem deposition, then deep-sea isotope stage 5 was at times the driest period of the last 130 k yrs. This is because at Aigamas hemicenote water levels were 40 m lower than today 129.9 ± 6.9 k yrs B.P. and 112 ± 5.3 k yrs B.P., and at Harasib they were 16 m below present at 107 ± 6.6 k yrs B.P. In the case of the Namibian caves, the submerged speleothems provide important information about past groundwater levels. In other areas (such as Florida) groundwater levels are influenced strongly by changes in sea level. In these low-lying areas the ages of submerged speleothems provide data on past sea levels rather than information on past water balance characteristics. Numerous caves in dry areas, including Lechuguilla in the southwestern U.S.A. (Bunnell, 1996), have submerged speleothems suggesting that regional groundwater levels were lower in the past. As demonstrated by the Namibian evidence, the ages of submerged speleothems can provide valuable information on more arid periods during the Quaternary.

Another aspect of speleothem age data worth mentioning is that, for individual speleothems, the beginning and terminal ages of growth are probably environmentally significant. Speleothems do not just start growing. Usually, the growth is triggered by a climatic/environmental change that either starts water dripping or stops it. The change may be an increase or decrease in annual rainfall or rainfall intensity, a change in the vegetation cover which affects water balance, or a change in the soil cover, or a combination of these. Speleothems can start or stop growing in a random fashion but in most cases growth is affected by environmental change. At Echo Cave in South Africa, for example, a core from a column in Bridal Chamber shows that the speleothem stopped growing shortly after 107 k yrs B.P., began again about 47 k yrs B.P., and stopped finally around 43 k yrs B.P. This record clearly suggests that there were significant environmental perturbations at 107, 47 and 43 k yrs B.P., growth probably indicating an increase in rainfall and cessation of growth a decrease in rainfall (Brook et al., 1997). Holmgren et al. (1995) found that a stalagmite in Lobatse II Cave in Botswana grew from 51-43 and from 27-21 k yrs B.P. The period 43-27 k yrs B.P., bounded by two major hiatuses in stalagmite growth, produced discontinuous speleothem deposition probably under drier conditions. Holmgren et al. also argue for drier conditions prior to 51 and after 21 k yrs B.P. The evidence suggests significant shifts to wetter conditions at around 51 and 27 k yrs B.P. and shifts to drier conditions at 43 and 21 k yrs B.P.

On their own, therefore, speleothem age data can often provide significant and lengthy records of environmental change in the world's arid areas. These records can form the basis for other, more specific studies of climatic change. The examples discussed above illustrate the usefulness of this research approach. Many, if not most, of the world's deserts have caves in them or in close proximity. Scientists have yet to tap the environmental information they contain.

Sediment Sequences

This paper does not attempt to discuss in detail the palaeoenvironmental data that can be obtained from cave clastic sediment sequences. However, because speleothems are frequently a part of such sequences, brief mention will be made of them. After deposition, speleothems may be re-dissolved because of rising groundwater levels or because drip waters become undersaturated with respect to calcite. When the surfaces of re-dissolved speleothems are scalloped and/or pass smoothly from carbonate to bedrock, or from carbonate to cemented fill, re-solution was probably caused by floodwaters moving slowly through the cave. In this case, re-solution suggests an increased water flow to the cave and therefore the possibility of increased precipitation in the past. The above morphological charactersteritics are absent when re-solution results from the inflow of more acid dripwaters. This is most likely to occur due to the combined effects of an increase in rainfall intensity coupled with a thinner soil and a sparser vegetation cover. All are an indication of drier climatic conditions near the cave.

In desert and semi-arid environments, wetter periods of the past were probably times of increased vegetation cover, rock weathering and soil accumulation. Conditions were probably suitable for speleothem growth because the thicker soils and denser vegetation cover would lead to higher levels of soil CO_2, and therefore more dissolution of the carbonate bedrock. Plant roots probably held the soil together so there was little soil erosion and therefore little inflow of clastic sediment into underground cavities. As a result, clean, white speleothems were deposited. With the onset of much drier conditions, the vegetation cover declined and soil became increasingly susceptible to soil erosion. During the infrequent intense storms that characterize arid and semi-arid climates, the loose soil was washed into the caves, gradually filling passages some almost to the ceiling, and burying many speleothems. During times of less-intense rain, these clastic fills were cemented by carbonate precipitated from seepage waters. A return to wetter conditions initially caused large amounts of water to flow into the caves eroding both clastic fill and speleothems. In many caves it is apparent that waters were ponded or perched on the clastic fills and that laminated sediments were deposited in them. Erosion was rarely complete and remnants of cemented fills are frequently still preserved on cave walls. Eventually, soil accumulation at the surface allowed the development of a dense vegetation cover and conditions suitable for extensive speleothem growth returned.

Evidence for events similar to those described above is to be found in many dryland caves. Brook et al. (1997), for example, note that in Ladder Chamber in Echo Cave, South Africa, where annual precipitation is about 450 mm, there are clastic fills of at least two ages and speleothems of fives ages (Fig. 5a). The oldest fill consists of coarse gravel that was later cemented by flowstone with two dates from different samples of > 350 k yrs B.P. (Ech 27 and 31). Later, stalagmite Ech 30 was deposited 187.6 k yrs B.P. and wall flowstones Ech 33 and Ech 62 at 131.3 and 36.2 k yrs B.P., respectively. After this last date, Ladder Chamber filled with sediment, which buried many of the older speleothem formations. The influx of sediment probably followed a drying of the climate that reduced the plant cover and left the soil extremely susceptible to erosion. This fill is laminated, each lamination having coarse and fine bands, suggesting deposition in standing water after periodic runoff into the cave. As flowstone Ech 34, dated to 13.9 k yrs B.P., is draped over the eroded surface of this fill near Second Ladder, much of this sediment must have already been eroded into lower cave passages by Late glacial times.

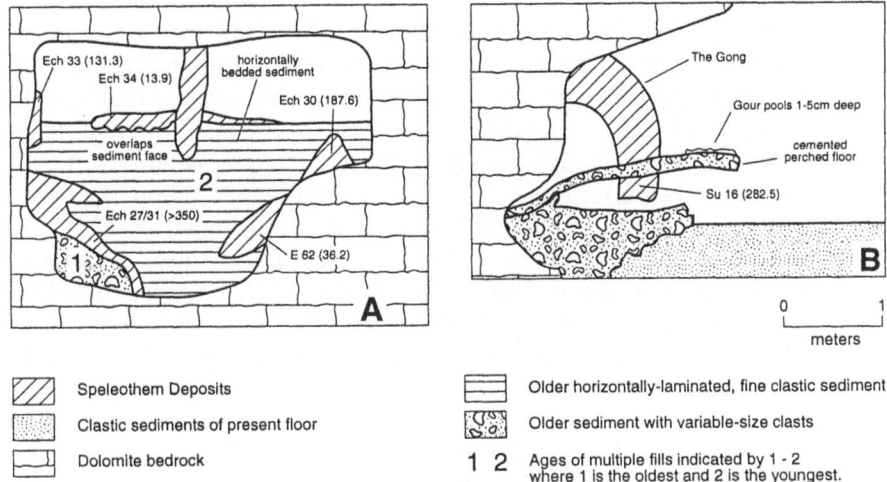

Fig. 5: Schematic diagrams of sediment sequences in Ladder Chamber, Echo Cave (A) and Entrance Hall, Sudwala Cave (B), South Africa (after Brook et al., 1997). Speleothem ages in k yrs B.P. are given in parentheses.

In Sudwala Cave, South Africa (precipitation 900 mm/yr), there is also evidence of a complex history (Brook et al., 1997). For example, in the Entrance Hall of this cave there is a large, heavily re-dissolved curtain called ``the Gong'', which is approximately 1.5 m in diameter and 3 m high. The basal 1.4 m of the curtain was formerly buried by clastic fill that has since been largely eroded. The fill surface was cemented by carbonate and remnants of the old floor are now preserved around the base of the Gong, 2 m above the present floor of the passage (Fig. 5b). A sample (Su 16) from the Gong, taken 51 cm above the present floor and 1.5 m below the ancient fill surface, gave an age of 282.5 ± 51.4 k yrs B.P. suggesting that this massive flowstone was probably deposited during deep sea oxygen isotope Stage 8 and was later buried by fill during a dry phase of climate when soil above the cave became unstable. The later erosion of the fill suggests that wetter conditions followed.

In the above examples from Echo and Sudwala Cave in South Africa, speleothem ages provide important chronological data for the interpretation of sediment fill sequences. As these sequences contain evidence of past drier and wetter periods, being able to date them accurately allows development of regional environmental histories.

Pollen Spectra

Rich pollen-bearing deposits like lake beds and peats are relatively scarce in many arid and semi-arid regions (Scott, 1984). Therefore, even though there is clear evidence that many deserts have undergone significant environmental changes in the past, poor pollen preservation and the small number of sites with data mean that there is still great uncertainty concerning the precise pattern of these changes. An important discovery in recent years is that some cave speleothems contain rich macrofossil and microfossil assemblages including gastropod shells, pollen and spores, diatoms, plant phytoliths, and plant cuticles, that can provide a wealth of palaeoenvironmental data (Bastin, 1978, 1982; Bastin et al., 1977, 1982; Bastin and Schneider, 1984; Bastin and Gewelt, 1986; Brook et al., 1987, 1990a, b, 1996; Brook and Nickman, 1996; Burney et al., 1994). Furthermore, studies by Bastin and Gewelt (1986), Brook et al. (1990a, b) and Burney and Burney (1993) confirm that pollen grains and spores in speleothems do provide reliable information on the general composition of the vegetation near the cave at the time of speleothem deposition.

In terms of desert research, the studies of speleothem pollen and spore spectra in Brook et al. (1990a, b) and Burney et al. (1994) are important. The work at Matupi Cave in northeastern Congo by Brook et al. (1990a, b) has shown how speleothem pollen spectra can provide information on the former extent of arid and semi-arid climates into areas that are much wetter today. Matupi Cave, at 1.5°N, 36°E, is developed in the Mt. Hoyo block located southwest of Lake Mobutu Sese Seko. The cave is at 1,100 m elevation on the western shoulder of the Western Rift Valley. Today, vegetation near the cave is species-rich equatorial rain forest, the mean annual temperature is 23°C and annual precipitation is about 1,687 mm. Pollen spectra from a Late Pleistocene stalagmite (U-series age 14.8 k yrs B.P.) and a modern active stalactite were found to be very different. The former was dominated by pollen of grassland plants, whereas these pollen types were relatively scarce in the modern spectrum. Significant traces of pollen types associated with high-elevation montane environments in eastern Africa were also better represented in the Pleistocene spectrum. On the other hand, the Pleistocene assemblage was virtually devoid of pollen types associated with the mid-elevation mesic forest characteristic of the site today and lacked many of the pollen types associated with humid tropical environments with human disturbance, vegetation mosaics, extensive riparian habitat, or other situations that include extensive forest edge or openings in the forest. These types are well represented in the modern stalactite as is *Pinus* pollen, a twentieth-century introduction to sub-Saharan Africa, confirming that the modern spectrum is indeed derived from the modern environment.

The pollen spectrum for the Late Pleistocene stalagmite indicates that around 15 k yrs B.P. the vegetation near Matupi Cave was a savanna grassland and that high montane vegetation may have existed closer to the site than at present. Moderately warm, dry conditions are indicated for this period with temperatures lower than today. Faunal remains of glacial and late glacial age from an archaeological excavation in one of the entrances to Matupi Cave, indicating a savanna grassland close to the cave, support the pollen evidence (Van Noten, 1977, 1982). Significantly, pollen were scarce and poorly preserved in the sediments excavated by Van Noten, but were extremely well preserved in the speleothems studied by Brook et al. The glacial and Late glacial fauna and pollen spectra from Matupi Cave are evidence that this area was not part of an ice-age refugium for montane equatorial forest as previously thought. Instead, it appears that semi-arid conditions extended into this area during what must have been a drier period as compared to the present.

Brook et al. (1990b) present spectra for four speleothems from Hayla Cave in northern Somalia, one modern and the others with U-series ages of 176.5 ± 27, 11.8 ± .4 and 10 ± 1.5 k yrs B.P. The pollen assemblages allow several cautious conclusions to be drawn. First, the modern spectrum seems to reflect the modern conditions with xeric plant communities adapted to high temperatures, ruderal plants responding to human disturbance, and exotic introductions. This is further proof that pollen trapped in speleothems can provide evidence about vegetation near the cave. Secondly, the spectra for the late glacial-early Holocene formations contain many dry-adapted plants associated with deciduous bushland and wooded grassland, but, in comparison to the modern spectrum from the cave, these indicate cooler environments than today. Also, a higher pollen diversity and the pollen of a few mesic types together with high values of *Typha* in one speleothem suggest less arid conditions than those of today. Across much of eastern Africa there is evidence of wetter conditions in Late glacial-Early Holocene times, further confirmation that speleothems can provide palaeovegetation data for arid and semiarid areas. A final conclusion from the work in northern Somalia is that speleothems of considerable age can also contain well-preserved pollen. Brook et al. (1990b) present a pollen spectrum for a speleothem U-series dated to 176.5 ± 27 k yrs B.P. The pollen were in sufficient quantity to allow a count of 503 grains. They revealed a cooler, wetter environment than now during a part of deep-sea oxygen isotope Stage 6. The information from northern Somalia suggests that speleothem pollen spectra could produce reliable records of vegetation change in presently arid and semi-arid areas for the last few hundred thousand years.

Speleothem pollen spectra from a column in Drotsky's Cave have provided the first record of Holocene vegetation change for Botswana, where the ubiquitous surface sands are not conducive to pollen preservation. Burney et al. (1994) examined pollen in a horizontal, 61 cm core drilled from a 2.63 m high column located 25 m from the southwest entrance to the cave. In the laboratory the core was cut lengthwise vertically and horizontally into quadrants. Eight subsamples of 20-40 gm were cut from the core for U-series dating and fifteen subsamples from 7.2-17.2 gm for pollen analysis (Fig. 6). Surface contaminants were removed from the samples by forced-air cleaning followed by a 5-minute wash in running deionized and filtered water. The carbonate matrix was digested with 12 N HCl, followed by processing of the residue according to conventional palynological methods described in Faegri and Iverson (1975). Growth layers exposed in the vertical and horizontal sections of the core revealed that it grew by the coalescence of at least two stalagmites, and that growth had been much greater at the distal end of the core, that is in the direction of drilling (Fig. 6). The sections through the core also revealed a number of layers of clastic sediment deposited on the column surface by either wind or by water percolating into the cave after heavy rain. Burney et al. (1994) note that two clastic layers at the proximal end of the core at 1.4 and 3.1 cm correlated with clastic layers at 58.4 and 49.5 cm at the distal end of the core. These correlations were later confirmed by U-series dating, allowing dates and pollen samples from opposite sides of the core to be placed in a single chronological sequence (Fig. 7).

Not surprisingly, pollen preservation was poor relative to most lake and bog sediments, with crumpled indeterminates comprising 12-34 % of the raw sum at each level. A total of 31 pollen and spore types were identified mostly to genus or family. Ages of pollen samples were determined by assuming constant sedimentation rates between dated parts of the core. Burney et al. (1994) identified three pollen zones (Fig. 7). The Zone 1 spectra, which on the ^{14}C timescale date to the beginning of the Holocene (ca. 10-7 k yrs B.P.), indicate dry conditions, with a grassland near the site and a few dry-adapted trees and shrubs in the landscape. Grasses are at their highest level in this zone and the Combretaceae, many species of which are characteristic of somewhat wetter savannas, are represented by their lowest values. In Zone 2, dating to approximately 7-6 k yrs B.P., a slight decline in grasses and an increase in Combretaceae suggest a somewhat wetter episode. Pollen spectra in Zone 3, from about 6-3 k yrs B.P., indicate a trend towards a generally wetter climate than existed in Zones 1 and 2 but there was some climatic variability. The periods 6-5 and 4-3 k yrs B.P. have less grass pollen and more Combretaceae and may therefore have been wetter that the intervening period 5-4 k yrs B.P. The highest values for *Combretaceae* and the lowest for grasses in the entire Holocene record are found in the interval 4-3 k yrs B.P., suggesting that this may have been an exceptionally wet period.

The conclusions arrived at by Burney et al. (1994) from speleothem pollen spectra are in broad agreement with other palaeoenvironmental

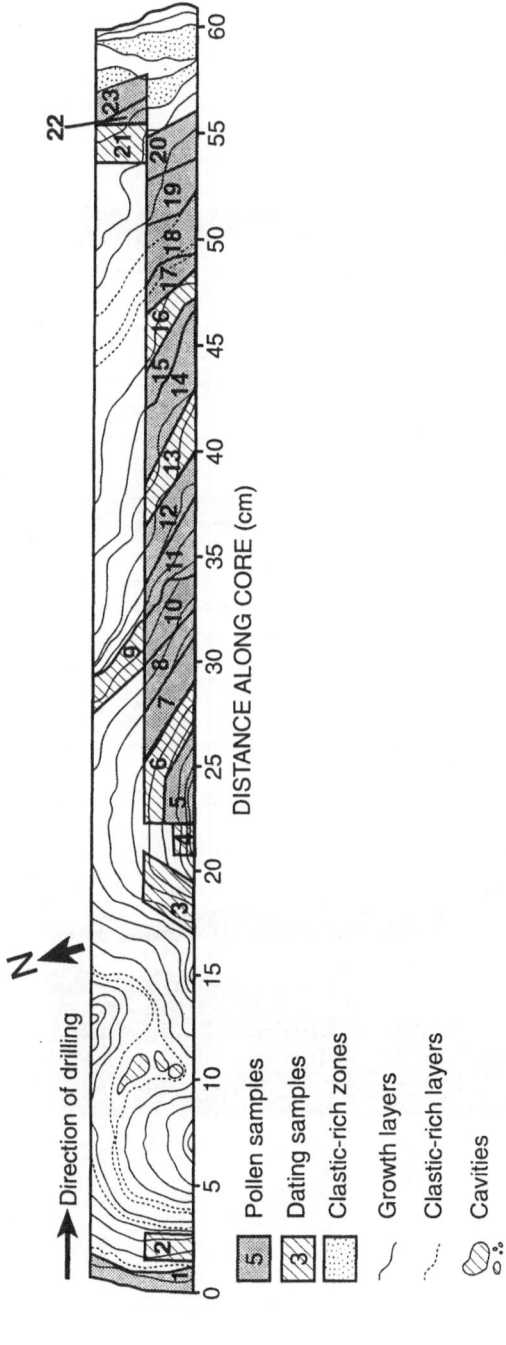

Fig. 6: Horizontal section through a 5 cm-diameter core DC87-2 drilled from a column in Drotsky's Cave, Kalahari Desert, Botswana. Growth layers show that the column formed by the coalescence of two stalagmites. Locations of samples taken for U-series dating and for pollen analysis are shown (after Burney et al., 1994).

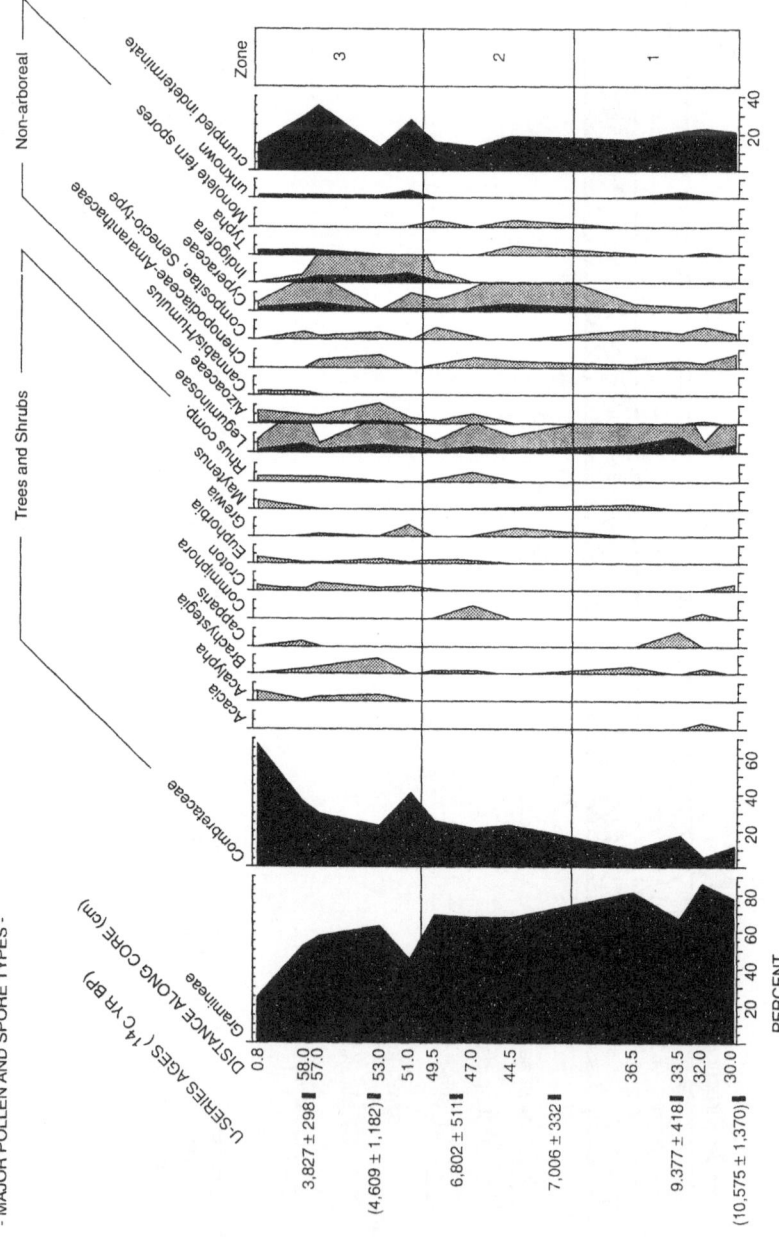

Fig. 7: Pollen diagram for spelaeothem core DC87-2 from Drotsky's Cave, Botswana. Grey shading indicates 5x exaggeration. Pollen sum based on all types except crumpled indeterminate. 'Leguminosae' refers to pollen types from this family, excluding *Acacia*, *Indigofera* and *Brachystegia* (modified after Burney et al., 1994).

data for the region. As Figs. 2 and 3 show, for example, speleothem age data for the summer rainfall zone of southern Africa indicate that the period 10-7 k yrs B.P. was generally drier than the interval 7-3 k yrs B.P. In particular, submerged speleothems from depths of 3-13 m in three flooded Namibian caves point to a significant dry interval in the region. The speleothem record also indicates a short interval of relative dryness around 4 k yrs B.P. with increased wetness shortly after and shortly before this time. This closely parallels the pollen data. Despite these seemingly informative results, Burney et al. (1994) add a word of caution for those using speleothem pollen spectra in environmental reconstruction. They point out that speleothem carbonate normally accretes at rates perhaps one to two orders of magnitude slower than peats and lake sediments. This means that pollen spectra for a speleothem sample of several grams are likely to represent more temporal smoothing than most conventional pollen-bearing sediments. Each pollen spectrum should therefore be viewed as representing the accumulation of many years of low-level pollen deposition, and environmental changes of brief duration (perhaps < 1 k yrs) may be missed or blurred in the record.

Despite problems in interpreting speleothem pollen spectra, it is clear that these can provide reliable information on past vegetation over potentially hundreds of thousands of years. Furthermore, pollen appear to be far better preserved in speleothems in desert caves than in the clastic sediments of these caves, suggesting that they are an under-utilized source of palaeoenvironmental data for these regions.

Stable Isotopes

Background

Isotopes are atoms whose nuclei contain the same number of protons but a different number of neutrons. Although, in general, the isotopes of an element behave almost identically, since chemical behavior is governed by the electron structure of the atom, differences in isotopic mass do lead to subtle but significant differences in the behavior of the isotopes during natural processes. Since the magnitude of "isotope effects" is proportional to the relative mass differences between isotopes, significant isotope variations in nature are limited to the light elements H, C, N, O, and S. As well as being relatively abundant, they are the most important elements in biological systems, and participate in most of the important geochemical reactions in nature. Isotope studies of speleothems have focused on O, C, and H. In the case of oxygen, the isotopes of interest are ^{16}O (99.763%)

and ^{18}O(0.1905%), for hydrogen ^1H(99.984%) and ^2H or D (0.016%), and for carbon ^{12}C (98.89%) and ^{13}C (1.11%) (Bowen, 1991).

The absolute abundances of the heavier isotopes (i.e. ^{18}O, ^{13}C and D), or the absolute values of the isotopic ratios ($^{18}O/^{16}O$, $^{13}C/^{12}C$ and D/H), are difficult to determine with sufficient accuracy for geochemical applications, so that differences are measured between sample isotopic ratios and those of standards. Important in this is that differences can be measured directly far more precisely than absolute values in multiple collecter mass spectrometers. As a consequence, relative differences in isotopic ratios are generally used for reporting stable isotope abundances and variations. The reporting notation is called the "δ-value" which is defined as:

$$\delta(x) = [(R_x - R_{std})/R_{std}] \times 100$$

where R_x is the isotopic ratio of the sample ($^{18}O/^{16}O$, $^{13}C/^{12}C$, D/H) and R_{std} is the corresponding ratio in a standard. The δ-value is the relative difference in the isotopic ratio (expressed as the heavy, rare isotope versus the light, abundant isotope) between the sample and the standard in parts per thousand or per mil. As the δ-value depends upon the isotope ratio of the standard it can either be positive ($R_x > R_{std}$) or negative ($R_x < R_{std}$). There are two standards used in speleothem isotope studies, PDB and SMOW. PDB, a belemnite (*Belemnitella americana*) from the Cretaceous Peedee Formation in South Carolina, is generally used for oxygen and carbon analyses of carbonates. The SMOW standard (Standard Mean Ocean Water) is a hypothetical water close to average ocean water and is usually used for oxygen and hydrogen isotopic studies of water. To relate the $\delta^{18}O$ values of calcite on the PDB and SMOW scales, the following expressions from Friedman and O'Neil (1977) are used:

$$\delta^{18}O_{SMOW} = 1.03086 \, \delta^{18}O_{PDB} + 30.86, \tag{a}$$

and

$$\delta^{18}O_{PDB} = 0.97006 \, \delta^{18}O_{SMOW} - 29.94. \tag{b}$$

In other words PDB calcite is +30.86 on the SMOW scale.

Oxygen Isotopes

When calcite is precipitated from solution more of the heavier isotopes of oxygen and carbon are deposited because of fractionation which is more effective at lower temperatures. Hence, calcite deposited from cold waters will have higher ^{18}O and ^{13}C than that deposited from warm waters of the same chemistry. When the amount of fractionation is determined by the ambient temperature alone, this is termed "equilibrium fractionation".

When it is determined by temperature and evaporation, and/or the rapid loss of CO_2 from solution, it is termed "kinetic fractionation". If a speleothem is deposited in isotopic equilibrium, its isotopic composition can yield the temperature of deposition if the isotopic characterstics of the seepage water are known (Schwarcz, 1986; Schwarcz and Yonge, 1983; Schwarcz et al., 1976). Temperatures in caves are generally constant throughout the year and are equal to the mean annual air temperature. Thus, temperatures of deposition derived from speleothems provide information on mean annual temperatures above the cave. Talma and Vogel (1992) comment that the low growth rate of most speleothems means that a single sample of *ca.* 20 mg of speleothem carbonate covers a few decades of deposition rather than a few years so that derived temperatures are sufficiently smoothed provide only decadal variations.

The most commonly used relationships for determining palaeotemperatures from speleothem data are:

$$t = 16.9 - 4.2(\delta^{18}O_c - \delta^{18}O_w) + 0.13(\delta^{18}O_c - \delta^{18}O_w)^2$$

after Craig (1965),   (c)

and

$$t = 16.9 - 4.38(\delta^{18}O_c - \delta^{18}O_w) + 0.10(\delta^{18}O_c - \delta^{18}O_w)^2$$

afer O'Neil et al. (1969),   (d)

where t is temperature in °C, and $\delta^{18}O_c$ and $\delta^{18}O_w$ are the oxygen isotopic characteristics of the speleothem and the water with respect to the same standard, usually PDB or SMOW (see equations (a) and (b)).

In order to use these relationships to estimate the temperature of formation of a speleothem, it is first necessary to determine if the calcite was deposited in isotopic equilibrium with the water, and secondly, the oxygen isotopic composition of the water must be known. Hendy (1971) suggested that isotopic equilibrium can be established if (i) $\delta^{18}O_c$ is constant along a single growth layer, and (ii) there is no correlation between $\delta^{18}O_c$ and $\delta^{13}C_c$ along the same growth layer. Variation of $\delta^{18}O_c$ along a growth layer, particularly a continuous increase, suggests that deposition may have been influenced by evaporation of water, rather than by slow loss of CO_2. Correlation of $\delta^{18}O_c$ with $\delta^{13}C_c$ along a growth layer indicates that a kinetic isotope effect such as rapid loss of CO_2 is occurring rather than the equilibrium distribution processes that affect $\delta^{13}C_c$ in a different manner to $\delta^{18}O_c$.

In their study of a stalagmite in Cango Caves, South Africa, Talma and Vogel (1992) tested for isotopic equilibrium deposition at 86 and 240 cm from the tip of a 2.7 m high formation. In both sets of measurements along growth layers the relatively flat tops and the steeply sloping flanks of the stalagmite showed different precipitation regimes. They found that on the tops $\delta^{18}O_c$ varied by only 0.3 per mil. while $\delta^{13}C_c$ was virtually constant. Along the flanks of the speleothem, however, $\delta^{18}O_c$ increased by up to 1.4 per mil. in the direction of water flow and correlated with an increase in $\delta^{13}C_c$. Talma and Vogel argue that the absence of correlation on the flat tops indicates that there was no kinetic fractionation there during carbonate formation. They suggest that the water film was much thicker and the release of CO_2 much slower so as to ensure that isotopic equilibrium existed between the liquid and solid phases. In contrast, along the flanks of the stalagmite ^{18}O enrichment is approximately 0.47 times that of ^{13}C enrichment indicating kinetic fractionation. Talma and Vogel ascribe this enrichment to the rapid release of CO_2 from the thin water layer on the flanks of the stalagmite. They conclude that samples from the flat top of the formation should, therefore, produce reliable palaeotemperature data while samples from the flanks will not.

The locality in the cave where the sample is collected can be shown to be suitable for equilibrium deposition if a speleothem forming presently at the site is in isotopic equilibrium with the water from which it is being deposited. Typically, such sites are far from the cave mouth, in passages which are blocked at one end, and exhibit relative humidities of close to 100% (Schwarcz, 1986). Talma and Vogel (1992) used this approach in their study in Cango Caves. They calculated the present temperature of deposition using equation (c) above by analysing stable isotopes in calcite from the top of the stalagmite. Their estimate of 17.7 °C was very close to the measured temperature of 17.5 °C. These findings suggest that for palaeotemperature studies it is preferable to have a whole stalagmite rather than a drilled core and that samples should be taken along the central growth axis of the formation. Where formations that are being deposited at present are available for investigations, it is possible to verify such calibrations so as to provide more confidence on values obtained from ancient material.

The main difficulty when calculating temperatures from the ^{18}O content of stalagmites is lack of knowledge of the isotopic composition of the water during calcite deposition. The ^{18}O content of present drip water can readily be measured but the ^{18}O of past drip water is usually unknown. One approach to solving this problem has been to analyse the fluid inclusions in the calcite lattice of the speleothem on the assumption that these preserve the characteristics of the original meteoric waters that precipitated it. As $\delta^{18}O_w$ could have been altered by oxygen exchange with the calcite during the time following deposition, δD_w is generally measured and $\delta^{18}O_w$ estimated from the general relationship for most meteoric waters (Craig, 1961) i.e.,

$$\delta D_w = 8 \, \delta^{18}O_w + 10. \qquad\qquad (e)$$

Once $\delta^{18}O_c$ and $\delta^{18}O_w$ have been determined for points along the growth axis of a speleothem, variations in temperature can be calculated. According to Gascoyne (1992), fluid inclusion analysis in palaeo-temperature determination has not been used as frequently as expected. Gasoyne suggests that this is because of analytical difficulties and because numerous studies of δD and $\delta^{18}O$ in modern precipitation have shown that both the slope and intercept in equation (e) above may vary depending on location (e.g. Fritz et al., 1987), with time for a given location such as between glacial and interglacial times (e.g. Harmon and Schwarcz, 1981). For these reasons the application of fluid inclusion analysis to absolute palaeotemperature determination has been limited and the results are often ambiguous (Gascoyne, 1992). As a result, there has been a tendency to interpret variations in $\delta^{18}O_c$ in terms of temperature change with or without quantification of the magnitude of the change. As the $\delta^{18}O$ of modern calcite in a cave is likely to represent values characteristic of warm conditions, isotopically heavier calcite is normally interpreted as being indicative of colder conditions. As Gascoyne (1992) points out, the position of maximum and minimum δ-values in an isotopic profile of a fossil deposit, relative to that of modern calcite, immediately indicates the sense of the climatic trends seen in the profile.

In general, therefore, isotopic equilibrium deposition requires that CO_2 release from the carbonate-oversaturated drip water is relatively slow. This is usually the case in a poorly ventilated cave where the CO_2 content of the cave air is close to the partial pressure of CO_2 in the drip water, and the relative humidity, which controls the amount of evaporation, is close to 100%. In arid and semi-arid areas, many caves are extremely dry today and, although they contin speleothems, it may appear unlikely that carbonate deposition was in isotopic equilibrium with the precipitating fluids. However, what must be understood is that past conditions in the cave may have been very different from those of the present. Frumkin et al. (1994), for instance, investigated stalagmites from Soreq Cave and two other caves revealed by quarrying and construction activities around Jerusalem in Israel. In all three they found limited variation of $\delta^{18}O$ along speleothem growth layers indicating equilibrium deposition. This suggests that the Jerusalem caves formerly remained at higher humidity by continuous recharge of seepage waters and that the speleothems could be useful for palaeoclimate studies.

Even in drylands, poorly-ventilated caves may have speleothems suitable for palaeoclimate studies. This will be the case if a sufficient plant growth during the wet season occurs to provide drip waters to humidify the cave and to release CO_2. The CO_2 is originally picked up in the soil and carried to the cave in percolation water. Its release into the cave atmosphere maintains a high level of CO_2 in the cave air, which prevents the rapid release of CO_2 from the carbonate-depositing drip waters that can result in kinetic fractionation. I have visited caves in the Kalahari Desert of Botswana, including Drotsky's and Bone Cave, that even today have humidities in some passages close to 100% and CO_2 levels elevated well above atmospheric values. In the Otavi Mountain Land of Namibia, southeast of Etosha Pan, there are numerous, poorly ventilated caves, such as Dragons Breath, that have large water bodies in their lower passages so that humidities are nearly 100% and also CO_2 levels that, on occasion may be a danger to cave explorers (Brook et al., 1996). These caves are also eminently suitable for speleothem palaeoclimate studies. Furthermore, even in dryland caves where there is no speleothem deposition today, conditions in the past may have been very different. Large, relict speleothems indicate more available water in the past, and many were probably laid down before the cave was opened to the surface to any significant degree. Old speleothems may, therefore, have been deposited in conditions of isotopic equilibrium that no longer exist today because of a drier climate and a better ventilated cave environment.

A recent example where $\delta^{18}O$ of speleothem carbonate has been used on its own to derive relative palaeotemperature data is the study by Holmgren et al. (1995) who examined oxygen and carbon isotope variations in a stalagmite from Lobatse II Cave in Botswana (Fig. 8). They argue that at least on the top of the stalagmite (following the arguments of Talma and Vogel, 1992) the carbonate was precipitated in isotopic equilibrium with the precipitating fluid. However, they report that on the flanks of the formation deposition may have been influenced by kinetic effects. Holmgren et al. (1995) note that the $\delta^{18}O$ of calcite layers ranged from -7.3 to $+3.4$ per mil. An enrichment in $^{18}O_c$ (less-negative $\delta^{18}O$ values) is interpreted as a shift toward a colder climate and so Holmgress etal. argue for a change to warmer conditions at 51-43 k yrs B.P. and to colder at 27-21 k yrs B.P. as inferred from an increase in $\delta^{18}O_c$ of 0.35 - 0.4 per mil. They argue that the few data between 50 - 49 k yrs B.P. probably record the warmest time during speleothem growth with minor warming trends at 46 and 27-26 k yrs B.P. Colder conditions are suggested at about 22 k yrs B.P. (Fig. 8). Holmgren et al. (1995) suggest that their oxygen isotope record indicates a temperature change of about 2°C, although they do not say how they arrived at this estimate.

In a 1974 study, Talma et al. examined the stable isotope characteristics of speleothems in Wolkberg Cave, South Africa, where the annual precipitation is about 600 mm and the mean annual temperature 16.5°C. By assuming that groundwater during the last ice age had an ^{18}O content 0.8 per mil lower than today, measurements of $\delta^{18}O_c$ in speleothems with ^{14}C ages of 29.6 and 19.8 k yrs B.P. were used in equation (c) above to estimate a last ice age temperature 8-9.5°C lower than now.

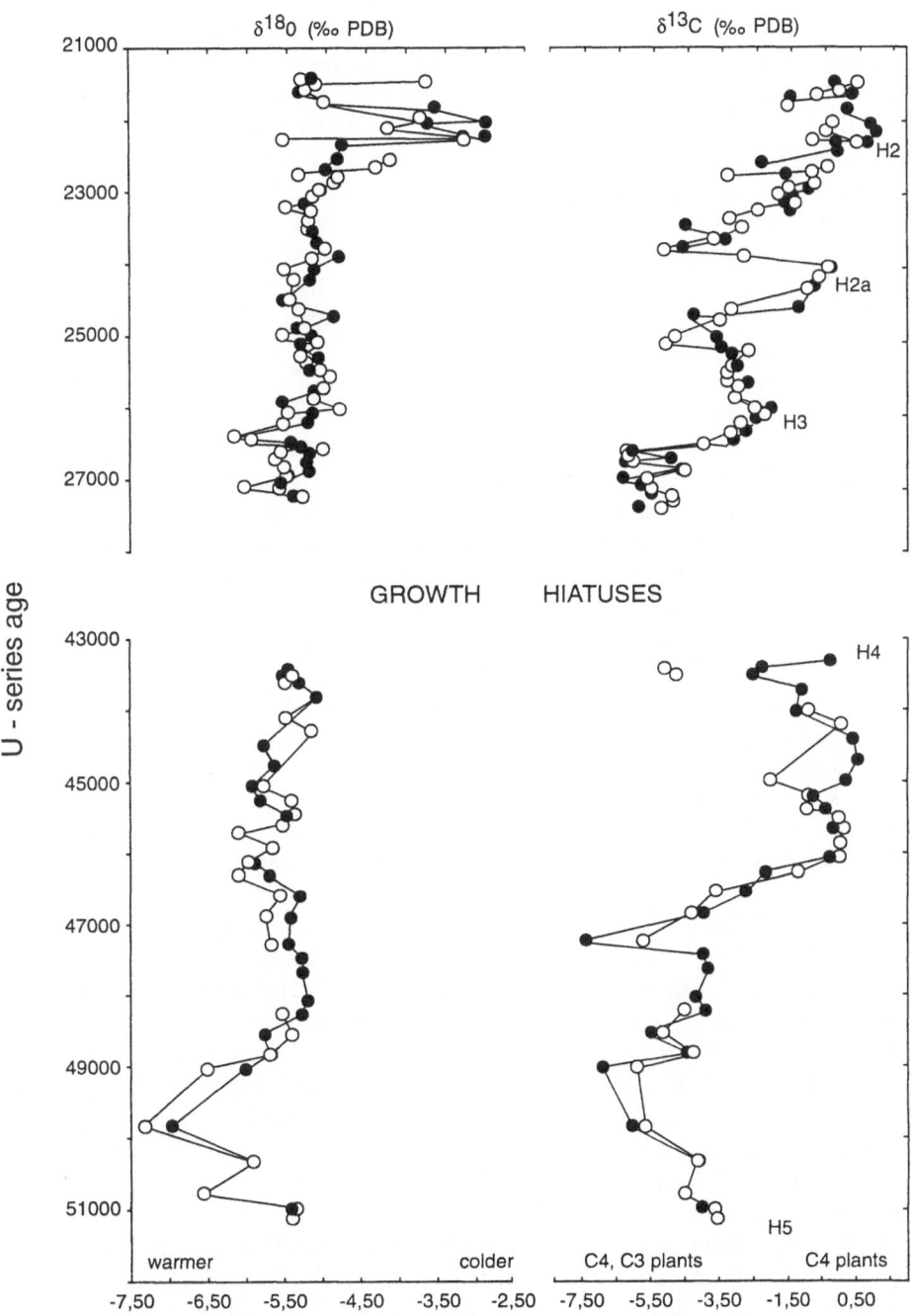

Fig. 8: $\delta^{18}O_c$ and $\delta^{13}C_c$ time series for a stalagmite from Lobatse II Cave, Botswana for the periods 51-43 and 27-21 k yrs B.P. (after Holmgren et al., 1995).

In their study of a 2.7 m high stalagmite in Cango Caves, South Africa, Talma et al. (1992) solved the problem of determining the isotopic characteristics of ancient cave drip waters by using the ^{18}O content of confined and dated groundwater from the Uitenhage artesian aquifer located 350 km east of the caves. The groundwater data were used to generate a meteoric drip water record for the last 30 k yrs by subtracting 0.95 per mil from the groundwater values, this being the difference between the youngest groundwater and present-day drip water in the cave (Fig. 9). Temperatures were calculated for the periods 30 - 13.8 and 5-0 k yrs B.P. using the Craig (1965) equation (c) given above (Fig. 10). No temperature data were obtained for 13.8-5 k yrs B.P. as the stalagmite did not grow during this time interval. Results showed that temperatures during the Last Glacial Maximum were, on average, about 6°C lower than those of today, with peaks up to 7°C lower. An important finding of the Talma and Vogel (1992) study was that the $\delta^{18}O_c$ data for the stalagmite did not show much systematic change between the glacial and postglacial periods. However, calculation of the carbonate deposition temperatures with the aid of the ^{18}O estimates of drip water indicated a highly regular temperature pattern in the past consisting of more or less constant temperatures during the past 5 k yrs and temperatures 4-7°C lower from 30-13.8 k yrs B.P. (Fig. 10). The temperature record shows a slow temperature decrease from 30 k yrs B.P. with minimum temperatures between 19 and 17 k yrs B.P. Thereafter, the temperature increased up to 13.8 k yrs B.P. when stalagmite growth ceased. On resumption of growth at 5 k yrs B.P., the temperature varied essentially within +1 and −2°C relative to present-day temperatures with generally lower temperatures between 5 and 2.5 k yrs B.P. and slightly higher values around 2 k yrs B.P.

In 1988, Winograd et al. published a 250 k yrs isotopic record for the Great Basin desert of the U.S.A. by studying a core (DH-2) from a calcite vein recovered from 21 m below water level in Devils Hole, a 1.5 m-wide open fault zone at the Ash Meadows spring in Nevada. The core was obtained by scuba divers using a submersible air-powered coring machine. The Ash Meadows oasis is the principal discharge of the Ash Meadows drainage basin, which covers about 12,000 square km, and which is underlain by 100 m to more than 1,000 m of Palaezoic carbonates. It is located about 115 km west-northwest of Las Vegas. To depths of 130 m below water table, which is 15 m below land surface, this open fissure is lined with a layer of dense mammary calcite more than 30 cm thick that precipitated from the calcite-saturated groundwater. Thirteen alpha-spectrometric, U-series ages were used to develop a chronology for the 140 mm thick vein deposit showing that deposition occurred from 308 ± 10.2 to 51 ± 2.8 k yrs B.P. The $\delta^{18}O_c$ record was developed by analysis of

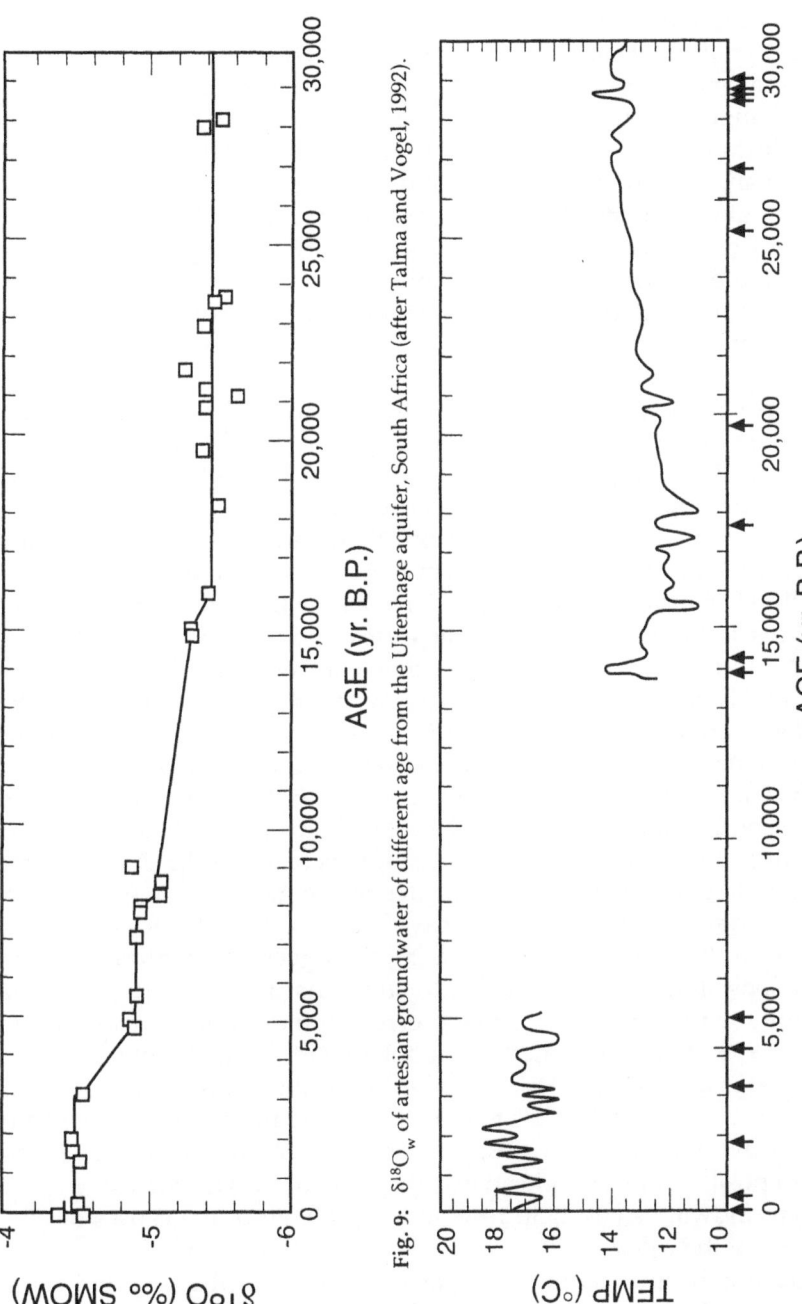

Fig. 9: $\delta^{18}O_w$ of artesian groundwater of different age from the Uitenhage aquifer, South Africa (after Talma and Vogel, 1992).

Fig. 10: Palaeotemperature record to 30 k yrs B.P. from $\delta^{18}O_c$ of speleothem calcite form Congo Caves and $\delta^{18}O_w$ of Uitenhage groundwater, Cango Caves, South Africa (modified after Talma and Vogel, 1992). The arrows indicate the ages of samples taken for dating.

110 samples from the vein calcite. The water in Devils Hole is slightly supersaturated with respect to calcite, and Winograd et al. (1988) report that calcite precipitated on most calcite crystals placed in the hole. Winograd et al. (1988) argue that, because only winter-spring precipitation appears to contribute to recharge, the $\delta^{18}O_c$ variations in DH-2 probably reflect changes in winter-spring air temperature and the source of air masses bringing the precipitation. They also consider that the record reflects changes in the $\delta^{18}O$ of the world's oceans between full glacial and peak interglacial times. Therefore, they interpret the heaviest isotopic values of the DH-2 time series as indicative of relatively warm weather and the lightest values as relatively cold weather. By comparing the DH-2 record with the marine and ice (Vostok) isotopic records. Winograd et al. (1988) argue that the last interglacial stage (marine stage 5) began before 147 ± 3 k yrs B.P., some 17 k yrs earlier than indicated by the marine record and 7 k yrs earlier than the ice core record (Fig. 11). Based on this they present the controversial argument that the indirectly dated marine ^{18}O chronology may need revision and that orbital forcing may not be the principal cause of the Pleistocene ice ages.

Following up on their earlier work, Winograd et al. (1992) present a 500 k yrs $^{18}O_c$ record (ca. 565-60 k yrs B.P.) derived from a 36 cm calcite core (DH-11) recovered from a point 30 m below the water table in Devils Hole. This is based on 285 $\delta^{18}O_c$ analyses and 21 TIMS U-series ages (Fig. 12). As in their previous paper, they conclude that the Devils Hole data are inconsistent with the Milankovitch hypothesis that orbitally controlled variations in solar insolation play a direct role in Pleistocene climate change. They argue that the hypothesis fails to predict the timing of deglaciations during the period 500-100 k yrs B.P. and that the observed increase in the duration of glacial cycles from about 80 to 130 k yrs during this time suggests that climate shifts were aperiodic. As Gascoyne (1992) points out, however, this controversial suggestion conflicts with a wealth of data from fossil reefs, cave calcites, peat bogs etc., which generally support the hypothesis. Gascoyne suggests that interlaboratory comparisons should be undertaken to show that the age differences are not simply a function of systematic error in the dating. Edwards and Gallup (1993) have presented a model where ^{230}Th dissolved in the groundwater at Devils Hole is adsorbed onto the walls of the cave and incorporated into the vein calcite thus increasing the apparent age of the deposit. Gascoyne (1992) also suggests analysis of fluid inclusions in the calcite deposit would provide an excellent record of past changes in $\delta^{18}O$ of the groundwaters and help resolve the problem of the causes of $\delta^{18}O$ variations in the calcite.

Artesian and other springs are not uncommon in the arid and semi-arid regions of the world. These could, if the waters are supersaturated

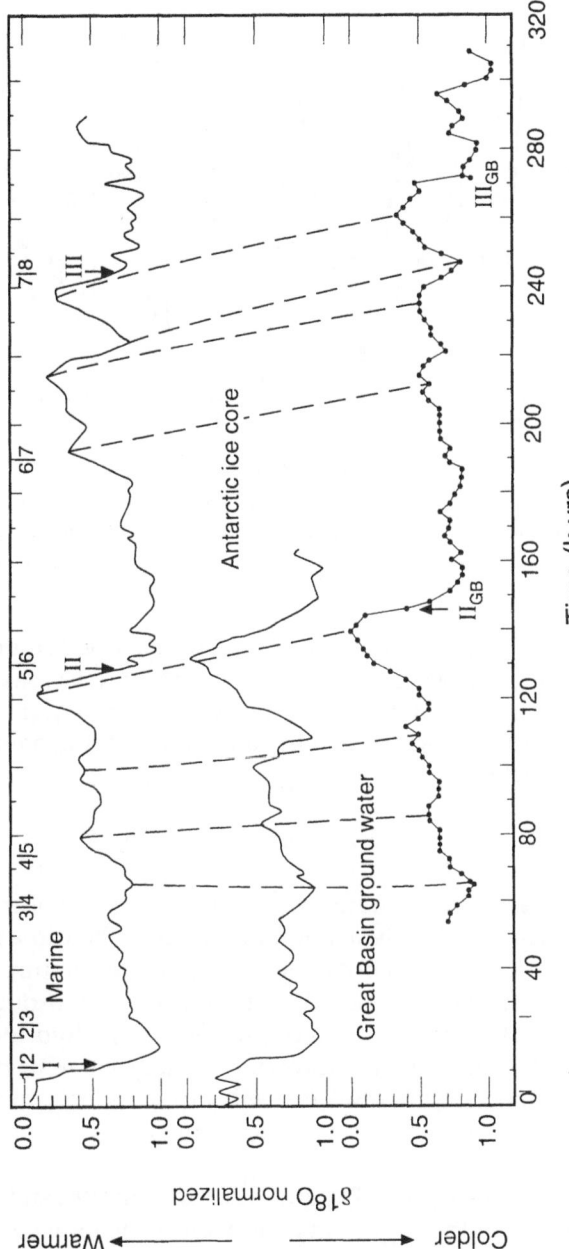

Fig. 11: Comparison of marine, Antarctic ice cap, and Devils Hole DH-2 δ¹⁸O records for the last 300 k yrs B.P. Numbers below the upper margin are isotope stage boundaries of the marine chronology. Points labeled I, II and III denote terminations (major deglaciations) in the marine record; II$_{CB}$ and III$_{GB}$ denote terminations suggested by the DH-2 record for the Great Basin. Vertical and sloping dashed lines connect major peaks and troughs believed to be correlative (after Winograd et al., 1988).

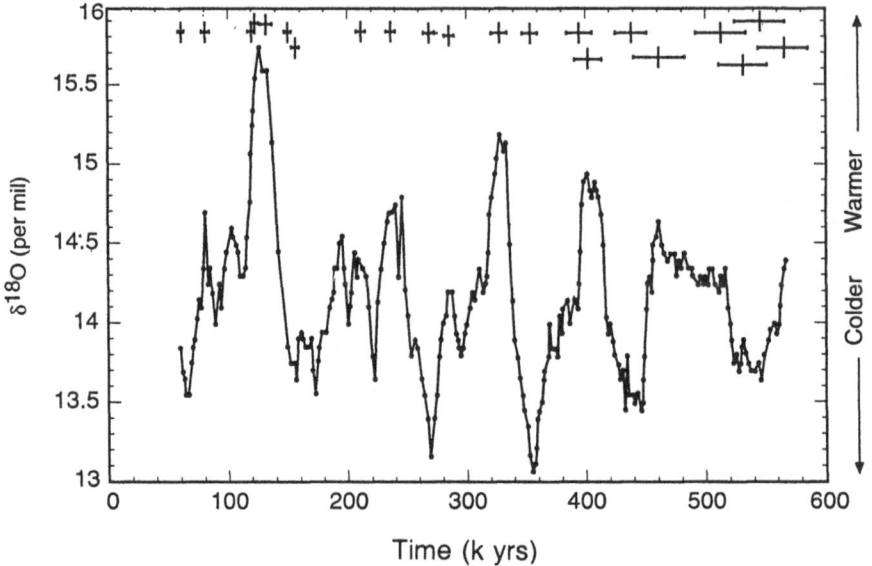

Fig. 12: Variations in $\delta^{18}O_c$ through Devils Hole core DH-11 over the last 560 k yrs B.P.
The distribution of TIMS U-series ages is shown by vertical lines below the
upper margin; error bars (2σ) are shown by horizontal lines (after Winograd et
al., 1992).

with respect to calcite, have subaqueous carbonate precipitates that
preserve a long record of past conditions in the recharge areas of the
aquifers. To determine if valuable deposits exist may require future
expeditions to the world's driest areas to include scuba diving equipment
as part of their baggage!

Hydrogen Isotopes

In addition to their use in determining the ^{18}O characteristics of meteoric
waters from speleothem fluid inclusion studies as described in equation
(c) above, δD_w of fluid inclusions can, under suitable circumstances,
provide important information about the air masses that brought the
precipitation. Little work has been done on this aspect of fluid inclusion
studies but the potential could be enormous. For example, in arid zones
the isotopic composition of groundwaters will likely reflect only major
recharge events, and so should preserve the isotopic chartacteristics of
the precipitation.

It must be remembered that the δD_w and $\delta^{18}O_w$ of rain depend largely
on the origin and history of the vapour mass as well as the local conditions
during rain-out. Rain which originates from a vapour mass with a short
oceanic history will have an isotopic composition which resembles ocean
water much more than rain from a vapour mass which had a long history

of rain-out and/or vapour addition over the continents. In Oman, for example, precipitation in the Dhofar Mountains and plains of the south has a southwest monsoon origin. The monsoon is most active from May to September over the Arabian Sea so that water vapor has only a very short residence time in the atmosphere. According to Fritz and Clark (1986), there is a narrow spread of isotopic values which provides a clear signature for monsoon rain contribution to ground water aquifers in the southern region. Southwest monsoon rains cluster around sea water values (i.e. δD_w and $\delta^{18}O_w$ are close to zero on the SMOW scale). By contrast, depressions from the Mediterranean and Red Sea bring rainfall of the continental type to northern Oman in winter, most frequently in January and February. These rains are depleted in both D and ^{18}O with δ-values typically less than –10 per mil. and –3 to –4 per mil. respectively (Fritz and Clark, 1986). Although no work has been done as yet, it seems highly likely that these major diferences in δD and $\delta^{18}O$ of atmospheric precipitation are recorded in speleothem fluid inclusions. If in the past the southwest monsoon extended further to the north than it does today, the evidence should be preserved in speleothems deposited at the time of this rainfall. Fluid inclusion waters should preserve a δD value that is close to the SMOW value. In fact, our studies indicate that a massive bank of travertine was deposited in Majlis al Jinn Cave in northern Oman 128 ± 11 k yrs B.P. As the cave is presently dry and dusty with little or no speleothem deposition, this suggests that during deep-sea isotope substage 5e the southwest monsoon extended further north than it does today bringing increased rainfall to central Oman. The proof of this may be preserved in fluid inclusions in the travertine.

Carbon Isotopes

The stable carbon isotope characteristics of speleothem carbonate can sometimes be used to infer past vegetation characterstics above a cave. This is particularly true in semi-arid areas where both C_3 and C_4 plants may be present. Dissolved carbon in the seepage waters that deposit speleothems is derived from: (i) atmospheric CO_2, (ii) decomposing organic matter in the soil and CO_2 respired by plants, and (iii) the carbonate bedrock. The $\delta^{13}C$ value of well-mixed air is about – 6 per mil.. The degree of ^{13}C enrichment in the organic matter produced by the vegetation above the cave depends on the photosynthetic pathway used by the plant. Plants using the C_4 or Hatch-Slack pathway generate cellulose with a modal $\delta^{13}C$ of –13 per mil., approximately 6 per mil. lighter than atmospheric CO_2. C_3 or Calvin-cycle plants produce carbon compounds with a modal $\delta^{13}C$ of –27 per mil. (Cerling, 1984). These isotopic characteristics are in large part transmitted to the CO_2 respired through the plant roots, and to the CO_2 formed by the oxidation of organic matter in the soil. However, because of their different atomic weights, $^{12}CO_2$ and $^{13}CO_2$ diffuse from the soil to the atmosphere at different rates, so that soil CO_2 is often isotopically heavier than soil organic matter and respired CO_2. Cerling (1984) has calculated limits of -22.2 per mil. and -8.5 per mil. for the $\delta^{13}C$ of soil CO_2 beneath pure C_3 and pure C_4 biomasses, where the original $\delta^{13}C$ of the biogenic CO_2 was -27 per mil. and -13 per mil. respectively.

Waters percolating through soil equilibriate with soil CO_2 and then dissolve limestone, which has a carbon isotopic composition of about +1 per mil. In closed-system dissolution of the limestone, the water first equilibriates with the CO_2 and is then brought into contact with the limestone; there is no further exchange with the CO_2 reservoir. The bicarbonate in the ground water is derived from equal parts of biogenic CO_2 and limestone. This means that beneath a pure C_4 biomass, with a soil CO_2 $\delta^{13}C$ of -8.5 per mil. and a limestone $\delta^{13}C$ of +1 per mil., the bicarbonate in the groundwater will have a $\delta^{13}C$ of about -4.3 per mil. Beneath a pure C_3 biomass with a soil CO_2 $\delta^{13}C$ of -22 per mil. the groundwater bicarbonate will have a $\delta^{13}C$ of -11 per mil.. If during the dissolution of the limestone the system remains open to the soil CO_2 reservoir, there is isotopic fractionation between the dissolved bicarbonate and the gaseous CO_2 amounting to about $+8$ per mil. at 20°C (Talma et al., 1974; Emrich et al., 1970; Hoefs, 1973; Salomons and Mook, 1986). Therefore the $\delta^{13}C$ of bicarbonate in the groundwater will be about -0.5 and -14 per mil. for pure C_4 and C_3 biomasses, respectively. Calcite deposited under isotopic equilibrium from such water is enriched in ^{13}C by about 2 per mil. at 20°C (Emrich et al., 1970).

Therefore, under a C_3 biomass (soil CO_2 $\delta^{13}C$ of -22 per mil.), the first speleothem calcite deposited in isotopic equilibrium from seepage waters is likely to have a $\delta^{13}C$ ranging from -12 to -9 per mil., depending on whether open or closed system conditions prevailed. Beneath a C_4 biomass (soil CO_2 $\delta^{13}C$ of -8.5 per mil.), the first calcite deposited is likely to have a $\delta^{13}C$ in the range -2.3 to $+1.5$ per mil.. Under non-equilibrium conditions and at lower temperatures than 20°C, the $\delta^{13}C$ values will be more positive. Thus shifts in the $\delta^{13}C$ of speleothem calcite can result from changes in the composition of the biomass (C_3 or C_4) above the cave.

A number of studies in arid and semi-arid areas have utilized $^{13}C/^{12}C$ ratios of speleothem carbonate to speculate about past vegetation characteristics. From studies at Wolkberg Cave in South Africa (mean annual temperature 16.5°C, annual precipitation 600 mm), Talma et al. (1974) concluded that the relatively high ^{13}C content of the inorganically dissolved carbon in groundwater and in the cave speleothems suggests that C_4, Hatch-Slack type plants contribute noticeably to the carbon dioxide in the soils in the vicinity of the cave. Talma and Vogel (1992) used carbon isotopes to speculate on vegetation changes since 47 k yrs recorded in a 2.7 m high stalagmite in Cango Caves, South Africa. Following Talma and Netterberg (1983) and Van der Merwe and Vogel (1984), they note that in many semiarid areas the main cause of $\delta^{13}C$ variation in the environment is the extent of C_4 grass present in the local vegetation. Some 40% of the present vegetation cover near Cango Caves is C_4 grasses. Talma and Vogel (1992) argue that calcite formed under open system conditions by dripwaters that have passed through soil supporting a 40% C_4 grass cover would have a ^{13}C value of -7 per mil. They note that the most recent calcite on the speleothem they examined, which was active when collected, was -6.4 per mil. suggesting that the ^{13}C content of the stalagmite does reflect the amount of C_4 vegetation cover at the surface. Measurements of ^{13}C of the stalagmite carbonate showed that from 47 k yrs to 13 k yrs B.P. little C_4 grass was present immediately above the cave. From 5 k yrs B.P. onwards C_4 grasses invaded the area reaching a maximum at 2 k yrs B.P. (Fig. 13).

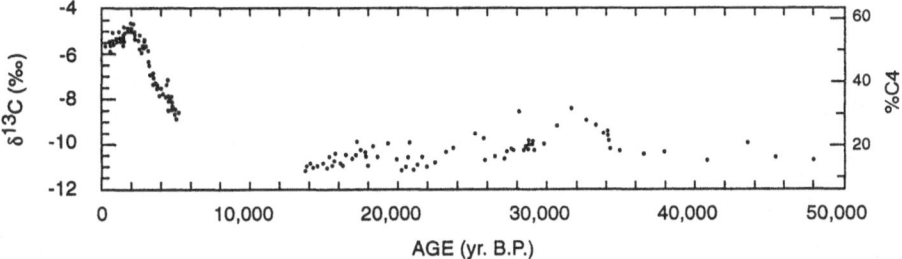

Fig. 13: $\delta^{13}C_c$ of speleothem calcite at Cango Caves, South Africa to 50 k yrs B.P. and presumed changes in the C_4 vegetation cover (after Talma and Vogel, 1992).

This approach has also been used by Brook et al. (1990a) in their study of palaeoenvironmental change at Matupi Cave in Congo. Today the vegetation near the cave is an equatorial forest made up primarily of C_3 plants so that actively growing speleothems should have a $\delta^{13}C$ of -12 to -9 per mil. as Table 1 shows, values for 3 modern formations (MAT 5, 14 and 16) range from -12.4 to -7.9 per mil. suggesting that speleothem $\delta^{13}C$ does indeed register the type of vegetation at the cave. The flat upper surface of MAT-12, deposited 6.0 ± 0.4 k yrs B.P., had a ^{13}C of -8.7 per mil. and that of MAT-23 dated to 14.8 ± 0.8 k yrs B.P. a value of -5.9 per mil. The ^{13}C contents of speleothems MAT-11 and MAT-13, 40.1 ± 3.5 and 50.3 ± 2.6 k yrs old, respectively, were -9.4 and 13.1 per mil. (Table 1). These analyses suggest that *ca.* 50 k yrs, *ca.* 40 k yrs and *ca.* 6 k yrs B.P. the vegetation at Matupi Cave was dominated by C_3 plants as it is today. However, the $\delta^{13}C$ content of MAT-23 suggests that in Late glacial times the vegetation was somewhat different with C_4 grasses making up about 50% of the vegetation cover. Pollen recovered from MAT-23 indicate a savanna grassland while that recovered from MAT-11 is typical of a high montane vegetation. This evidence is consistent with a 50% C_4 grass cover in Late glacial times and a C_3-dominated vegetation *ca.* 40 k yrs B.P.

Table 1: U-series age and ^{13}C data for speleothems from Matupi Cave, Zaire (after Brook et al., 1990a).

Sample ID	Age k yrs B.P.	$\delta^{13}C$ (per mil.)
MAT-5	modern	−12.4
MAT-14	modern	−11.7
MAT-16	modern	−7.9
MAT-12	6 ± 0.4	−8.7
MAT-23	14.8 ± 0.8	−5.9
MAT-11	40.1 ± 3.5	−9.4
MAT-13	50.3 ± 2.6	−13.1

In a study of a Late Pleistocene stalagmite from Lobatse II Cave in Botswana, which was deposited in the periods 51-43 k yrs and 27-21 k yrs B.P., Holmgren et al. (1995) used ^{13}C content to infer the importance of C_4 plants near the cave in the past. The ^{13}C values ranged between +1 and −6.5 per mil. demonstrating that at times of speleothem growth C_4 plants were the dominant biomass as they are today. However, lower values at 51-47, 28-27, ca. 25 and ca. 23.5 k yrs B.P. suggest a change in climate significant enough to allow for some C_3 vegetation (Fig. 8). Holmgren et al. (1995) believe that the region was probably influenced by both summer and winter rainfall at these times; presently most of the rainfall comes in the summer months. Higher $\delta^{13}C$ values at 46-43, ca. 26, ca. 24.5 and 23-21 k yrs B.P. suggest a drier climate with less winter rain and a vegetation dominated by C_4 plants, particularly grasses.

Baskaran and *Krishnamurthy* (1993) argue that in some cases speleothem $\delta^{13}C$ may be a proxy for atmospheric CO_2 $\delta^{13}C$. They note that in the period 1920-1990 the $\delta^{13}C$ of speleothem calcite from a cave in Texas, U.S.A., deposited in isotopic equilibrium, declined by −0.032 per mil. per year. They point out that this is in good agreement with estimates from tree rings, ice cores, C_4 plants, and direct measurements. One problem to using this approach to determine long-term records of atmospheric $\delta^{13}C$ is that, as we have seen above, speleothem $\delta^{13}C$ is to a large extent determined by the percentage of C_4 plants in the vegetation above the cave. If vegetation characteristics can be determined by other methods, such as pollen analysis, then this method may have potential. In any case, arid and semi-arid areas, that often have a mix of C_4 and C_3 plants which can change rapidly with the vagaries of climate, may not be the ideal ecosystems in which to conduct this kind of research.

Annual Layers

In recent years it has been determined that cave speleothems may have annual layers that could provide high-resolution climate data. An early reference to annual layers was by Broecker et al. (1960) who found them in a travertine covering a human femur from Moaning Cave in California. The deposit was discovered in 1951 and contained 1,206 identifiable layers with an additional 200 obscured. The layers were visible using a petrographic microscope and averaged about 0.1 mm thick. Radiocarbon dating confirmed that the layers were annual. Cazzoli et al. (1988) describe laminations in cylindrical speleothems deposited over nylon thread in the Acquafredda gypsum cave near Bologna, Italy. These were clearly visible through a petrographic microscope. The nylon thread was left in the cave in 1969. Although the speleothems had only been growing for 18 years when examined, they had hundreds of laminations which Cazzoli et al. (1988) related to intra-annual precipitation variations in the vicinity of the cave. Average deposition on the speleothems was 0.33 mm/yr. This study clearly indicates that speleothem laminae visible through a petrographic microscope may record intra-annual climatic events of extremely short duration.

Brook et al. (1992), Chen (1992) and Railsback et al. (1994) report annual layers in a stalagmite, 40 cm high and 18 cm wide at the base, from Drotsky's Cave in northwestern Botswana. The stalagmite was cut along its length and thin sections were made from the top to the base. As with the travertine from Moaning Cave in the U.S.A., and the cylindrical speleothems from the Acquafredda gypsum cave in Italy, the layers were visible through a petrographic microscope. They were found to consist of a basal unit of calcite deposited during the wet season and an upper unit of aragonite laid down during the following dry season. The layers of calcite and aragonite could be identified on the basis of crystal morphology (Fig. 14). Layer counts and radiocarbon ages indicate that the calcite-aragonite pairs are annual, representing about 1,500 years of deposition. In the upper 27 cm of the speleothem, annual layer thickness averaged about 0.36 mm. Rainfall at Drotsky's Cave is estimated at 475 mm/yr based on the average of three stations: Shakawe, Ghanzi and Maun; 150, 200 and 230 km away. Rainfall is highly seasonal. October through March is considered to be the wet season and April to September the dry season. Based on data for Shakawe, Ghanzi and Maun, in most years more than 90% of the annual rainfall at Drotsky's Cave falls in the wet season months.

According to Railsback et al. (1994), each annual cycle of calcium carbonate precipitation begins with a relatively intense fluid flow, sometimes sufficient to dissolve some of the underlying aragonite before

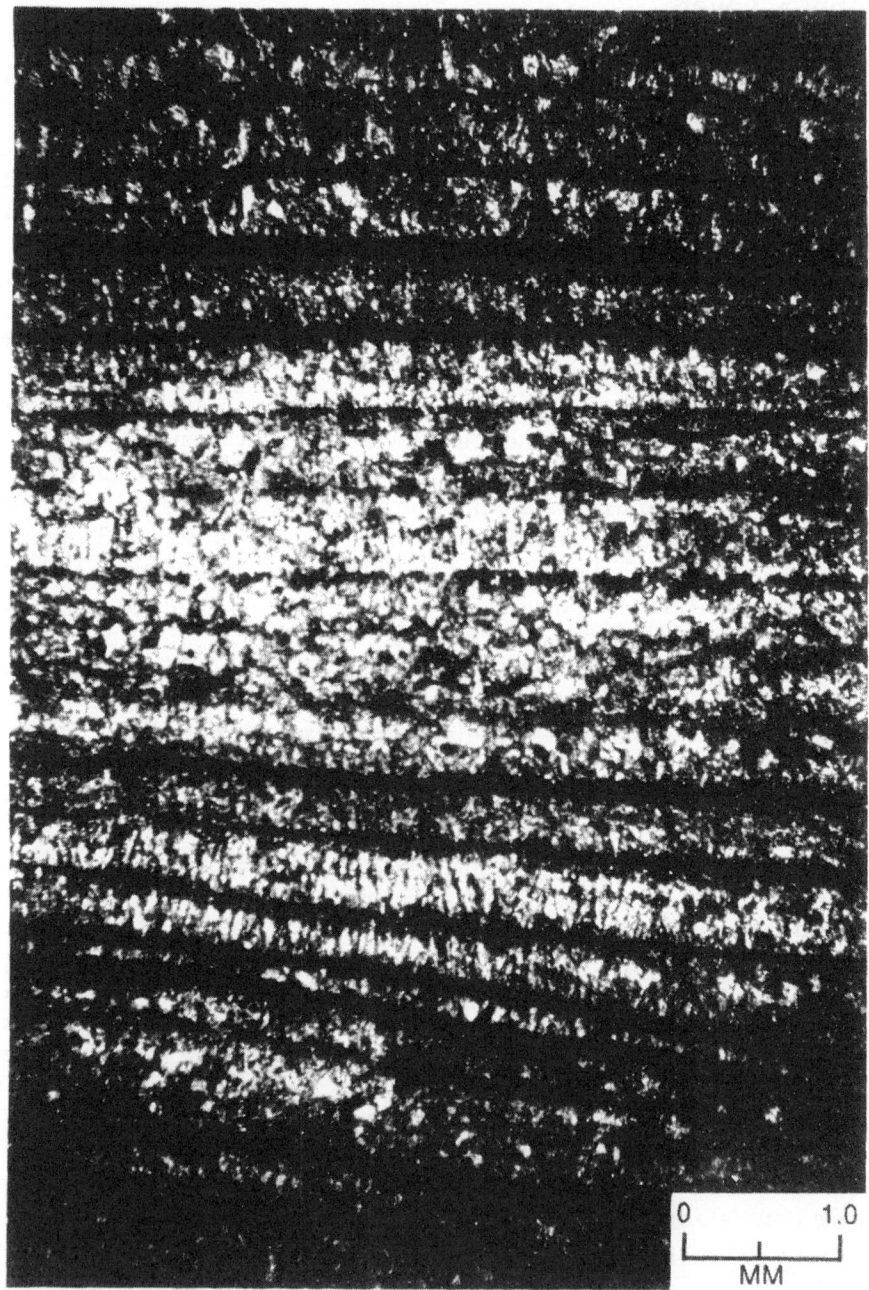

Fig. 14: Photomicrograph of annual growth layers in stalagmite D-18 from Drotsky's Cave, Kalahari Desert, Botswana. The generally thicker, lighter band in each layer is calcite and the thinner, darker band is made up of clusters of radiating aragonite needles.

precipitation of calcite takes place. Calcite precipitation under a thick fluid layer allows euhedral crystals to form at first, but thinning of the fluid to a film permits only flatly terminated calcite crystals by season's end. As fluid flow diminishes, increasing evaporation, increasing Mg/Ca ratios in the fluid, and perhaps increasing temperature, combine to cause aragonite precipitation to begin, particularly on the sides of the speleothem. In some years, fluid flow diminishes to the point that dust accumulates on aragonite surfaces before the onset of the next year's precipitation.

Importantly, comparison of the uppermost layers of the speleothem, which looked to be active when collected during the 1987 dry season, with meteorological data beginning in the 1920s, has shown a statistically significant correlation between annual layer thickness and annual precipitation. In a study of the most recent 600 layers in the deposit, covering the period AD 1374-1973, Chen (1992) isolated periodicities at 10, 20, 30, 40, 75, and 150 years (Fig. 15). The 10 and 20 year periodicities

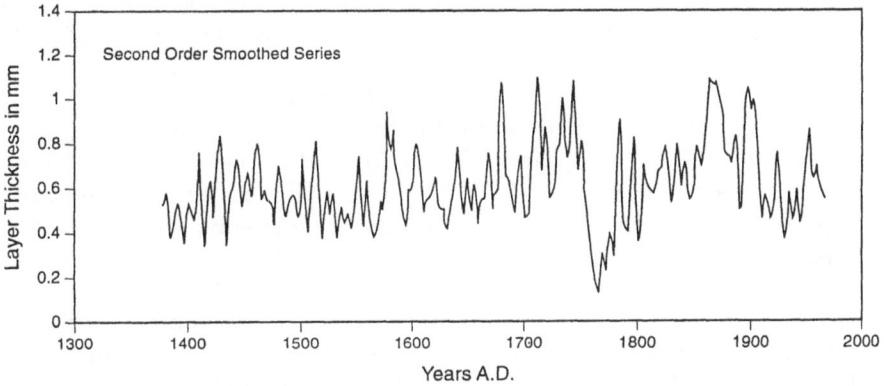

Fig. 15: Six hundred year time series of stalagmite annual layer thickness AD 1374-1973. Stalagmite D-18, Drotsky's Cave, Botswana. The series has been smoothed twice using a binomial filter to emphasize longer-term trends in the data (after Chen, 1992).

were also apparent in rainfall data for stations near the cave. Spectral models used to predict rainfall variability beyond 1973 to AD 2100 accurately predicted the drought of the 1980s and suggest that from AD 1973-2100 precipitation will only be above normal in the period AD 2010-2060 (Fig. 16). If annual layer thickness in some speleothems is a proxy measure of yearly rainfall, these formations could provide high-resolution records of precipitation for perhaps thousands or even tens of thousands of years. What is more, as Chen (1992) has shown, studies of periodicities in the layer (precipitation) record, with the formulation of spectral models, can allow speculations on likely precipitation in the future. To be able to

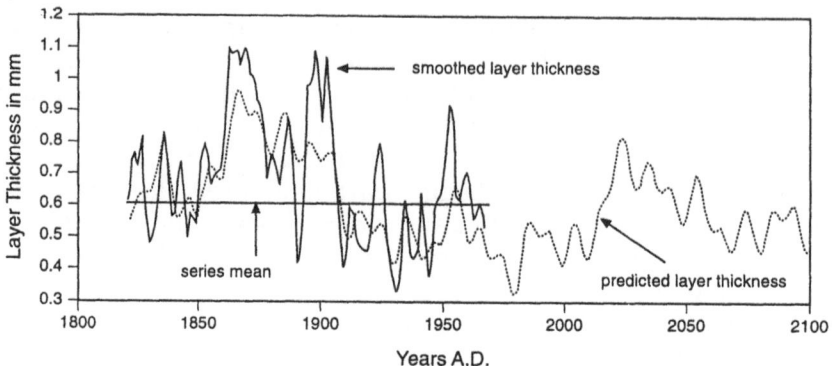

Fig. 16: Measured stalagmite annual layer thicknesses AD 1820-1973 and values for AD 1820-2100 predicted using a spectral model. Stalagmite D-18, Drotsky's Cave, Botswana (after Chen, 1992).

predict future droughts or times of plentiful rainfall in the world's semi-arid areas is an exciting possibility. Certainly, such predictions, if reliable, would be of inestimable value to long-range planners and those interested in sustainable development. In a study of a second stalagmite from Drotsky's Cave, also active when collected, Sheen (1996) found a clear relationship between layer thickness and rainfall.

Most speleothems show no visible evidence of annual layering when viewed through a petrographic microscope. However, when illuminated by ultraviolet light, many display luminescent banding parallel to growth layering. The luminescence is caused by organic substances (fulvic and humic acids) in the carbonate (Lauritzen et al., 1986). High precision TIMS U-series and AMS ^{14}C dating have shown that often this banding has a period of about one year (eg. Baker et al., 1993) but that in some cases the bands are resolvable to seasons or even days (e.g. Shopov et al., 1989, 1990, 1991, 1994). That many bands have a period of about one year indicates that there is a well-defined annual cycle of deposition in many vadose-zone speleothems. This cycle is probably a response to hydrological events in the recharge area of the cave. As Shopov et al. (1994) point out, luminescence banding is a record of fluctuations in the abundance of fulvic and humic acids occluded in the growing speleothem. Fulvic acids are produced by photosynthesis and are released through plant roots. They are readily soluble and probably enter speleothem feed waters preferentially during the growing season. Humic acids are products of organic decomposition in the soil and dissolve at a slower pace. Together, the two may indicate productivity in the overlying soil and plant cover and therefore may be a proxy measure of palaeoclimate as it affected these variables.

Shopov et al. (1990) were able to identify 35,000 layers in a cave flowstone from Bulgaria by examining variations in luminescence in the carbonate. Spectral analysis of luminescence intensity revealed climate/ solar activity periodicities at 1, 2, 11-12, 22, 95, 180, 300, 400, 600, 900, 1200, 2300 and 3350 years. As solar activity, such as sunspot frequency, and climate appear to be related (e.g. Lund, 1987), luminescence banding in speleothems may one day provide detailed climatic data if transfer functions between luminescence and climate variables can be developed. Clear relationships between luminescence in a speleothem from Coldwater Cave in Iowa, U.S.A., and atmospheric CO_2 production since AD 1000, and between luminescence in a speleothem from Jewel Cave, South Dakota, U.S.A., and the ^{18}O record of planktonic foraminifera in deep-sea core V28-238, suggest that a great deal of palaeoenvironmental information can be acquired from annual bands in cave speleothems (Shopov et al., 1994).

Little work has been conducted on luminescence banding in speleothems from currently arid or semi-arid areas although preliminary data from a Botswana stalagmite suggest that it is present at least in some formations (Brook and Railsback, unpublished data). Certainly, luminescence banding is likely to be more pronounced in areas of greater vegetative activity where fulvic and humic acids are produced in abundance. However, speleothems growing in semi-arid areas with a pronounced dry season may have the greatest potential for palaeoclimatic research. This is because speleothems are more likely to have layers that are visible petrographically if drip rates to the speleothem, and therefore the chemical characteristics of fluids on the speleothem surface, vary substantially from wet to dry season. In areas of more abundant precipitation, drip rates to speleothems and fluid chemistry may not vary enough to affect the crystal and chemical characteristics of carbonate precipitated from solution.

Railsback et al. (1994), for example, suggest that the conditions most likely to produce calcite/aragonite annual layers in speleothems are a semi-arid climate, and a highly seasonal climate with a distinct wet season. They also note that a dolomitic bedrock, or one with a high magnesium content, is more likely to induce aragonite deposition during the dry season than a pure calcium carbonate limestone. This is because higher concentrations of magnesium ions in solution increase the likelihood that Ca/Mg ratios will reach a level suitable for aragonite rather than calcite deposition. They point out that these are the conditions prevailing near Drotsky's Cave in Botswana and at Moaning Cave in California where more than 90% of the annual precipitation of 400-500 mm comes in 8 months of the year. If some speleothem annual layers are visible both petrographically and through variations in luminescence, as are those at Drotsky's Cave in Botswana, this should allow detailed studies of the intra- and inter-annual climatic factors that affect luminescence.

It is now clear that caves in the arid and semi-arid areas of the world may contain speleothems that preserve proxy, high-resolution (at least annual) palaeoclimatic data over periods of several to several tens of thousands of years. Furthermore, as Chen (1992) has shown, it may be possible to use these data to develop models that can predict annual climatic variables into the 21st century.

Conclusions

Cave sediments in deserts, particularly speleothems, can be accurately dated by a variety of methods to at least 500 k yrs B.P. and with somewhat less accuracy using the palaeomagnetic reversal time scale to many millions of years ago. Speleothems can potentially provide relative or even absolute palaeotemperature data, *via* stable oxygen and hydrogen levels in their carbonate and fluid inclusions, if carbonate deposition occurred in isotopic equilibrium. The δD_w of fluid inclusion waters may fingerprint the origin of the airmass that brought ground water to the area at the time of speleothem growth. Finally, the $\delta^{13}C$ content of the speleothem carbonate can indicate the percentage cover of C_4 plants above the cave at the time of speleothem deposition.

In caves where there is no or very limited growth of speleothems today, large, relict forms are an indication of greater availability of water in the past. Lengthy records of wet and dry periods can be developed by studying periods of growth or no-growth of speleothems in an area. In coastal desert areas, where speleothems in caves are now submerged beneath the sea, these tell of times when the sea level was substantially lower than now and even indicate the magnitude of sea level decline. Where speleothems in caves far from the sea, or well above the present sea level, are submerged by ground water they tell of drier times in the past when the water table was much lower than now. In such cases, the ages of submerged speleothems are an accurate way of determining when desert dunes in nearby drier areas are likely to have been active and when nearby rivers, lakes and ponds are likely to have dried up.

Speleothems also trap airborne and waterborne particulate matter including inorganic dust, pollen, diatoms, charcoal and bat guano. The pollen in particular can give a reliable picture of vegetation at the time of speleothem growth at least as long as the cave was open to the air so that bats and airborne pollen could enter. It appears from our studies that most pollen in speleothems derives from air rather than water and so an opening to the surface is a prerequisite. Pollen has been recovered from cave travertines more than a million years old.

Finally, it appears that some speleothems can contain an annual record of deposition. Annual layers can be visible petrographically where there is a seasonal change in crystal character or a change from calcite to aragonite deposition, for example. Sometimes, dust is introduced during the dry season, again leading to differentiation between seasonal deposits. In other cases, annual layers are only visible when the speleothem carbonate is subjected to ultraviolet radiation which causes humic and fulvic acids in the carbonate to luminesce. As flow of organic acids to a speleothem may be seasonal, bands of variable luminescence may be apparent when the speleothem is exposed to ultraviolet light. Pairs of bands may form an annual couplet. Based on my experience, annual layers that are visible petrographically are more likely to form in dry areas with a highly seasonal precipitation of around 400-500 mm. They are also more likely to form in dolomites and magnesian limestones than in pure limestones. This is because drying of the speleothem during the pronounced dry season is likely to bring about ionic concentrations and Ca/Mg ratios that are conducive to the deposition of aragonite rather than calcite. Aragonite generally forms radiating bundles of crystals that are easily distinguishable from calcite under the microscope. Importantly, annual layers in speleothems from arid areas are likely to contain information on past precipitation as layer thickness may be related to annual precipitation. Elsewhere, luminescent banding in speleothems may provide information on variations in solar radiation and general climatic conditions of temperature and precipitation. This is because these influence plant growth and therefore the production of organic acids that can be leached to the subsurface and ultimately deposited in caves. Annual layers in cave speleothems promise to provide information on cyclic variations in climatic variables over long periods of time and may ultimately provide data that allow prediction of climates during the 21st century.

Solution caves and speleothems are present in most of the world's deserts, yet remain little studied. The wealth of data that can be obtained from speleothems suggests that more work is needed on this neglected desert landform. Speleothems could yet produce excellent high-resolution records for deserts going back thousands to tens of thousands of years.

Acknowledgements

Much of the work referenced here in which the author participated was funded by the National Science Foundation, the National Oceanic and Atmospheric Administration, the National Geographic Society, the University of Georgia Research Foundation and an Africa Regional Research Fulbright Award. Work in Somalia was conducted in collaboration with the Somali Academy of Sciences and Arts, that in

Botswana with the Botswana National Museum, in Namibia with the National Museum of Namibia and in Congo in cooperation with the Institut Zairois pour la Conservation de la Nature. Work in the Sultanate of Oman was undertaken with the support of the Ministry of Water Resources and was greatly facilitated by the assistance of John Kay, Training Advisor. I would also like to thank the following for their assistance with the various research projects: Steven Brandt (Somalia); Eugene Marais (Namibia); Alec Campbell, John Cooke, Paul Shaw and Larry Robbins (Botswana); Noel Boaz (Zaire); and Jim Cowart and Brooks Ellwood (U.S.A.).

References

Baker, A., Smart, P.L., Edwards, R.L. and Richards, D.A. (1993). Annual growth banding in a cave stalagmite. *Nature*, 364: 518-520.

Bard, E., Hamelin, B., Fairbanks, R.G. and Zindler, A. (1990). Calibration of the ^{14}C timescale over the past 30,000 years using mass spectrometric U-Th ages from Barbados corals. *Nature*, 345: 405-410.

Baskaran, M. and Krishnamurthy, R.V. (1993). Speleothems as proxy for the carbon isotope composition of atmospheric CO_2. *Geoph. Research Letters*, 20: 2905-2908.

Bastin, B. (1978). L'analyse pollinique des stalagmites: Une nouvelle possibilité d'approche des fluctuations climatiques du Quaternaire. *Ann. Soc. Géol. Belg.*, 101: 13-19.

Bastin, B. (1982). Premier bilan de l'analyse pollinique de stalagmites holocénes en provenance de grottes Belges. *Rev. Belg. Geogr.*, 106: 87-97.

Bastin, B., Dupuis, C. and Quinif, Y. (1977). Preliminary results of the application of Quaternary geological methods to speleogenetic studies of a Belgian cave. *Proc. of the 7th International Congress of Speleology*, pp. 24-28, Sheffield, U.K.

Bastin, B., Dupuis, C. and Quinif, Y. (1982). Étude microstratigraphique et palynologique d'une croûte stalagmitique de la Grotte de la Vilaine Source (Arbre, Belgique): Méthodologie et résultats. *Rev. Belg. Géogr. Phys. Quat.*, 106: 109-120.

Bastin, B. and Schneider, A.M. (1984). Palynologie. In: *Le Karst Belge. Köln. Geogr. Arb.*, 45: 87-93.

Bastin, B. and Gewelt, M. (1986). Analyse pollinique et datation ^{14}C de concrétions stalagmitiques holocénes: Apports complémentaires des deux méthodes. *Géographie Physique et Quaternaire*, 40: 185-196.

Bluszcz, A., Goslar, T., Hercman, H., Pazdur, M.F. and Walanus, A. (1988). Comparison of TL, ESR and ^{14}C dates of speleothems. *Quat. Sci. Revi.* 7: 417-421.

Bowen, R. (1991). *Isotopes and Climate*. Elsevier, London.

Broecker. W.S., Olson, E.A. and Orr, P.C. (1960). Radiocarbon measurements and annual rings in cave formations. *Nature*, 185: 93-94.

Brook, G.A., Keferl, E.P. and Nickmann, R.J. (1987). Paleoenvironmental data for N.W. Georgia, U.S.A., from fossils in cave speleothems. *Inter. J. Speleology*, 16: 69-78.

Brook, G.A., Burney, D.A. and Cowart, J.B. (1990a). Paleoenvironmental data for Ituri, Zaire, from sediments in Matupi Cave, Mt. Hoyo. *Virginia Museum of Natural History, Memoir* 1: 49-70.

Brook, G.A., Burney, D.A. and Cowart, J.B. (1990b). Desert paleoenvironmental data from cave speleothems with examples from the Chihuahuan, Somali-Chalbi, and Kalahari deserts. *Palaeogeography, Palaeoclimatology, Palaeoecology*, 76: 311-329.

Brook, G.A., Railsback, L.B., Cooke, H.J., Chen, J. and Culp, R.A. (1992). Annual growth layers in a stalagmite from Drotsky's Cave, Ngamiland: Relationships between growth layer thickness and precipitation. *Botswana Notes and Records,* 24: 151-163.

Brook, G.A. and Nickmann, R.J. (1996). Evidence of late Quaternary environments in north-west Georgia from sediments preserved in Red Spider Cave. *Physical Geography,* 17(5): 465-484.

Brook, G.A., Cowart, J.B. and Marais, E. (1996). Wet and dry periods in the southern African summer rainfall zone during the last 300 k yrs from speleothem, tufa and sand dune age data. *Palaeoecology of Africa,* 24: 147-158.

Brook G.A., Cowart, J.B., Brandt, S.A. and Scott, L. (1997). Quaternary climatic change in southern and eastern Africa during the last 300 ka: The evidence from caves in Somalia and the Transvaal region of South Africa. *Zeitschrift für Geomorphologie* N.F. Suppl.-Bd. 10: 15-48.

Bunnell, D. (1996). Diving the lakes of Lechuguilla. *NSS News,* 58(8): 206-210.

Burney, D.A. and Burney, L.P. (1993). Modern pollen deposition in cave sites: Experimental results from New York State. *New Phytologist,* 124: 523-535.

Burney, D.A., Brook, G.A. and Cowart, J.B. (1994). A Holocene pollen record for the Kalahari Desert of Botswana from a U-series dated speleothem. *The Holocene,* 4: 225-232.

Cazzoli, M.A., Forti, P. and Bettazzi, L. (1988). L'accrescimento di alabastri calcarei in grotte gessose: Nuovi dati dall'inghiottitoio dell'Acquafredda (3/ER.BO). *Estratto da Sottoterra,* 8: 16-23.

Cerling, T.E. (1984). The stable isotopic composition of modern soil carbonate and its relationship to climate. *Earth and Planetary Science Letters,* 71: 229-240.

Cerling, T.E. and Hay, R.L. (1986). An isotopic study of paleosol carbonates from Olduvai Gorge. *Quater. Res.* 25: 63-78.

Chen, J. (1992). Climate variations in Botswana over the last 600 years: Evidence from annual growth layers in a stalagmite from Drotsky's Cave, Ngamiland, Botswana. M.A. Thesis, Department of Geography, University of Georgia, Athens, Georgia, 125 pp.

Cooke, H.J. and Verhagen, B.T. (1977). The dating of cave development—an example from Botswana. *Proc. the 7th Inter. Congress of Speleology,* pp. 122-124, Sheffield, U.K.

Craig, H. (1961). Standard for reporting concentrations of deuterium and oxygen-18 in natural waters. *Science,* 133: 1833-1834.

Craig, H. (1965). The measurement of oxygen isotope palaeotemperatures. In: *Stable Isotopes in Oceanographic Studies and Palaeotemperatures,* E. Trongiorgi (ed.) pp. 161-182. Spoleto, CNR, Lab. Geol. Nucl., Pisa.

Edwards, R.L. and Gallup, C.D. (1993). Dating of the Devils Hole calcite vein. *Science,* 259: 1626-1627.

Emrich, K., Ehhalt, D.H. and Vogel, J.C. (1970). Carbon isotope fractionation during the precipitation of calcium carbonate. *Earth and Planetary Science Letters,* 8: 363-371.

Faegri, K. and Iverson, J. (1975). *Textbook of Pollen Analysis.* Hafner, New York.

Franke, H.W. and Geyh, M.A. (1971). Radiokohlenstoff-Analysen an Tropfsteinen. *Umschau,* 71: 91-92.

Friedman, I. and O'Neil, J.R. (1977). Compilation of stable isotope fractionation factors of geochemical interest. *In:* M. Fleischer (ed.), *Data of Geochemistry, Sixth Edition. U.S. Geological Survey Professional Paper* 440-KK.

Fritz, P. and Clark, I. (1986). *Origin and Age of Groundwater in Oman: A Study of Environmental Isotopes.* Report PAWR 86-7, Cansult Limited in association with Gartner Lee International Inc. for The Sultanate of Oman, Public Authority for Water Resources.

Fritz, P., Drimmie, R.J., Frape, S.K. and O'Shea, K. (1987). The isotopic composition of precipitation and groundwater in Canada. *International Symposium on the Use of Isotope Techniques in Water Resources Development,* IAEA-SM-299, pp. 539-550.

Frumkin, A., Schwarcz, H.P. and Ford, D.C. (1994). Evidence for isotopic equilibrium in stalagmites from caves in a dry region: Jerusalem, Israel. *Israel J. Earth Sci.* 43: 221-230.

Gascoyne, M. (1984). Twenty years of uranium-series dating of cave calcites: A review of results, problems and new directions. *Studies in Speleology*, 5: 15-30.

Gascoyne, M., (1992). Palaeoclimatic determination from cave calcite deposits. *Quater. Sci. Rev.* 11: 609-632.

Gascoyne, M., Benjamin, G.J., Schwarcz, H.P. and Ford, D.C. (1979). Sea-level lowering during the Illinoian Glaciation: Evidence from a Bahama "blue hole". *Science*, 205: 806-808.

Gascoyne, M. Ford, D.C. and Schwarcz, H.P. (1981). Late Pleistocene chronology and paleoclimate of Vancouver Island determined from cave deposits. *Canadian J. of Earth Sci.* 18: 1643-1652.

Gascoyne, M., Ford, D.C. and Schwarcz, H.P. (1983a). Rates of cave and landform development in the Yorkshire Dales from speleothem age data. *Earth Surface Processes and Landforms*, 8: 557-568.

Gascoyne, M., Schwarcz, H.P. and Ford, D.C. (1983b). Uranium-series ages of speleothem from north-west England: Correlation with Quaternary climate. *Philosophical Transactions of the Royal Society of London*, B301: 143-164.

Grün, R. (1989). Electron spin resonance (ESR) dating. *Quaternary International*, 1: 65-109.

Harmon, R.S., Ford, D.C. and Schwarcz, H.P. (1977). Interglacial chronology of the Rocky and Mackenzie Mountains based on 230Th/234U dating of calcite speleothems. *Canadian J. Earth Sci.* 14: 2543-2552.

Harmon, R.S., Schwarcz, H.P. and Ford, D.C. (1978). Late Pleistocene sea level history of Bermuda. *Quater. Res.*, 9: 205-218.

Harmon, R.S. and Schwarcz, H.P. (1981). Changes of ^2H and ^{18}O enrichment of meteoric water and Pleistocene glaciation. *Nature*, 290: 125-128.

Harmon, R.S., Land, L., Mitterer, R.M., Garrett, P., Schwarcz, H.P. and Larson, G.J. (1981). Bermuda sea level during the last interglacial. *Nature*, 289: 481-483.

Harmon, R.S., Mitterer, R.M., Kriausakul, N., Land, L., Schwarcz, H.P., Garrett, P., Larson, G.J., Vacher, H.L. and Rowe, M. (1983). U-series and amino-acid racemization geochronology of Bermuda: Implications for eustatic sea-level fluctuation over the past 250,000 years. *Palaeogeography, Palaeoclimatology, Palaeoecology*, 44: 41-70.

Hendy, C.H. (1971). The isotopic geochemistry of speleothems I: The calculation of the effects of different modes of formation on the isotopic composition of speleothems and their applicability as paleoclimatic indicators. *Geochimica et Cosmochimica Acta*, 35: 801-824.

Hennig, G.J., Bangert, U., Herr, W. and Freundlich, J. (1980). Uranium series dating of calcite formations in caves: Recent results and a comparative study on age determinations via $^{230}Th/^{234}U$, ^{14}C, TL and ESR. *Revue Archaeometry*, 4: 91-100.

Hennig, G.J., Grün, R. and Brunnacker, K. (1983). Speleothems, travertines, and palaeoclimates. *Quater. Res.* 20: 1-29.

Hoefs, J. (1973). *Stable Isotope Geochemistry*. Springer-Verlag, Berlin.

Holmgren, K., Karlén, W. and Shaw, P.A. (1995). Paleoclimatic significance of the stable isotopic composition and petrology of a Late Pleistocene stalagmite from Botswana. *Quater. Res.*, 43: 320-328.

Holmgren, K., Lauritzen, S.-E. and Possnert, G. (1994). $^{230}Th/^{234}U$ and ^{14}C dating of a late Pleistocene stalagmite in Lobatse II Cave, Botswana. *Quaternary Geochronology (Quaternary Science Reviews)*, 13: 111-119.

Ivanovic, M. and Harmon, R.S. (eds) (1982). *Uranium Series Disequilibrium: Applications to Environmental Problems*. Clarendon Press, Oxford, 571 pp.

Jones, D.L., Brock, A. and McFadden, P.L. (1986). Palaeomagnetic results from the Kromdraai and Sterkfontein hominid sites. *South African J. Sci.* 82: 160-163.

Kai, A., Miki, T. and Ikeya, M. (1988). ESR dating of teeth, bones and eggshells excavated at a palaeolithic site of Douara Cave, Syria. *Quater. Sci. Rev.*, 7: 503-507.

Kashiwaya, K., Atkinson, T.C. and Smart, P.L. (1991). Periodic variations in Late Pleistocene speleothem abundance in Britain. *Quater. Res.* 35: 190-196.

Latham, A.G., Schwarcz, H.P., Ford, D.C. and Pearce, G.W. (1979). Palaeomagnetism of stalagmite deposits. *Nature* 280: 383-385.

Latham, A.G., Schwarcz, H.P., Ford, D.C. and Pearce, G.W. (1982). The palaeomagnetism and U-Th dating of three Canadian speleothems: Evidence for the westward drift, 5.4-2.1 k yrs B.P. *Canadian J. Earth Sciences*, 19: 1985-1995.

Latham, A.G., Schwarcz, H.P. and Ford D.C. (1986). The paleomagnetism and U-Th dating of Mexican stalagmite, DAS2. *Earth and Planetary Science Letters*, 79: 195-207.

Lauritzen, S.-E., Ford, D.C. and Schwarcz, H.P. (1986). Humic substances in speleothem matrix-paleoclimate significance. *Proceedings of the 9th International Congress of Spelaeology* 2: 77-79, Barcelona, August 1986.

Lauritzen, S.-E., Haugen, J.E., Løvlie, R. and Gilje-Nielsen, H. (1994). Geochronological potential of isoleucine epimerization in calcite speleothems. *Quater. Res.* 41: 52-58.

Li, W.X., Lundberg, J., Dickin, A.P., Ford, D.C., Schwarcz, H.P., McNutt, R. and Williams, D. (1989). High-precision mass spectrometric uranium-series dating of cave deposits and implications for paleoclimitic studies. *Nature*, 339: 534-536.

Lund, B.G.A. (1987). Solar cycles and floods. *South African J. Sci.* 83: 669-670.

O'Neil, J.R., Clayton, R.N. and Mayeda, T. (1969). Oxygen isotope fractionation in divalent metal carbonates. *J. of Chemical Physics*, 51: 5547-5558.

Railsback, L.B., Brook, G.A., Chen, J., Kalin, R. and Fleisher, C.J. (1994). Environmental controls on the petrology of a Late Holocene speleothem from Botswana with annual layers of aragonite and calcite. *J. Sedimentary Res.* A64: 147-155.

Richards, D.A., Smart, P.L. and Edwards, R.L. (1994). Maximum sea levels for the last glacial period from U-series ages of submerged speleothems. *Nature*, 367: 357-360.

Rink, W.J., Schwarcz, H.P., Grün, R., Yalcinkaya, I., Taskiran, H., Otte, M., Valladas, H., Mercier, N., Bar-Yosef, O. and Kozlowski, J. (1994). ESR dating of the last interglacial Mousterian at Karain Cave, southern Turkey. *J. Archaeological Sci.* 21: 839-849.

Salomons, W. and Mook, W.G. (1986). Isotope geochemistry of carbonates in the weathering zone. In: *Handbook of Environmental Geochemistry* P. Fritz and J–C. Fontes (eds.) 2, 239-269, Elsevier, Amsterdam.

Schmidt, V.A. (1982). Magnetostratigraphy of sediments in Mammoth Cave, Kentucky. *Science*, 217: 827-829.

Schmidt, V.A., Jennings, J.N. and Haosheng, B. (1984). Dating of cave sediments at Wee Jasper, New South Wales, by magnetostratigraphy. *Australian J. Earth Sci.* 31: 361-370.

Schwarcz, H.P. (1986). Geochronology and isotopic geochemistry of speleothems. In *Handbook of Environmental Isotope Geochemistry: The Terrestrial Environment*, J–C. Fontes, P. Fritz (eds.) 8, 271-303, Elsevier.

Schwarcz, H.P., Harmon, R.S., Thompson, P. and Ford, D.C. (1976). Stable isotope studies of fluid inclusions in speleothems and their paleoclimatic significance. *Geochimica et Cosmochimica Acta*, 40: 657-665.

Schwarcz, H.P. and Yonge, C.J. (1983). Isotopic composition of paleowaters as inferred from speleothem and its fluid inclusions. In: *Paleoclimates and Palaeowaters: A Collection of Environmental Isotope Studies*, R. Gonfiantini (ed.), pp. 115-133, IAEA.

Scott, L. (1984). Palynological evidence for Quaternary palaeoenvironments in southern Africa. In: *Southern African Prehistory and Paleoenvironments* R.G. Klein (ed.), pp. 65-80, Balkema, Rotterdam.

Sheen, S.-W. (1996). *Cyclic variations in southern African rainfall and water surplus, the Southern Oscillation, and annual layers in a Botswana speleothem.* M.A. Thesis, Department of Geography, University of Georgia, Athens, 219 pp.

Shopov, Y.Y., Dermendjiev, V. and Buyukliev, G. (1989). Investigation on the old variations of the climate and solar activity by a new method — LLMZA of cave flowstone from Bulgaria. *Proc. of the 10th International Congress of Speleology*, Budapest, August 1989, pp. 95-97.

Shopov, Y.Y., Dermendjiev, V. and Buyukliev, G. (1990). Methods for research of the solar activity in the past and flowstone luminescent records of the solar activity. *Studia carsologica*, 2: 139-149.

Shopov, Y.Y., Dermendjiev, V. and Buyukliev, G. (1991). A new method for dating natural materials with periodical macrostructure by autocalibration and 1st application for study of the solar activity in the past. *Proceedings of the International Conference on Environmental Changes in Karst Areas - I.G.U. - U.I.S.* - Italy, September 1991. Quaderni del Dipartimento di Geografia, Università di Padovà, 13: 17-22.

Shopov, Y.Y., Ford, D.C. and Schwarcz, H.P. (1994). Luminescent microbanding in speleothems: High-resolution chronology and paleoclimate. *Geology*, 22: 407-410.

Spalding, R.F. and Mathews, T.D. (1972). Submerged stalagmites from caves in the Bahamas: Indicators of low sea stand. *Quaternary Research* 2: 470-472.

Talma, A.S., Vogel, J.C. and Partridge, T.C. (1974). Isotopic contents of some Transvaal speleothems and their palaeoclimatic significance. *South African J. Sci.*, 70: 135-140.

Talma, A.S. and Netterberg, F. (1983). Stable isotope abundances in calcrete. In: *Residual Deposits*, R.C.L. Wilson (ed.) Geological Society of London Special Publication 11: 221-233. Blackwell, Oxford.

Talma, A.S. and Vogel, J.C. (1992). Late Quaternary paleotemperatures derived from a speleothem from Cango Caves, Cape Province, South Africa. *Quater. Res.* 37: 203-213.

Van der Merwe, N.J. and Vogel, J.C. (1984). Recent carbon isotope research and its implications for African archaeology. *African Archaeological Review*, 1: 33-56.

Van Noten, F. (1977). Excavations at Matupi Cave, *Antiquity*, 51: 35-40.

Van Noten, F. (1982). The Stone Age in the north and east. In: *The Archaeology of Central Africa*, F. Van Noten (ed.) Akademische Druck-und Verlagsanstalt, Graz.

Vogel, J.C. (1983). ^{14}C variations during the Upper Pleistocene. *Radiocarbon*, 25: 213-218.

Winograd, I.J., Szabo, B.J., Coplen, T.B. and Riggs, A.C. (1988). A 250,000-year climatic record from Great Basin vein calcite: Implications for Milankovitch theory. *Science*, 242: 1275-1280.

Winograd, I.J., Coplen, T.B., Landwehr, J.M., Riggs, A.C., Ludwig, K.R., Szabo, B.J., Kolesar, P.T., and Revesz, K.M. (1992). Continuous 500,000-year climate record from vein calcite in Devils Hole, Nevada, *Science*, 258: 255-260.

Present Flora as an Indicator of Palaeoclimate: Examples from the Arabian Peninsula

9

Shahina A. Ghazanfar[*]

ABSTRACT

Distribution patterns of the flora, both endemic and relict taxa in the Arabian Peninsula, can serve as indicators of past environments. In particular, the floras of Southwest Arabia (Yemen and Dhofar) and northern Oman, which consist of African and Iranian elements respectively, can be used to reconstruct past vegetation types and serve as proxy indicators of past climatic regimes. Similarities between the present floras of the northern mountains of Oman and that of Baluchistan, indicate that plant species migrated across the Arabian Gulf from their centres of origin in Asia and established themselves there possibly during pluvial times. The restricted distribution of these species to higher altitudes, where the climate is still equable, suggests that they are climatic relicts in an arid land. The presence of species of African origin in both northern Oman and southwest Arabia suggest that a period (or periods) of relatively wetter climate than known today occurred, when these species were perhaps more widespread.

Introduction

The age and degree of isolation of neighbouring regions is often reflected in the similarities and dissimilarities of their flora and the distribution patterns of related taxa: relict and disjunctive species often reveal past floristic ties. In the Arabian Peninsula, the proximity of southern Arabia to Africa and the presence of African floral elements in the floras of sourth-western Arabia indicate that floral movements have occurred from Africa to Arabia. Similarly, the proximity of northern Oman to Iran and SW Pakistan (Baluchistan) and a strong representation of the Iranian elements in the Oman flora indicate that plants migrated from SW Iran and Pakistan across the Arabian Gulf (Kürschner, 1986; Kürschner, 1998). It has been suggested that interchange of species between Africa and Asia through the Middle East probably took place during the Miocene when there were land bridges between the two areas (Braithwaite, 1987).

[*]Department of Biology, Sultan Qaboos University, Muscat, Sultanate of Oman.

During the Late Miocene, Arabia was separated from Africa by the rifting of the Red Sea and the land bridge between the two regions ceased to exist. However, similarities between species and species communities in NE Africa and SW Arabia suggest that migrations also took place in relatively recent times, possibly during the Pleistocene and Holocene (Delany, 1989). There is evidence that land connections between Africa and Arabia and between southern Iran and northern Oman could have existed over the last 70,000 years. It is suggested that, for this period, world sea-levels were at least 30 m below the present, falling to -80 m and -125 m around 34,000 years BP and 53,000 years BP (Chappell and Thom, 1977) and that land connections might have been present during the last glaciation (20,000 -12,000 years BP) (Thunell et al., 1988) across the Arabian Gulf and the southern part of the Red Sea where seas are shallow.

During the Late Pliocene to Early Pleistocene a relatively humid phase prevailed over the Arabian Peninsula. Alluvial deposits were laid down, the vegetation consisted of savanna and forest, and lakes were present (Anton, 1984). During the Mid-Pleistocene, an arid phase started and the vegetation changed to a relatively xerophytic type. In the Late Pleistocene (35,000 to 17,000 years BP) there was another humid phase and some changes in the vegetation occurred (McClure, 1978; Sanlaville, 1992). From the Late Pleistocene to the Early Holocene there was yet another arid phase which lasted from about 17,000 to 11,000 years BP, followed by a humid phase from 11,000 to 6000 years BP. During this last humid phase there is evidence that lakes occurred and that the vegetation was more extensive than at the present time, although the species composition was more or less similar (Schulz and Whitney, 1986; El-Moslimany, 1983). Evidence from palaeontological studies in the Sahara suggest that in most localities the savanna type of vegetation was progressively replaced by xeric Saharan plant communities between 6000 and 4000 years BP (Schulz, 1987). Sanlaville (1992) suggests that, in Arabia the last humid phase lasted until about 4000 years BP and, since that time, an arid phase has prevailed over the Arabian Peninsula and neighbouring regions.

This paper examines the similarity between the present flora of the Arabian Peninsula, in particular that of SW Arabia and northern Oman, and that of NE Africa and SW Asia, as well as the presence and distribution of endemic and relict flora of Oman, in order to corroborate our knowledge of the palaeoclimate of the Arabian Peninsula.

Methods and Sources of Data

One of the most frequently employed techniques in palaeoenvironmental research is the reconstruction of past vegetation types. Analyses of palaeobotanical information, together with geomorphological and archaeological data, can be used for studying regional trends in environmental evolution that occur primarily as a response to changes in climatic factors.

Amongst the available plant-based techniques, the analysis of recent and fossil pollen and spore spectra is used extensively. In regions with a temperate climate, palynology is a powerful tool and is a major contributor to our knowledge of past environments, especially for the Quaternary. In arid zones, however, the combination of poor pollen production as a result of sparse vegetation, poor pollen preservation, relatively high rates of sediment accumulation, strong erosion, and the long distance transportation of pollen by winds, rivers and sea currents have made it difficult to use traditional palynological techniques and interpretation (Horowitz, 1992, and references therein). Nevertheless, research methods and palynological techniques are being revised for arid lands, with encouraging results (Horowitz, 1992).

Other methods used for the reconstruction of vegetation and past environments are based on floral relationships and distribution patterns of extant taxa. Floral relationships such as similarities of taxa are used to estimate the degree and length of isolation among locations (Raup and Crick, 1979; McCoy and Heck, 1987). Phytogeographic relationships and patterns of geographical distribution of related taxa, especially disjunctive distribution patterns are also used to study palaeogeographical developments within a region (Estrada-Loera, 1991; Kadmon and Pulliam, 1993; Lioubimtseva, 1995). Studies of speciation of endemic taxa and areas of high endemism have also been used to study regional biogeographic relationships.

Various indices have been used to compare floral and faunal lists in order to evaluate their similarities and differences. Perhaps one of the most widely used is the Simpson's Similarity Coefficient, in which $S = 100k/B$, where S = Simpson's Similarity Coefficient, k = number of taxa common to two samples A and B, and B = the total taxa found in the smaller sample. The coefficient varies from 0 to 100 and gives the percentage similarity. Jaccard's Similarity Coefficient in which $S_j = a/(a + b + c)$, where S_j = Jaccard's Similarity Coefficient, a = number of species in sample A and sample B (joint occurrences), b = number of species in sample B but not in sample A, and c = number of species in sample A but not in sample B (for details on measurements of similarity see Krebs, 1989). Several authors have criticized the accuracy of these and other such indices and have proposed alternative indices of similarities (Raup and Crick, 1979 and references therein).

Percentage similarity of genera and species between two regions can also be used to estimate similarities between regions (Ghazanfar, in press.).

This can be estimated by calculating the percentage of the total number of genera in each region and the number of genera common to pairs of regions. The percentage similarity between two regions can be calculated by dividing the total number of genera in each region by the number of genera common to both and standardising it as a percentage (Table 2).

Table 1: Regional floras for the Arabian Peninsula, NE Africa, Pakistan and SW Iran. Generic and species descriptions of floras are arranged under families.

Region/Country	Reference
Bahrain	Cornes and Cornes, 1989
Kuwait	Daoud and Al-Rawi, vol. 1, 1985; Al-Rawi, vol. 2, 1987
Oman	Ghazanfar, 1992a; Miller and Morris, 1988 (Dhofar); Mandaville, 1985 (northern Oman)
Qatar	Batanouny, 1981
Saudi Arabia	Migahid, 1978; Collenette, 1988; Mandaville, 1990 (Eastern Saudi Arabia)
United Arab Emirates	Western, 1989; Karim, 1992a, b; 1993, 1994
Yemen	Gabali and Al-Gifri, 1990; Boulos, 1985; Baltter, 1914
NE Africa	Thulin (ed.), 1993 (Somalia); El-Amin, 1990 (Sudan, trees and shrubs); Andrews, 1950 (Sudan)
Pakistan	Nasir and Ali (eds.), 1974-1994; Burkill, 1909 (Baluchistan)
SW Iran	Leonard, 1981-1989

Table 2: Total number of genera (diagonal) and common genera (in parentheses) with percentage similarity between regions. Percentage similarity between two regions is calculated by dividing the total number of genera in each region by the number of genera common to both. (Ghazanfar, in press.).

	S. Yemen	Dhofar	C. Oman	N. Oman	UAE	Baluchistan	Deserts of Iran
S Yemen	254	(214) 48%	(77) 30%	(173) 68%	(121) 48%	(161) 63%	(77) 30%
Dhofar	(214) 74%	449	(78) 17%	(234) 52%	(159) 35%	(181) 40%	(118) 26%
C Oman	(77) 72%	(78) 73%	107	(80) 74%	(81) 76%	(64) 59%	(58) 54%
N Oman	(173) 44%	(234) 61%	(80) 21%	386	(173) 45%	(155) 40%	(152) 39%
UAE	(121) 57%	(159) 75%	(81) 38%	(173) 82%	212	(137) 65%	(119) 56%
Baluchistan	(161) 43%	(209) 56%	(64) 17%	(155) 42%	(137) 37%	368	(181) 49%
Deserts of Iran	(77) 27%	(118) 42%	(58) 21%	(152) 55%	(119) 52%	(181) 65%	278

Multivariate analyses of distributional (presence/absence) data can also be used effectively to study patterns of relationship between regions and ordination techniques can also be applied to measure the relationships between regions (Fig. 2). It must always be borne in mind that, as with other measures used in the reconstruction of palaeoenvironments, present-day floral indicators can only be used successfully when corroborated by other proxy evidence.

Regional floras and checklists are good sources of generic and species lists for the investigation of phytogeographic relationships. Several checklists and regional floras are available for Arabia, NE and E Africa and SW Asia. Table 1 provides a list of the relevant references for the flora and vegetation of the Arabian Peninsula, NE Africa, Pakistan and SW Iran (Ghazanfar and Fisher, 1998 and references there in).

Flora of the Arabian Peninsula and its Relationships

The floras of SW Arabia (SE Yemen and Dhofar) and N Oman are composed of a number of phytogeographical elements which reflect

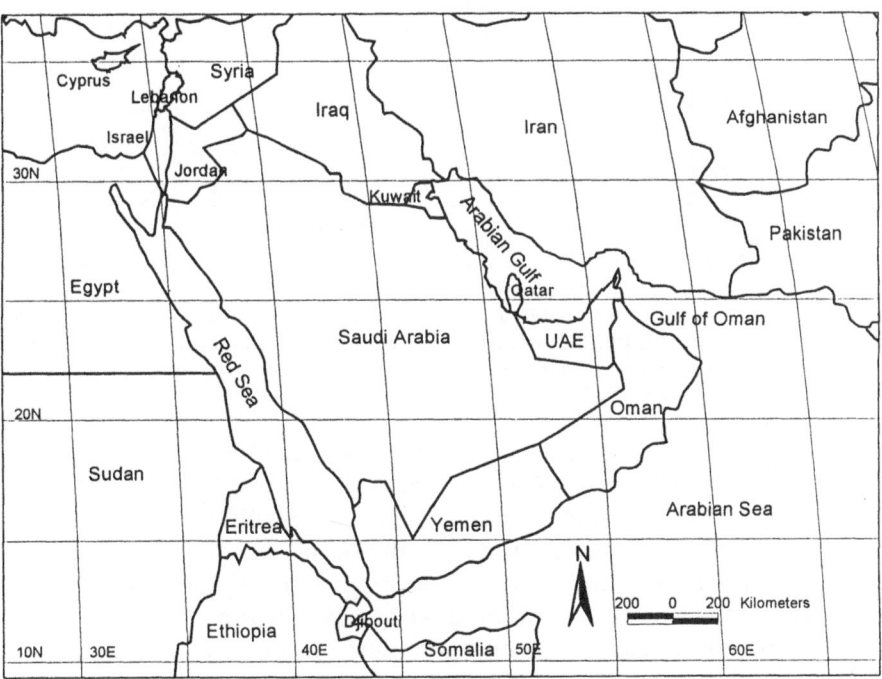

Fig. 1: The Arabian Peninsula and its neighbours.

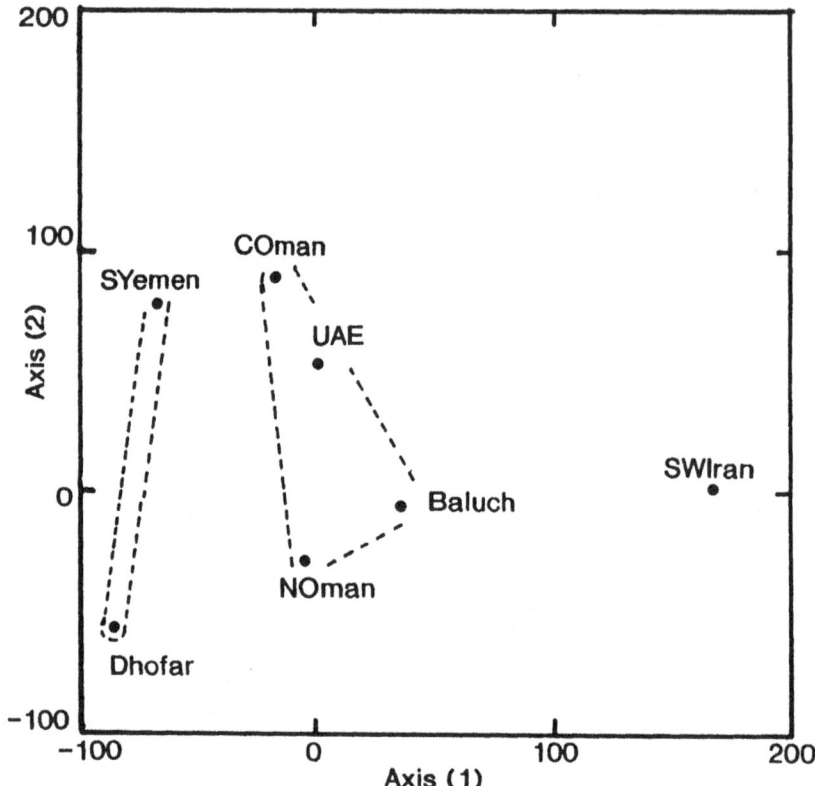

Fig. 2: Multivariate analysis of 703 plant genera plotted for seven sites (central Oman, N Oman, Dhofar, S Yemen, UAE, Baluchistan and SW Iran). Axis 1 of the ordinate extracts the general geographical locations of the regions from west to east and also separates the three groups delimited by classification. Axis 2 separates the regions from south to north from generally montane areas to plains (Ghazanfar, in press).

different affinities and migration routes from NE Africa in the south and SW Iran and Pakistan in the north (Lavranos, 1975; Mandaville, 1984; Kürschner, 1986; 1998, Ghazanfar, 1991a, b; 1992a, b) (Fig. 1).

The SW region of Arabia with the escarpment mountains of SW Saudi Arabia, Yemen and Dhofar, forms an integral part of Africa, both in geological and biogeographical terms. The majority of the species occurring below altitudes of 1500 m in southern Arabia also occur on the African side of the Red Sea and the Gulf of Aden in NE and E Africa (White and Leonard, 1991). These include several species of *Acacia* such as *A. asak, A. edgeworthii, A. etbaica, A. hamulosa* and *A. oerfota, Boswellia sacra, Cadaba baccarinii, Caesalpinia eriantha,* species of *Commiphora* such as *C. foliacea, C. gileadensis, C. habessinica, C. myrrha* and *C. schimperi,* species of *Lannea,*

Maytenus, Premna, Rhigozum, Turraea and *Livistonia*. Several Afromontane genera such as *Anagyris, Ceratonia, Juniperus, Iris, Myrsine, Myrtus, Pavetta, Rhamnus* and *Teclea* reach the SW region of Saudi Arabia, Yemen and S Oman. It has been suggested that the taxa present in the Ethiopian-Somalian region and SW Arabia migrated from Africa and became established at a time when the climate was relatively more moist than today and the desert areas between N. and S. Arabia were narrower (Lavranos, 1975, Kürschner, 1998).

The phytogeographic relations of Oman are interesting since the country comes under the influence of Asian elements in the north and African elements in the south. The flora can be classified into four major biogeographic zones which extend from NE Africa and SW Asia into Oman. The phytochorology as outlined by White (1983) for the vegetation of Africa, can be extended into Oman. These zones are: Somali-Masai zone which extends from NE tropical Africa into Dhofar; Saharan and Arabian regional subzones (of the Saharo-Sindian regional zone) which extend from N. Africa into Central Oman; and the Nubo-Sindian zone which extends from Afghanistan, Baluchistan and SW Iran into N. Oman (Fig. 3; for details see Ghazanfar, 1992b). The composition of plant communities and the morphology and distribution of the endemic and regionally endemic species indicates close floristic relations between the floras of these regions. Qualitative analysis was used to assess the similarity of the flora (at the generic level) of three regions of Oman: northern Oman, central Oman and Dhofar with each other and with those of Baluchistan (SW Pakistan), United Arab Emirates, Republic of Yemen (South) and the desert regions of S. Iran (Ghazanfar, in press).

On comparing Oman with its neighbours, the highest number of genera were recorded in Dhofar, (southern region of Oman), followed by N Oman, Baluchistan, S. Yemen, the deserts of S. Iran, the United Arab Emirates and central Oman. Generic similarity is highest between the floras of Dhofar and S. Yemen and between the floras of N. Oman, UAE and Baluchistan, and lowest between the floras of S. Yemen and the deserts of Iran. Over 50% similarity of genera exists between the desert floras of Iran and N. Oman, but only 27% between the deserts of Iran and S. Yemen.

Central Oman shares more than 70% of its total genera with N. Oman, UAE, Dhofar and S. Yemen; reciprocally, N. Oman shares only 21%, UAE 38%, Dhofar 17% and S. Yemen 30% of its total genera with central Oman. The reason for this disparity is the impoverished flora of the hyper arid region of central Oman: the species which occur there are the common hardy species which are widespread throughout the arid regions of SW Asia. These common species belong to genera such as *Aerva, Anabasis, Calotropis, Cenchrus, Capparis, Cressa, Euphorbia, Fagonia, Heliotropium, Ochradenus, Panicum, Suaeda, Tephrosia, Tribulus* and *Zygophyllum*.

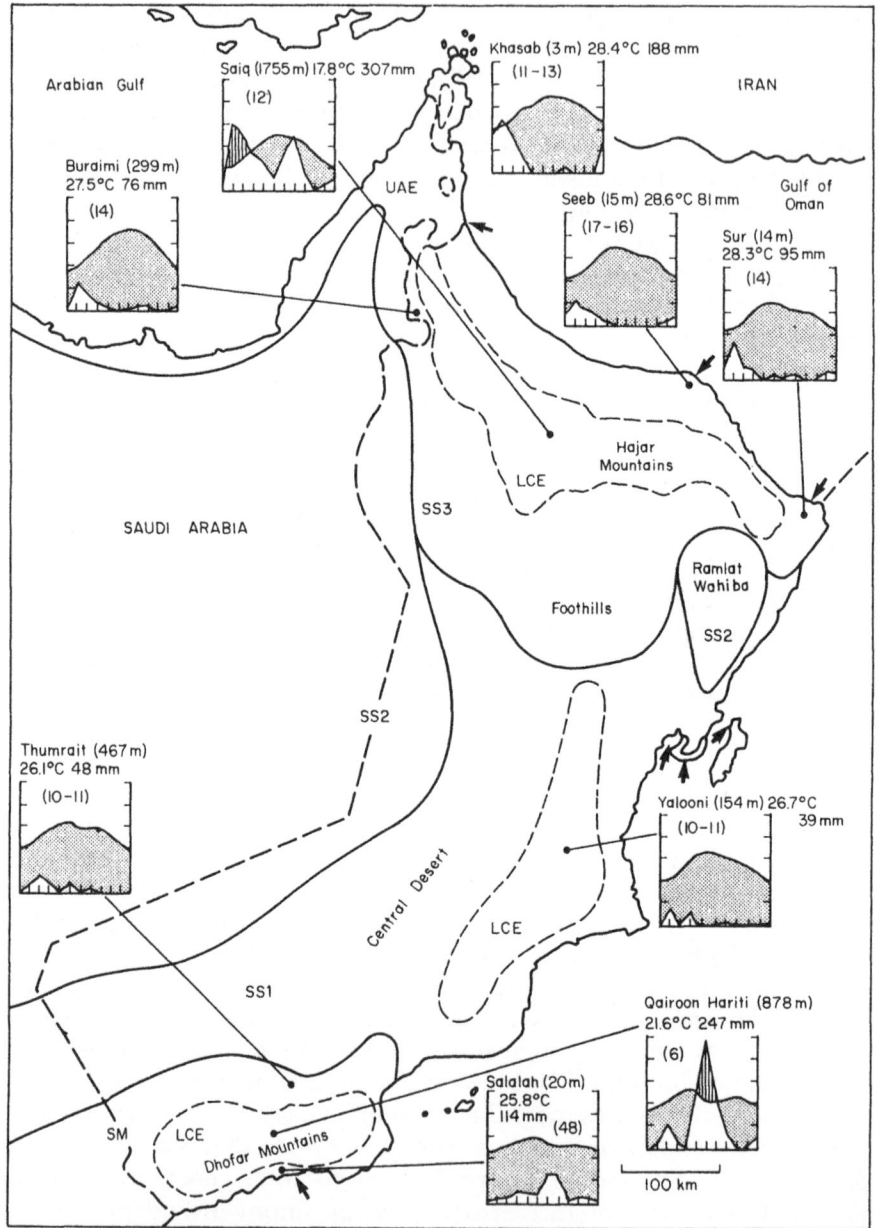

Fig. 3: Biogeographic regions of Oman. SM, Somali-Masai regional centre of endemism; SS1, Saharan regional subzone; SS2, Arabian subzone; SS3, Nubo-Sindian local centre of endemism. LCE. Local centre of endemism; → Azonal vegetation. (In each climate diagram the left ordinate is mean temperature (0-50°C), the right is mean precipitation (0-100 mm), and the abscissa is months, January-December) (after Ghazanfar, 1992b).

Not surprisingly, the floras of N. Oman and UAE are closely related (82% generic similarity). They occur in the same geomorphic region and the same climatic regime. The similarity between N Oman and Baluchistan (c. 40 %) consists mostly of species and plant communities present on the northern mountains of Oman and the hills of SW Baluchistan and SW Iran. It is suggested that between 30,000 and 20,000 years BP (humid phase), montane plant communities such as the *Juniperus exselsa* subsp. *polycarpos, Helianthemum lippii, Ephedra pachyclada* community and associated species such as *Cymbopogon schoenanthus, Lonicera aucheri* and *Segeretia spiciflora* were established in N. Oman (Mandaville, 1984; Kürschner, 1986; Ghazanfar, 1992b), crossing the Arabian Gulf from Baluchistan and SW Iran. Distribution of the flora probably declined during the last arid phase (between 11,000 and 4000 years BP) and became restricted to areas of favourable climate at the higher altitudes in the mountains of N. Oman and the Musandam peninsula. During the last humid phase, which lasted until 4000 years BP, sea levels rose rapidly in the Arabian Gulf and changes were recorded both in the vegetation (attested by the presence of mangroves) and human habitation of coastal areas (Sanlaville, 1992 and references therein). This phase was shorter and probably less humid than the previous phases, as indicated by the development of sabkhas rather than lakes. The flora and fauna that existed were similar in terms of species richness to those of the present but were more extensive and luxurious. Evidence of a wetter climate in N Oman is supported by the presence of species such as *Caralluma flava, Commiphora gileadensis, Dalechampia parvifolia, Heliotropium fartakense* and *Limonium* cf. *axillare* in the foothills and rocky coastal areas of the eastern Hajar mountains. There, these species are restricted in their distribution, being present in sheltered wadis and in a few locations on the coasts. However, these species are commonly distributed throughout Dhofar, sustained, like the other vegetation, by the moisture received from the annual monsoon season. These and other species such as *Ceratonia oreothauma* subsp. *oreothauma* on the summit areas of the eastern Hajar mountains, are now relics confined to a few localities in N. Oman where there is relatively more moisture and lower temperatures (Ghazanfar, 1994).

The flora of N. and central Oman also contains several species which are E. African (Sudanian floristic elements) in origin. These species, such as those of *Acacia, Ziziphus* and *Rhazya* are considered Sudanian relics that may have migrated during the Late Pliocene through wadis flowing from W. Arabia into E. Arabia (Mandaville,1984). The present distribution of these species in Oman, especially those of *Acacia*, suggests that only the more hardy species were able to become established there. More than 50 species are reported for E. Africa (Coe and Beentje, 1991), of which 10 species occur in Oman; of these, only three species (*A. tortilis, A. ehrenbergiana* and *A. leucodermis*) occur in central and N. Oman, while others (such as *A. asak*, *A. etbaica*, *A. oerfota* and *A. hamulosa*) occur only on the escarpment mountains of Dhofar (S. Oman) where more moisture is available.

The floras of Yemen (north and south) and Oman are the richest in endemic species (Table 3). In Oman, the southern region of Dhofar has the highest number of endemic species. The influence of the south-west monsoon in S Arabia sustains a flora which is broadly E African in origin and is reminiscent of the upland regions of the African savanna. Several species are regionally endemic, being present both in S. Yemen and Dhofar. Unlike many of the endemic species in N. and central Oman which are restricted in distribution, several of the endemic species in Dhofar are common and widespread. Species such as *Anogeissus dhofarica, Dyschoriste dalyi, Euphorbia dhofarica* and *Jatropha dhofarica* form an important component of the woodland vegetation. The present flora appears to be a restricted representative of a flora which formerly extended over a wider area. It is likely that, after the last pluvial phase (c. 4000 years BP), as the climate became drier and the influence of the monsoon decreased, cosmopolitan species common in semi-arid regions became established. These species, represented by genera such as *Alhagi, Cornulaca, Ficus, Leptadenia, Portulaca, Teucrium* and *Sonchus*, are present in the drier areas of Dhofar, S. Yemen, N. Oman and Baluchistan.

Table 3: Endemic flora of the Arabian Peninsula
(data from Miller and Nyberg, 1991).

Countries	Species	Endemics
Saudi Arabia	3020	34 (2%)
Yemen	2750	137 (10.5%)
Oman	1204	57 (5%)
Qatar	312	0 (0%)
Bahrain	280	0 (0%)
Kuwait	280	0 (0%)
UAE	268	0 (0%)

The desert flora of Iran is shared by two phytochorologies, the Nubo-Sindian and the Irano-Turanian (Leonard, 1981-1989). The northern mountains of Oman fall within the Nubo-Sindian phytochorology, but the influence of the Irano-Turanian phytochorology is absent. Genera such as *Amygdalus, Anthemis, Berberis, Bassia, Ranunculus, Scabosia, Stipa* and *Verbascum*, which are Nubo-Sindian floristic elements, occur in the northern mountains but Irano-Turanian floristic elements such as *Acer, Chesneya, Gamanthus, Trachynia* (which occur in Central Asia and Turkey) are not represented.

Conclusion

Several methods are used to reconstruct palaeoenvironments. Plant-based methods based on similarities of taxa, endemism and distribution of relict taxa can serve as indicators of past climates. Species similarities between the floras of SW Arabia and NE Africa and those of N. Oman and Baluchistan and SW Iran and their present-day distribution indicates a relatively humid phase in the climate of Arabia when plant species could have migrated and become established. Endemic and relict taxa and their distribution indicate past floristic ties between regions which can be used to corroborate our knowledge of palaeoenvironments.

Acknowledgements

My sincere thanks to Martin Fisher for many helpful suggestions and to the Department of Biology, Sultan Qaboos University for providing research facilities.

References

Al-Rawi, A. (1987). *Flora of Kuwait*. Vol.2. University of Kuwait.

Andrews, F.W. (1950). *The Flowering Plants of Sudan*. London, UK.

Anton, D. (1984). Aspects of geomorphological evolution; palaeosols and dunes in Saudi Arabia. In: *Quaternary Period in Saudi Arabia* 2 A.R. Jado and J.G. Zolt (eds.) pp. 275-296. Springer-Verlag, Vienna.

Batanouny, K.H. (1981). *Ecology and Flora of Qatar*. Centre for Applied and Scientific Research, University of Qatar, Doha.

Blatter, E. (1914). Flora of Aden. *Records Botanical Survey of India*, 8: 1-518.

Boulos, L. (1988). A contribution to the flora of South Yemen (PDRY). *Candollea* 43: 549-585.

Braithwaite, C.J.R. (1987). Geology and palaeogeography of the Red Sea region. In: *Red Sea* A.J. Edwards and S.M. Head (eds.), pp. 22-44. Pergamon Press, Oxford.

Burkill, I.H. (1909). *A working list of the flowering plants of Baluchistan*. Calcutta, India. Superintendent Government Printing, India.

Chappell, J. and Thom, B. (1977). Sea levels and coasts. In: *Sunda and Sahul* J. Allen, J. Golson and R. Jones (eds.) Academic Press, New York.

Coe, M. and Beentje, H. (1991). *A Field Guide to the Acacias of Kenya*. Oxford University Press.

Collenette, S. (1985). *An Illustrated Guide to the Flowers of Saudi Arabia*. Scorpion Publishing Ltd. London.

Cornes, M.D. and Cornes, C.D. (1989). *The Wild Flowering Plants of Bahrain, An illustrated Guide*. Immel Publishing, UK.

Daoud, H.S. and Al-Rawi, A. (1985). *Flora of Kuwait*. Vol. 1, University of Kuwait.

Delany, M.J. (1989). The zoogeography of the mammal fauna of southern Arabia. *Mammal Rev.* 19: 133-152.

El-Amin, H.M. (1990). *Trees and Shrubs of Sudan*. Ithaca Press, UK.

El-Moslimany, A.P. (1983). History of climate and vegetation in the eastern Mediterranean and the Middle East from the Pleniglacial to Mid-Holocene. Ph. D. Dissertation, University of Washington.

Estrada-Loera, E. (1991). Phytogeographic relationships of the Yucatan Peninsula. J. Biogeogr. 18: 687-697.

Gabali, S.A. and Al-Gifri, A.N. (1990). Flora of South Yemen - Angiospermae. A provisional checklist. *Feddes Repert.* 101: 373-383.

Ghazanfar, S.A. (1991a). Floristic composition and the analysis of vegetation of the Sultanate of Oman. *Flora et Vegetatio Mundi* 9: 215-227.

Ghazanfar, S.A. (1991b). Vegetation structure and phytogeography of Jabal Shams, an arid mountain in Oman. *J. Biogeogr.* 18: 299-309.

Ghazanfar, S.A. (1992a). Annotated catalogue of the vascular plants of the Sultanate of Oman and their vernacular names. *Scripta Botanica Belgica*, vol. 2. Jardin botanique national de Belgique.

Ghazanfar, S.A. (1992b). Quantitative and biogeographic analysis of the flora of the Sultanate of Oman. *Global Ecology and Biogeogr. Letters* 2: 189-195.

Ghazanfar, S.A. (1994). *Novitates* from the flora of the Sultanate of Oman. *Edinb. J. Bot.* 51: 59-63.

Ghazanfar, S.A. (in press). *The Influence of Paleoclimate on lte Phytogeography of Oman* Proceedings of the V International Symposium on Plant Life of Southwest and Central Asia, 18-23 May, Tashkent, Uzbekistan.

Ghazanfar, S.A. and Fisher, M. (eds.). (1998). *Vegetation of the Arabian Peninsula.* 362 pp, Kluwer Academic Press, The Netherlands.

Horowitz, A. (1992). *Palynology of Arid Lands.* Elsevier, Amsterdam.

Kadmon, R. and Pulliam, H. R. (1993). Island biogeography: effect of geographical isolation on species composition. *Ecology* 74: 977-981.

Karim, F.M. (1992a). New records of the flora of the UAE (Part: 2). *Arab Gulf. J. Scient. Res.* 10: 105-115.

Karim, F.M. (1992b). New records of the flora of the United Arab Emirates (Part 2). *Arab Gulf J. Scient. Res.* 10: 105-115.

Karim, F.M. (1993). New records of the flora of the United Arab Emirates (Part 3). *Arab Gulf J. Scient. Res.* 11: 391-401.

Karim, F.M. (1994). New records of the flora of the United Arab Emirates (Part 4). *Arab Gulf J. Scient. Res.* 12: 109-118.

Krebs, C.J. (1989). *Ecological Methodology.* Harper & Row, Publishers, New York.

Kürschner, H. (1986). Omanisch-Makranische Disjunktionen Ein Bietrag zur Pflanzengeographischen Stellung und zu den florengenetischen Bezeihungen Omans. *Bot. Jahrb. Syst.* 106: 541-562.

Kürschner, H. (1998) Biogeography and Introduction to Vegetation. pp. 64-98 in *Vegetation of the Arabian Penissula.* S.A. Ghazanfar and M. Fisher (eds.) Kluwer Academic Press, the Nertherlands.

Lavranos, J.J. (1975). Note on the temperate element in the flora of the Ethio-Arabian region. *Boissiera* 24: 67-69.

Leonard, J. (1981-1989). *Contribution a l'etude de la flore et de la vegetation des deserts d'Iran.* Fasc. 1-10. Jardin botanique national de Belgique.

Lioubimtseva, E. U. (1995). Landscape changes in the Saharo-Arabian area during the last glacial cycle. *J. Arid Environments* 30: 1-17.

Mandaville, J.P. (1984). Studies in the Flora of Arabia XI: Some Historical and Geographical aspects of a Principal Floristic Frontier. *Notes Royal Bot. Gard. Edinb.* 42(1): 1-15.

Mandaville, J.P. (1985). A Botanical Reconnaissance in the Musandam Region of Oman. *J. Oman Studies* 7: 9-28.

Mandaville, J.P. (1990). *Flora of Eastern Saudi Arabia.* Kegan Paul Int., London and New York.

McClure, H.A. (1978). Ar Rub'Al Khali. In: *Quaternary Period in Saudi Arabia* 1 S.S. Al-Sayari and J.G. Zolt (eds.), vol. 1 pp. 252-263. Springer-Verlag, Vienna.

McCoy, E. D. and Heck, K. L. J. (1987). Some observations on the use of taxonomic similarity in large-scale biogeography. *J. Biogeogr.* 14: 79-87.

Migahid, A. M. (1978). *Flora of Saudi Arabia.* vols. 1 and 2. Riyadh University, Saudi Arabia.

Miller A.G. and Morris, M. (1988). *Plants of Dhofar: The southern region of Oman, traditional economic and medicinal uses.* Office of the Advisor for Conservation of the Environment, Diwan of Royal Court, Sultanate of Oman.

Miller, A.G. and Nyberg, J.A. (1991). Patterns of endemism in Arabia. *Flora et Vegetatio Mundi* 9: 263-279.

Nasir, E. and Ali, S.I. (eds.). (1972-1994). *Flora of Pakistan.* Fasc. 1-193. Karachi University, Pakistan.

Raup, D. M. and Crick, R. E. (1979). Measurement of faunal similarity in palaeontology. *Palaeontology* 53: 1213-1227.

Sanlaville, P. 1992. Changements climatiques dans la Péninsule Arabique durant le Pléistocéne Supérieur et l'Holocéne. *Paléorient.* 18: 5-26.

Schulz, E. (1987). Holocene palaeoenvironments in the central and southern Sahara. In: *Current Research in Affrican Earth Sciences* Matheis and H. Schandelmeier (eds.). A.A. Balkema, Rotterdam.

Schulz, E. and Whitney, J.W. (1986). Upper Pleistocene and Holocene lakes in the An Nafud, Saudi Arabia. *Hydrobiologica* 143: 175-190.

Thulin, M. (ed.) (1993). *Flora of Somalia,* vol. 1. Royal Botanic Gardens Kew, UK.

Thunell, R.C., Locke, S.M. and Williams, D.F. (1988). Glacio-eustatic sea-level control on Red Sea salinity. *Nature* 334: 601-604.

Western, A.R. (1989). *The Flora of the United Arab Emirates, an Introduction.* United Arab Emirates University.

White, F. (1983). *The vegetation of Africa, a descriptive memoir to accompany the UNESCO, AETFAT, ANSO vegetation map of Africa.* UNESCO, Paris.

White, F. and Leonard, J. (1991). Phytogeographical links between Africa and Southwest Asia. *Flora et Vegetatio Mundi* 9: 229-246.

10

Anthracology (Charcoal Analysis) in Investigation of Desert Margin Shifts: A Case Study of Southern Sahara

H.N. Barakat[1] and C. Rolando[2]

ABSTRACT

The Sahara of North Africa is the largest desert on earth, and includes the driest parts of the world. However, it seems that, during the Holocene, this desert was less hostile to human occupation. Evidence for such favourable conditions is clear from the presence of a large number of Neolithic archaeological sites all over the Sahara and the Sahel. Anthracological studies (analysis of archaeological charcoals) in these sites allow us to reconstruct the vegetation in their vicinity and to infer the prevailing palaeoclimatic conditions, and, at a more general level, to infer fluctuations of the Saharan margin. This Chapter reviews the basic principles of anthracology used in palaeo-environmental reconstruction in arid regions including the sources and nature of archaeological charcoal, the methodology and techniques of retrieval, sampling and identification of the material, as well as outlining different approaches to the interpretation of results. Examples from recent research in the Sahara and sub-Saharan regions illustrate the application of anthracological analysis to the investigation of the prehistoric Saharan margin and underline the fact that study of sites with high spatial and temporal frequency is essential in order to delimit precisely the southern (and northern) margins of the Sahara for the time since the beginning of the Holocene.

Introduction

The application of the study of archaeological charcoal fragments for the reconstruction of the Holocene vegetation in Africa has developed dramatically over the past decade. In addition to reconstruction of vegetation, anthracological data have been used for general palaeoenvironmental reconstruction as well as for inferring palaeoclimatic conditions in tropical, subtropical (Neumann 1988, 1992; Otto 1993; Prior and Price 1985; Rolando 1992a, 1992b) and drier regions of Africa including the Sahara (Barakat 1995a; 1995b, Barakat in press; Neumann 1987a, 1987b, 1989a, 1989b).

[1]Faculty of Science, Cairo University, Giza Egypt.
[2]IMEP-Case 461-Fac. ST Jerome F 13397, MARSEILLE Cedex 13, France

The Sahara of North Africa is the largest desert on earth, and includes the driest parts of the world. However, increasing evidence has accrued which suggests that during the Holocene, this desert was less hostile to human occupation. Evidence for such favourable conditions is clearly provided by the presence of a large number of Neolithic archaeological sites all over the Sahara and the Sahel. At the Saharan margin, there is a barely perceptible shift in the plants, animals and physiographic characteristics. This margin fluctuates in response to climatic conditions. For instance, the surface area of the Sahara has become enlarged by 15% in just four years (1980-1984) (Tucker et al., 1991) and by 50-95% in 10,000 years (Faure et al., 1993). Conversely, the same desert shrank by 30-97% during the Holocene climatic optimum (Faure et al., 1993). By what methods can the desert margins be reconstructed?

To define more precisely such favourable habitation conditions and detect desert margin variations through the millennia, it is necessary to combine results from several disciplines related to biology, geology, hydrology and archaeology. In this chapter, we consider only those methods concerned with the study of the terrestrial plant remains viz. carpology, palynology and anthracology.

Carpology deals with the study of seeds and grains found in archaeological sites. This discipline is useful if such remains are common and well preserved. Unfortunately, this is not the case in Saharan sites, where macro-remains are scarce and often poorly preserved (with the exception of those having a ligneous endocarp (eg. *Celtis*)). The retrieval method is also important. Accordingly, carpology has so far played only a minor role in palaeoenvironmental reconstruction (Celles and Schulz 1983).

Pollen-analysis is the most commonly used tool for the reconstruction of palaeoenvironment and vegetation. During the past 20 years, a number of workers have contributed valuable palaeoenvironmental reconstructions for the southern Sahara and tropical Africa using palynology, including Cour and Duzer (1976), Haynes et al. (1989), Lezine (1987, 1989), Lezine et al. (1990), Maley (1977, 1981, 1983), Ritchie (1987, 1994), Ritchie et al. (1985), Ritchie and Haynes (1987) and Schulz (1988, 1991). Pollen analysis of Holocene lake deposits (Ritchie et al., 1985, Ritchie and Haynes 1987, Ritchie 1994) has been used to set up a model of vegetation change that envisages a 400 km (5″) northward shift of the vegetational belts, in which the present day desert region was occupied by a savanna and desert grassland during the period from 10,000 to 5,000 yrs BP.

Pollen grains are indeed numerous in Quaternary sediments. Moreover, they are often well preserved and can be identified at least to the genus level. However, a major problem in pollen analysis is the effect of long distance transport on the pollen grains. Studies of present pollen grains

(Cour and Duzer 1976) provide evidence for long distance transport (800-1000 km) of floristic elements. One result of this is that, in present desert conditions such as those in the Taoudeni sebkha, the pollen spectra give the impression of a Sahelian savanna.

Anthracology is concerned with the scientific study of charcoal. Charred wood fragments are rarely subjected to transport, often being found in situ. They are the most common plant remains in archaeological sites (Dimbleby 1978). Moreover, dating is carried out on the sample itself, so that anthracoanalysis offers the best level of spatio-temporal precision for palaeoenvironmental reconstruction. In the case of archaeological charcoals, it must be stressed that they yield information about local vegetation, related to peculiar local conditions, especially the presence of water without which human settlement would not be possible. The identification of charcoals found in the soil, whether of anthropogenic or natural origin (pedoanthracology, Thinon 1992), is a new and promising tool which has not yet been used in tropical areas. Neumann and Schulz (1987) and Neumann (1989a, 1989b, 1992) identified charred woods from Holocene lake sediments in the depression of Fachi, in the hyperarid Tenere. The Sudano-Sahelian species which formed a dense woodland at the edge of the lake were identified, and based on these the authors inferred northward shift of the Sahelian vegetation by at least 400 km. This model, based on anthracological analysis in Neolithic sites along a N-S axis in Egypt and northwestern Sudan (Neumann 1989b), assumes a northward shift of the vegetation zones by 4° latitude during the studied period (7000-5800 yrs BP).

This chapter reviews the basic principles underlying the use of anthracology in palaeoenvironmental reconstruction in arid regions, including the sources and nature of archaeological charcoal, the methodology and techniques of retrieval, sampling and identification of the material, as well as the different approaches to the interpretation of the results. Examples from recent research in Saharan and sub-Saharan regions illustrate the use of anthracological analysis in the investigation of the prehistoric Saharan margin.

Nature and Source of Archaeological Charcoal

The fact that charcoal is the most common plant material recovered in archaeological sites of all ages has often been emphasized in discussions of various aspects of palaeoethnobotanical research. Fortunately, charcoal was recognized at an early stage as the most suitable material for dating, especially in prehistoric sites, and has long been recovered by archaeologists. Charcoal is at least as abundant in prehistoric sites in arid and semiarid regions as it is in humid regions. In fact, archaeobotanists working in Saharan sites are often overwhelmed by the large quantities of charcoal even in sites lacking any other type of plant macro-remains (seeds, grains etc.).

Charcoal fragments found in archaeological sites are the result of the burning of wood. Sources of wood include trees, shrubs, and only rarely grasses and reeds. Wood may have been brought to the site for many different purposes including construction. Timber has been by far the most common constructional material. It has been used to make household implements, and has been the dominant fuel providing fire for cooking, pottery making, religious rituals as well as serving as a source of warmth and protection (Dimbleby, 1978). Wood has also been used for manufacturing implements of agriculture, hunting and war.

In all the above mentioned cases, charcoal fragments found in archaeological sites are the result of either intentional or accidental burning of wood. This is an important point in relating the taxa assemblage in the charcoal to the local flora. Charcoal fragments should be considered as true artifacts. The source of charcoal (wood) was consciously collected by people who undoubtedly made use of their knowledge and experience to choose the species appropriate for particular purposes. This element of selectivity is the most important factor in defining the charcoal assemblage. Other factors include the specific burning and preservation processes during which some taxa were turned into ash while others were perfectly preserved as charcoal. Other important factors were the temperature of the fire, the position of the wood within the fire (on top or at the base of the hearth), its moisture content, etc. The recovery techniques and sampling strategies are other factors determining the charcoal assemblage, as will be discussed later. These factors must be borne in mind by the excavator in order to avoid biased sampling.

Mostly charcoal recovered from archaeological sites in arid and semi-arid regions are found in household hearths. In rare cases it can also be found dispersed within sediments in the settlement's middens rather than the hearth (Barakat 1995a). In those middens, charcoals as well as other organic (bone and plant macro-remains) and non-organic (lithics and sherds) debris resulting from domestic activities were intentionally discarded from the surface of the settlement area and so accumulated over time to form a mound. The dispersion of charcoal is a result of the regular and recurrent burning of wood in household fires followed by the discarding of charcoal and ash in middens where it is further fragmented and mixed with other sediments. Such charcoal fragments, dispersed as they are within sediments, are a more interesting source of information on former phytogeographical, vegetational and climatological conditions than the in situ fragments found in hearths (Chabal 1988,

1992). Dispersed charcoal is usually qualitatively richer in the number of taxa. Moreover, the semi-quantitative approach (discussed below) necessary for the reconstruction of the vegetation is more applicable to dispersed charcoal than to material from intact hearths.

Archaeological Charcoal Analysis

Retrieval

The retrieval of charcoal and other plant macro-remains in archaeological sites is one of those aspects of palaeoethnobotanical research often dealt with in detail in specialised publications. The techniques of recovery are diverse, but for arid and semi-arid regions there seems to be one possible yet efficient technique, namely the dry sieving of the archaeological sediments. Dry sieving has been found to be the most suitable recovery technique for charcoal fragments that are too dry to be floated off and which frequently burst on contact with water. Moreover, the medium used for floating (water) is very often a rare commodity in Saharan sites. Dry sieving is carried out using standard soil analytical metal sieves of mesh sizes 2mm and 1mm, and, when smaller plant macro-remains (seeds and grains) are present, a 0.5mm mesh size sieve is used. Sieving is followed by manual sorting to pick out all charcoal fragments and other macro-remains using special soft tweezers.

In addition to sieving bulk samples, anthracologists often have to deal with samples that have been hand-picked by the archaeologist from particularly rich sediments where charcoal is visible during the excavation process (household hearths in situ). These are usually larger pieces which the excavator could not resist the temptation of extracting on the spot, for fear that they might otherwise be lost. Hand-picked samples need no further sorting but are often less interesting than bulk samples for palaeoenvironmental interpretation since, in addition to their being biased samples (excavators tending to choose larger pieces), they are often monospecific samples (one large piece broken into several smaller ones).

Sampling, On and Off-site

Sampling is a two-fold procedure: it begins in the field but often extends to the laboratory. Sampling procedures are best thought of for each site specifically and have to be adapted to each case in particular. However, a standardized sampling procedure should be chosen by the archaeologist-anthracologist in the field and this should ensure that all features have been sufficiently sampled from each excavated unit.

Laboratory sampling (or sub-sampling) is often necessary, particularly when extensive recovery procedures are being adopted. Sub-sampling should also be specifically adapted for each site or number of sites within the same region, and experimentation is often necessary before a sub-sample size is selected. Experiments are carried out in order to determine the minimum representative size category as well as the minimum representative sub-sample. This minimizes the risk of overlooking rare taxa or those whose charcoals fragment easily (for detailed discussion of subsampling procedures, see Smart and Hoffman 1988, van der Veen 1982, 1985, Chabal 1988).

Identification: Optical Devices and their Development

Although archaeological charcoals have been identified since the end of the 19th century in Europe, the long process of sample preparation and the mediocre performance of the old optical devices have meant that, in most cases, charcoal has been identified only at the level of the family, and less often at the genus level. This has long been a constraint on the discipline of anthracology. Nowadays, the systematic use of episcopic microscopy or SEM renders possible the routine identification of most charcoal fragments to the species level, provided that an exhaustive reference collection and accurate identification criteria are available.

Archaeological charcoal is satisfactorily identified only when compared with charred reference material. Over the years, the authors have collected samples of woody species of the south Saharan and Sahelo-Sudanian biogeographic areas. The wood samples were charred by embedding them in sand in a muffle furnace at 600°C. The charred wood fragments were then fractured manually along the three classic planes, transverse, tangential and radial, and immediately examined without further preparation. The anatomical features were examined under an incident light microscope equipped with a differential interference contrast device. The very high contrast and resolving power of the DIC (Differential Interferential Contrast) episcopic microscope made it possible to examine routinely most of the anatomical features of charred wood (at magnifications from 100 to 1000) and yielded a good view of the relief without using an SEM (Rolando and Thinon 1988, Thinon 1988).

Databases of the anatomical and ecological features of the woody species of the south Saharan and Sahelo-Sudanian region (Rolando 1992a, 1992b, Barakat in prep.) allow the specific identification of archaeological charcoal. The identification of such charcoal at the species level is indispensable for the reconstruction of the vegetation and the palaeoclimate in the vicinity of the sites from which they were retrieved. The anatomical features of samples constituting the reference collection are then coded according to an exhaustive list of characters, adapted to the description of charcoal and developed by Thinon (1992).

Given a suitable microscope and a good reference collection and/or the relevant database identification of archaeological charcoal becomes more or less routine. Problems occur when dealing with badly preserved fragments which are usually fragile and crumble at the touch, thereby rendering the fracturing process more difficult. Less problematic are smaller fragments especially if well preserved. Provided that the microscope is well equipped, anthracologists can identify fragments as small as 2 mm in diameter.

The identification procedure includes the fracturing of the archaeological charcoal fragment under a binocular (x20) into the three planes (cross section, tangential longitudinal and radial longitudinal sections) thus exposing the structural features. Examination of all three planes provides complementary data for the identification of the taxon (genus or species). At this stage, the use of a pre-coded database and computer-aided image analysis may prove very useful. Confirmation of the identified material should be carried out by comparing archaeological charcoal with modern charred wood of the expected species in the reference collection. In well preserved fragments, the woody structure of the archaeological sample matches remarkably well with that of the modern charred sample.

Presentation and Interpretation of the Results

Quantification in Anthracology

Presentation of the results of anthracological analysis must involve a certain degree of quantification. In archaeobotany, quantification is a controversial issue. Three approaches, as mentioned in Popper (1988), are currently in use by archaeobotanists and may be summarized as follows:

1. Absolute counts in the form of the raw numbers of each taxon per sample

These rarely provide an adequate measurement for archaeobotanical remains, so that standardisation by expressing them as ratios seems more realistic. The following factors determine the unsuitability of charcoal samples for absolute counts.

   i. Selection or rejection of certain species by the user (Godwin and Tansley 1941, Western 1971, Miller 1985) needs to be considered as a factor influencing species data. The amount of wood of a particular species brought on to a site does not necessarily represent its relative abundance in the vegetational community, although the availability of a taxon might influence its selection. Charcoal fragments are artifacts, having passed through the human filter: the anthracocoenosis (Thinon 1992) is thus a biased sample.

ii. Burning on the site. Different woody species do not burn equally well in open fires, some tending to become charcoal more readily than others. This may alter the proportions of taxa in the charcoal samples (Smart and Hoffman 1988). Moreover, the preservation of charcoal differs according to the species.

iii. Dead wood is often used for fires. Fallen and collected dead wood does not necessarily occur in the same proportions as that of the woody species growing in the vicinity.

The quantitative approach in the form of the number of charcoal fragments in tables should thus be exclusively used for the description of the results but not for their interpretation.

2. Ubiquity

Presence of a taxon among the samples from a site (expressed as percentage of the total number of samples) disregards the absolute count of the number of charcoal fragments of this taxon. Rather, it considers the number of samples in which the taxon appears among a group of samples belonging to the same site, so that the taxon is scored as present or absent in each sample regardless of its quantity. Thus, the ubiquity of a taxon does not question the quantity of wood burnt but deals with the frequency with which the wood taxon was brought to the site or used in a domestic hearth. The use of this approach requires an appropriate sampling procedure where a clear definition of the sample exists. This is in order to avoid splitting of a sample or grouping of several samples into a single one. Ubiquity analyses are useful within limitations for showing general trends when one has little control over the source of patterning in one's data. Ubiquity is also difficult to relate to the actual vegetation. Nevertheless, the presence of a taxon in all or most of the samples could be considered as evidence that it was fairly common in the woody vegetation around the site (Godwin and Tansley 1941). Despite this, its presence in a relatively small number of samples cannot be used as a measurement of its relative abundance in the flora (Smart and Hoffman 1989) especially if the number of samples is small.

3. Diversity

This approach is used to summarize data in order to describe a plant assemblage. It includes the total number of taxa in a sample and the abundance of each taxon. When the pattern involves well controlled data (comparing taxa between the same type of context at one site), one may want a more specific measurement (ratio-diversity).

The palaeoenvironmental interpretation of the results of the anthracological analysis in archaeological sites in arid regions poses some problems for the following reasons.

1. The taxa assemblage in archaeological charcoal often includes very few taxa compared to the contemporary surrounding vegetation (Smart and Hoffman 1988). Principal components of the regional flora may be completely missing.

2. Different phytogeographical regions often share common arboreal species, e.g. in north Africa the semi-desert scrubland contains many of the arboreal species growing in the desert region to the north. The difference is quantitative, in the form of the total cover, size and number of trees and shrubs.

3. The closest 'natural' vegetation types, with which the reconstructed Holocene vegetation may be compared, have been subjected to the human impact over long periods of time, causing important modifications in the plant cover, and sometimes leading to the introduction of species and the extinction of others.

Generally speaking, there are three levels of inference in the interpretation of the results of archaeoanthracology (Smart and Hoffman 1988, Ritchie 1994). These are as follows.

(a) *Phytogeographical inferences*—In this case, identified taxa are simply considered to have grown in the area. Such an approach is suitable for small as well as for large scale reconstructions, the taxa being grouped into geographical categories, and the nearest discrete geographical region which is floristically similar to the taxa assemblage being sought.

(b) *Vegetation and environment* —This approach assumes that the identified taxa indicate the plant communities growing in the area, and that the preferred habitats of the identified taxa indicate former environments in the area and around it. A vegetation type is sought in which a semi-quantitative translation of the taxa assemblages (taxa ubiquity) can be used. Such a reconstruction depends upon the relative stability of the regional vegetation through time, the fidelity of the identified taxa, the availability of ecological studies of modern vegetation analogues in the region and the difference between the modern environment and past conditions. The multidisciplinary approach seems a more appropriate tool for this level of inference, since the results from macro-remains, palynological, geomorphological and sedimentological analyses may be combined.

(c) *Palaeoclimatological inferences*—The level of inference using this approach remains imprecise for arid and hyperarid regions (Rognon
1980). Interpretation of the vegetation in climatic terms poses the
following several problems.

(i) Many vascular plants persist during arid phases in dormant
states and have a rapid response to sporadic rains.

(ii) The complicated water availability (controlled by the topogra-
phy, rainfall runoff and ground water recharge) – precipitaton
relationship in arid regions renders it difficult to relate the vegetation (a product of the available moisture) to the
actual precipitation.

Moreover, the results of anthracological analysis could be used for a
palaeoethnobotanical interpretation bearing on the possible uses of the
identified taxa other than in household hearths as firewood. It might also
be used in discussions of the human impact on the vegetation occupying
the vicinity of the Neolithic archaeological sites during the Holocene.
These should include two main aspects: the evidence from the
archaeobotanical data and the discussion of the probable settlement
patterns and activities to which the human impact on the surrounding
environment is directly related (Barakat 1995a, Neumann 1993).

Recent Anthracological Research in the Saharan and Sub-Saharan Regions Application to the Investigation of the Southern Saharan Margin

The work of Barakat (in the eastern Sahara) and Rolando (in the southern
and western Sahara) is specifically concerned with the southern Sahara
margins (Fig. 1).

In northern Niger, the Early Holocene site of Tin Ouaffadene (about
9200 years BP) yielded charcoal from under a compact diatomite layer
1.20 m thick (Rolando and Roset 1991). These charcoals, which were
found among a large quantity of animal bones, belonged to two species,
namely *Calotropis procera* and *Leptadenia pyrotechnica*. These two
Asclepiadacae are still used for building shelters and are likely to have
been used for this purpose on the site. Both plants characterize disturbed
places and sandy soils and, are very frequently found in the Sahara
desert and its margins.

In Mauritania (Rolando and Riser 1992), charred wood fragments were
collected on the site of Hassi el Def'a (Tagant) among potsherds and
broken pieces of ostrich eggshell near rockshelters featuring rock art. The
^{14}C dating of the charcoals indicates that they are 2830±60 years old. All
the charcoals studied have been identified as *Acacia raddiana*. Rare

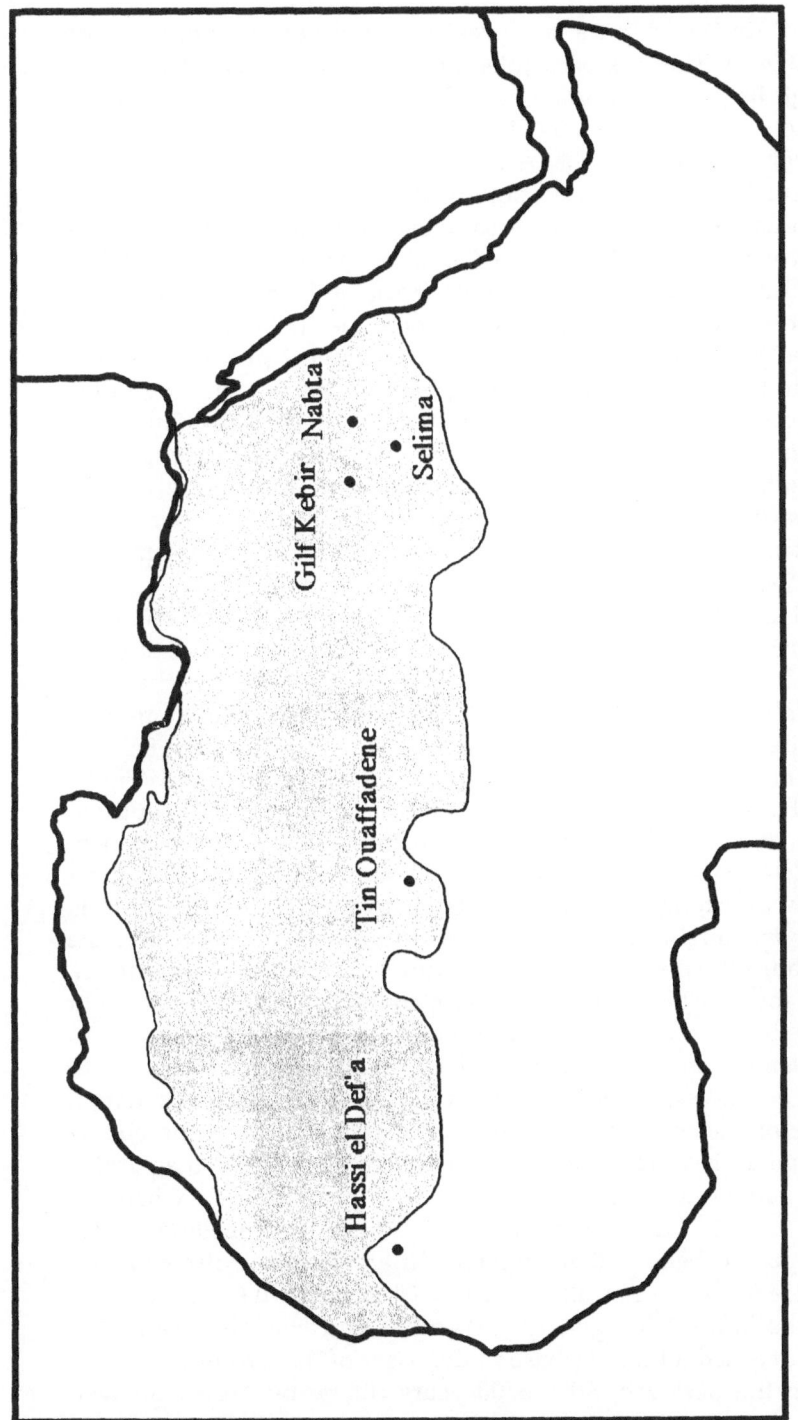

Fig. 1: The present day extension of the Sahara and location of the sites mentionned.

individuals of this species are still present, being very scattered in the vicinity of the site. The species is tolerant to long periods of drought so that, although its prefered rainfall range is about 700-800 mm, it can be found in places receiving between 50 and 1000 mm. Giraffes painted on the rockshelter walls indicate that *Acacia raddiana* was abundant, and that the trees were large and dense enough to provide giraffes with their required fodder. Study of charcoal samples from eight Neolithic sites extending over the whole Neolithic occupation (9000-6000 years BP) in the eastern Sahara (Nabta playa, Egypt) (Barakat 1995b, and in press) show that, at the beginning of the occupation phase (8960±110 ^{14}C years BP), arid conditions and sparser desert vegetation were prevalent. This is deduced from the fact that only *Tamarix* charcoals were identified in the anthracological assemblage. However, from slightly later sites (final early Neolithic), some 8000-7400 years old, charcoal specimens belonging to several other taxa (*Acacia raddiana, Acacia ehrenbergiana, Acacia nilotica, Ziziphus spina-christi, Cassia senna, Capparis decidua* and other *Capparidaceae*) were identified. Vegetation reconstructed on the basis of this assemblage of taxa represents a primitive type of oasis vegetation around a water body as well as less arid desert conditions. There is an indication of the return of hyperarid conditions by the end of the Neolithic period, through the dominance of *Tamarix* sp. and the presence of *Chenopodiaceae* charcoals in samples from late Neolithic sites. These results imply desert margin fluctuations during the Middle and Late Holocene in the eastern Sahara.

The present day limits of the Sahara are shown on the map (Fig.1). However, what is the most realistic basis for these limits? The southern limit of *Cornulaca monacantha*, the northern limit of the distribution of *Cenchrus biflorus* distribution, and the 150 mm isohyet are the most commonly used criteria, especially the latter to which the other criteria are related. One might also consider, as does Monod (1954), that the desert is characterized by a permanently restricted vegetation. Accordingly, the transition zone from restricted to diffuse vegetation might be used to delineate the southern margin of the Sahara desert.

Earlier attempts to reconstruct the past vegetation have tended to suggest that, on the assumption that the well-defined latitudinal zonation characteristic of the modern vegetation existed in the past, the climax plant communities have remained virtually unchanged and that a parallel shift of the climatic and vegetation belts took place so that when the isohyets moved northwards or southwards, so did the vegetation belts. We are inclined to believe that regional differences and discrepancies (Muzzolini 1985a, b) do not allow such a comprehensive synthesis. In addition to the information given by Wendorf and Hassan (1980) who, for instance, detected a humid phase in the oases of the western desert of Egypt during the period of 9000-6700 years BP, or by Neumann who describes a Sudanian flora in central Sahara at about 7000 years BP, (Neumann 1987b), there is evidence for several Middle Holocene (7500-6500 years BP) arid phases (Pachur and Kropelin 1989, Rognon 1987, Servant 1973, Servant and Servant-Vildary 1980). This also emphasizes the need to investigate a large number of sites of comparable age within a limited area and to apply a multidisciplinary approach using complementary methods.

The examples presented by the authors in this Chapter constitute the first step towards a palaeoenvironmental approach through anthraco-analysis in the African Sahara. These few scattered results show the necessity to study sites with a high spatial and temporal frequency so as to determine precisely the southern margins of the Sahara from the beginning of the Holocene. Moreover, a multidisciplinary approach, including carpology and pollen analysis, is indispensable for a sound and realistic reconstruction of the former vegetation.

Conclusion

The results of archaeological charcoal analysis (anthracology) in arid and semi-arid regions can be used for palaeoenvironmental reconstruction, and the reconstruction of the vegetation and palaeoclimate, as well as for the detection of fluctuations in desert margins. The development of the methodology integral to the discipline, such as high performance optical devices, the building up of exhaustive reference collections and databases including descriptions of the reference collection as well as computer assisted image analysis, has led to more precise identification of the material and has thus rendered the palaeoenvironmental reconstructions more realistic.

The examples discussed in this Chapter constitute preliminary results of the identification of charcoals from Neolithic sites in the eastern Sahara (Egypt) and the southern Sahara (northern Niger). The identified charcoals enhance our understanding of past fluctuations in the southern Saharan desert margins. They further emphasize the need to investigate a large number of sites of similar age within a limited area, and the value of the application of a multidisciplinary approach using various complementary palaeoecological tools.

Dedication

This Chapter is dedicated to Prof. M. Kassas and to Prof. Jean Paul Barry on the occasion of their 75th and 72nd birthdays, respectively.

References

Barakat, H.N. (in prep.). The anatomy of north-eastern African Acacia charcoals.

Barakat, H.N. (in press). Anthracological studies in the north-eastern african Sahara, methodology and preliminary results from the Nabta playa. Paper presented during the international symposium 'Interregional contacts in the later prehistory of northeastern Africa'. September 8-12, 1992, Poznan, Poland.

Barakat, H.N. (1995a). Middle Holocene vegetation and human impact in central Sudan: charcoals from the neolithic site at Kadero. *Vegetation history and archaeobotany*, 4: 101-108.

Barakat, H.N. (1995b). Charcoals from E-75-6, Nabta, Egypt. *Acta palaeobotanica*, 35: 163-166.

Celles, J.C. and Schulz, E. (1983). Pollens et macrorestes vegetaux. In: Petit-Maire N. et Riser J. (eds.) Sahara ou Sahel? quaternaire recent du Bassin de Taoudenni (Mali), Impremerie Lamy, Marseille : pp. 125-131.

Chabal, L. (1988). Pourquoi et comment prelever les charbons de bois pour la periode antique: les methodes utilisees sur le site de Lattes (Herault). Lattara, 1: 187-222.

Chabal, L. (1992). La representativite paleo-ecologique des charbons de bois archeologiques issus du bois de feu. Bull. Soc. Bot. Fr., 139 (2-3-4): 213-236.

Cour, P. and Duzer, D. (1976). Persistance d'un climat hyperaride au Sahara meridional et central'au courd'e l'Holocene. Rev. Geogr. phys. et Geol. dynamique, 17: 175-198.

Dimbleby, G.W. (1978). Plants and archaeology. Granada publ. ltd, England, 190 pp.

Faure, H., Branchu, Ph. and Ambrosi, J.P. (1993). Contribution de l'Afrique au cycle global du carbone depuis 18 000 ans. Wurzburger Geograpische Arbeiten, 87: 443-463.

Godwin, H. and Tansley, A.G., (1941). Prehistoric charcoals as evidence of former vegetation, soil and climate. J. Ecol., 29 (1): 117-126.

Haynes, C. V., Eyles, C.H., Pavlish, L.A., Ritchie, J. C. and M. Rybak., (1989). Holocene palaeoecology of the eastern Sahara: Selima Oasis. *Quat. Sc. Rev.*, 8: 109-136.

Lezine A.M. (1987). Palaeoenvironnements vegetaux d'Afrique nord-tropicale depuis 12000 BP. These Doc. en Sci, Univ. Aix-Marseille II.

Lezine, A.M. (1989). Vegetational palaeoenvironments of northwest tropical Africa since 12000 BP: pollen analysis of continental sedimentary sequence. Palaeoecology of Africa, 20: 187-188.

Lezine, A.M. Casanova, J. and Hilaire-Marcel, C. (1990). Across an early Holocene humid phase in western Sahara: pollen and isotope stratigraphy. Geol., 18: 264-267.

Maley, J. (1977). Palaeoclimates of central Sahara during the early Holocene. *Nature*, 269: 573-577.

Maley, J. (1981). Etudes palynologiques dans le bassin du Tchad et paleoclimatologie de l' Afrique nord-tropicale de 30000 ans a l'epoque actuelle. Trav. et documents de l' ORSTOM no. 129, ORSTOM Paris.

Maley, J. (1983). Histoire de la vegetation et du climat de 1 Afrique nord-tropicale au quaternaire recent. *Bothalia*, 14: 377-389.

Miller, N. F., (1985). Palaeoethnobotanical evidence for deforestation in ancient Iran: a case study of urban Malyan. *Journal of Ethnobiology*, 5 (1): 1-19.

Monod, Th., (1954). Modes contracte et diffus de la vegetation saharienne. In: Cloudsley-Thompson J. L. (ed.). Biology of deserts, London: 35-44.

Muzzolini, A. (1985a). Les pluviometries durant les humides Holocenes au Sahara. Cah. Ligures de prehistoire et de protohistoire. 2/3: 163-179.

Mozzolini, A. (1985b). Les climats au Sahara et sur ses bordures, du Pleistocene final a l' aride actuel. Empuries, 47, Barcelona, 8-27.

Neumann, K. (1987a). Middle holocene vegetation of Gilf Kebir, S W Egypt: a reconstruction. Palaeoecology of Africa, 18: 179-188.

Neumann, K. (1987b). Jebel Tageru: A contribution to the flora of the southern libyan desert. Journal of arid environ., 12: 27-39.

Neumann, K. (1988). Die Bedeutung von Holzkohluntersuchungen fur die Vegetationsgeschichte der Sahara- das Beispiel Fachi/Niger. Wurzburger Geogr. Arb., 69: 71-85.

Neumann, K. (1989a). Vegetationsgeschichte der Ostsahara im Holozan: Holzkohlen aus prahistorischen Fundstellen (mit einem Exkurs über die Holzkohlen von Fachi-Dogouboulo/Niger). Africa-praehistorica, 2: 341 pp.

Neumann, K. (1989b). Holocene vegetation of the Eastern Sahara charcoal from prehistoric sites. The African archaeological rev., 7: 97-116.

Neumann, K. (1992). Une flore soudanienne au Sahara central vers 7000 BP: les charbons de bois de Fachi, Niger. Bull. Soc. Bot. Fr., 139: 565-569.

Neumann, K. (1993). Holocene prehistoric economies in the Eastern Sahara. In: Geoscientific research in northeast Africa, Thorweihe et Schandelmeier, (eds.) pp. 609-611 Balkema, Rotterdam.

Neumann, K. and Schulz E. (1987). Middle holocene savanna vegetation in the central Sahara-preliminary report. Palaeoecology of Africa, 18: 163-166.

Otto, T. (1993). Phyto-archeologie de sites archeologiques de l'age du fer au Diamare, Nord-Cameroun: le site de Salak. Etude de bois et de graines carbonises. These de Doct. Sci., Universite de Montpellier.

Pachur, H. J. and Kropelin, S., (1989). L'aridification du Sahara oriental a l'Holoc ene moyen et superieur. Bull. Soc. geol. France, 8(1): 99-107.

Popper, V., (1988). — Selecting quantitative measurements in palaeoethnobotany. In: Current palaeoethnobotany, analytical methods and cultural interpretations of archaeological plant remains, C. A. Hastorf, and V. S. Popper (eds.) pp. 53-71, University of Chicago press.

Prior, J. and Price Williams, D. (1985). An investigation of climatic change in the Holocene epoch using archaeological charcoal from Swaziland, Southern Africa. J. Archaeol. Sci., 12: 457-475.

Ritchie, J.C. (1987). A Holocene pollen record from Bir Atrun, north-west Sudan. Pollen et spores, 29(4): 391-410.

Ritchie, J.C. (1994). Holocene pollen spectra from Oyo, northwestern Sudan. Problems of interpretation in a hyperarid environment. The Holocene, 4(1): 9-15.

Ritchie, J.C., Eyles, C.H. and Haynes, C. V. (1985). Sediment and pollen evidence for a middle Holocene humid period in the Eastern Sahara. Nature, 314: 352-355.

Ritchie, J.C. and Haynes, C. V. (1987). Holocene vegetation zonation in the Eastern Sahara. Nature, 330: 645-647.

Rognon, P. (1980). Pluvial and arid phases in the Sahara: the role of non-climatic factors. Palaeoecology of Africa, 12: 45-62.

Rognon, P. (1987). Les phases d'aridite du Pleistocene superieur et de l'Holocene au Sahara; arguments sedimentologiques. Palaeoecology of Africa, 18: 11-33.

Rolando, C. (1992a). Contribution de l'analyse anthracologique al'etude des paleoenvironnements Saheliens. These Doct. Sci., Université Aix-Marseille 3: 270 pp.

Rolando, C. (1992b). Identification des charbons d'Acacia saheliens de l'ouest africain. Etude preliminaire. Bull. Soc. Bot. Fr., 139: 255-263.

Rolando, C. and Riser, J. (1992). Application de l'analyse anthracologique au site neolithique de Hasi el Defa (Tagant, Mauritanie). Premiers resultats. C. R. Acad. Sci. Paris, t. 315, sarie II: 511-514.

Rolando, C. and Roset, J.P. (1991). Approche par l'analyse antracologique de la vegetation de Tin Ouaffadene (gisement archeologique de l'Holocene ancien, Niger nord-oriental). Geodynamique, 6(1): 87-91.

Rolando, C. and Thinon, M. (1988). Perspectives offertes par l'identification de petits fragments de charbons de bois. PACT 22, III. 3: 173-177.

Schulz, E. (1988). Der Sudrand der Sahara. Wurzburger Geogr. Arb., 69: 167-210.

Schulz, E. (1991). Holocene environments in the Central Sahara. Hydrobilogia, 214. 359-365.

Servant, M. (1973). Sequences continentales et variations climatiques: evolution du Bassin du Tchad au Cenozoique superieur, These Doct. Sci., Universite Paris VI.

Servant, M. and Servant-Vildary, S. (1980). L'environnement quaternaire du Bassin du Tchad. In: The Sahara and the Nile, M.A. Williams and H. Faure (eds), Balkema, Rotterdam.

Smart, T.L. and Hoffman, E. S. (1988). Environmental interpretation of archaeological charcoal. In: Current palaeoethnobotany, analytical methods and cultural interpretations of archaeological plant remains C. A. Hastorf and V. S. Popper (eds.) pp. 167-205, University of Chicago press.

Thinon, M. (1988). Utilisation de la microscopie episcopique interferentielle pour l'identification botanique des charbons de bois. PACT 22, III. 4: 179-188.

Thinon, M. (1992). L'analyse pedoanthracologique, aspects methodologiques et applications. These de doctorat es sciences, Universite Aix-Marseille III.

Tucker, C.J., Dregne, H.E. and Newcomb, W.W. (1991). Expansion and contraction of the Sahara Desert from 1980 to 1990. Science, 253: 299-301.

Veen, M. van der (1982). Sampling seeds. J. Archaeol. Sci., 9: 287-298.

Veen, M. van der (1985). Carbonised seeds, sample size and on-site sampling. BAR international series, 401: 99-107.

Wendorf, F. and Hassan, F.A. (1985). Holocene ecology and prehistory in the egyptian Sahara. In: The Sahara and the Nile; M.A. Williams and H. Faure (eds.), Balkema, Rotterdam.

Western, A.C., (1971). The ecological interpretation of ancient charcoals from Jericho. Levant 3, British school of archaeology in Jerusalem: 31-40.

11 Radiocarbon Dating of Arid Zone Deposits

M.J. HEAD[*]

ABSTRACT

The technique of radiocarbon dating has now been in existence for almost 50 years and its advent has resulted in major advances in the fields of archaeology and landscape evolution. However, the fact that most fossil materials form an open system with respect to environmental inorganic and organic carbon has been a distinct disadvantage in its use. The message that many published radiocarbon dates do not actually date the event they are supposed to has not been spread as widely as it should have been. There still seem to be users willing to build a story around a series of radiocarbon dates rather than first building a story around their own research and then testing the radiocarbon dates to see if they fit. The most straightforward aspects of obtaining radiocarbon dates are the sample collection and the measurement of the radiocarbon concentration. The other components of the system, such as interpretation of the collection site to correlate the relationship between sample and event, pretreatment of samples to remove components not associated with the event being dated, and final interpretation of results must never be regarded as straightforward.

This paper summarises the development of the measurement techniques used in radiocarbon dating and reviews many of the approaches that have been used in an attempt to isolate material representative of the event to be dated. There are still advances that can be made in the pretreatment and purification of samples for dating. This paper will provide an insight into possible strategies needed for this development.

Introduction

Outline of the Technique

Radiocarbon (^{14}C) is created in the upper atmosphere because of a natural nuclear reaction involving the capture of a cosmic ray produced

[*]Research School of Earth Sciences, The Australian National University, Canberra, ACT 0200, Australia. Present Address: State Key Laboratory for Loess and Quaternary Geology, Chinese Academy of Sciences, P.O. Box 17, Xi'an 710054, China.

neutron(n) by a nitrogen atom. This results in the emission of a proton(p), and the creation of a ^{14}C atom. The ^{14}C reacts with O atoms in the presence of neutrons to produce CO (Pandow et al., 1960). The ^{14}CO then undergoes an exchange reaction with CO_2, triggered by ultraviolet radiation from sunlight (Dorn et al, 1962). The $^{14}CO_2$ is distributed evenly throughout the global atmosphere. Plants assimilate ^{14}C during photosynthesis and animals eat plants. Thus all living terrestrial creatures maintain their ^{14}C input during life. $^{14}CO_2$ dissolves in the oceans and terrestrial water systems so that all creatures during their life continuously replenish their ^{14}C content. When the plants or animals die ^{14}C input ceases and the remaining ^{14}C atoms slowly revert to ^{14}N, with the emission of β radiation. This decay of ^{14}C occurs with a half-life, which for publication purposes is taken as 5568 ± 30 years, even though a more accurate value of 5730 ± 40 years was reported and accepted as the most probable value at the Fifth International Radiocarbon Conference held in Cambridge in 1962. The concentration of ^{14}C in atmospheric CO_2 is approximately 1 part in 10^{12} and the best estimate of the mean specific activity of ^{14}C in equilibrium with the biosphere is 13.56 ± 0.07 disintegrations per minute per gram (Karlén et al., 1966).

The ^{14}C age is based primarily on the assumption that the concentration of ^{14}C relative to ^{12}C remains constant in the atmosphere and that the balance between the terrestrial and oceanic reservoirs also remains constant. Hence the ^{14}C concentration relative to ^{12}C in a sample at the time of its deposition (death) is assumed to be the same as in a contemporary sample derived from the same environment. Libby et al., (1949) recognised that this assumption needed to be tested and later demonstrated that the conditions were fulfilled with a precision of approximately 10% (Arnold and Libby, 1949). First documented evidence of past variations of ^{14}C concentration in the atmosphere was presented by de Vries (1958), who measured the ^{14}C activity of two wood samples, one that had grown around 1700 AD and the other about 1500 AD. He found that both samples had a ^{14}C activity about 2% higher than that for 19th century wood. He also established the need to consider ^{14}C age determinations in ^{14}C years BP on a radiometric timescale and not an absolute timescale representing calendar years. The year 1950 AD was adopted as the standard reference year.

Since the work of deVries, much effort has gone into the preparation of ^{14}C calibration plots using high precision ^{14}C activity determinations on known-age tree-ring samples that now cover the period from about 12,000 years ago to the present. The periodicity of the peaks and troughs found in the plots has been related to climate variations (changes in the balance of CO_2 between reservoirs), variations in the earth's dipole moment and variations in solar activity. These variations and the techniques used to document them have been extremely well summarized by Stuiver and Pearson (1992); Becker (1992); Sonett (1992); Duplessy et al. (1992); and Lal (1992). These ^{14}C calibration plots have been published by Stuiver and Kra (1986) and Stuiver, Long and Kra (1993).

Anthropogenic dilution of the ^{14}C concentration in the atmosphere from burning of fossil fuels was first noticed by Suess (1955), who measured the ^{14}C activity of early 20th Century tree rings. A more detailed examination of this effect was published by Houtermans et al., (1967), showing that the dilution was of the order of 2%. Since the mid 1950's thermonuclear bomb testing has introduced a significant pulse of ^{14}C in the atmosphere. The maximum concentration of ^{14}C in the atmosphere occurred in the Northern Hemisphere in mid-1963 (more than twice the expected natural value) and in the Southern Hemisphere in late 1964 (almost 1.7 times the expected value). Since then, the concentration of ^{14}C has been gradually decreasing, helped by dilution with fossil fuel derived CO_2 (Levin et al., 1980, 1992). Figure 1 illustrates the bomb-induced ^{14}C variations in the atmosphere. As can be seen from the Figure, this bomb-introduced atmospheric ^{14}C has been an invaluable tool for tracer experiments such as the air-sea gas exchange rate of CO_2 (Levin et al.,1992). The bomb spike also makes it possible to obtain very precise (within 2-3 years) ^{14}C ages for short-lived materials that have died within the last 45 years.

Modern reference standards for ^{14}C age determinations initially consisted of known-age tree rings from the period between 1850 and 1890 AD. Since this material is rather scarce, two oxalic acid standards distributed by NIST (formerly NBS) are recognised as the primary modern reference standards (Stuiver and Polach, 1977; Klinedinst et al., 1994), namely SRM 4990B (Oxalic Acid I) and SRM 4990C (Oxalic Acid II).

Natural and Laboratory-induced Isotopic Fractionation

There are three isotopes of carbon found in nature: stable isotopes ^{12}C and ^{13}C, together with the radioactive isotope ^{14}C. The natural abundances of ^{12}C and ^{13}C are 98.89% and 1.11% respectively, while that of ^{14}C is one part in 10^{12}. Isotope equilibrium exchange reactions within the inorganic carbon system i.e., atmospheric CO_2 - dissolved bicarbonate - solid carbonate lead to an enrichment of ^{13}C and ^{14}C in carbonates, while kinetic isotope effects during photosynthesis concentrate the light isotope ^{12}C in the synthesised organic material (Hoefs, 1997). Fractionation of ^{14}C to ^{12}C is twice that of ^{13}C to ^{12}C (Craig, 1959). Isotope ratios of ^{13}C/^{12}C can be measured by mass spectrometer and are reported as an enrichment or depletion (δ^{13}C) with respect to the PDB standard. A representation of

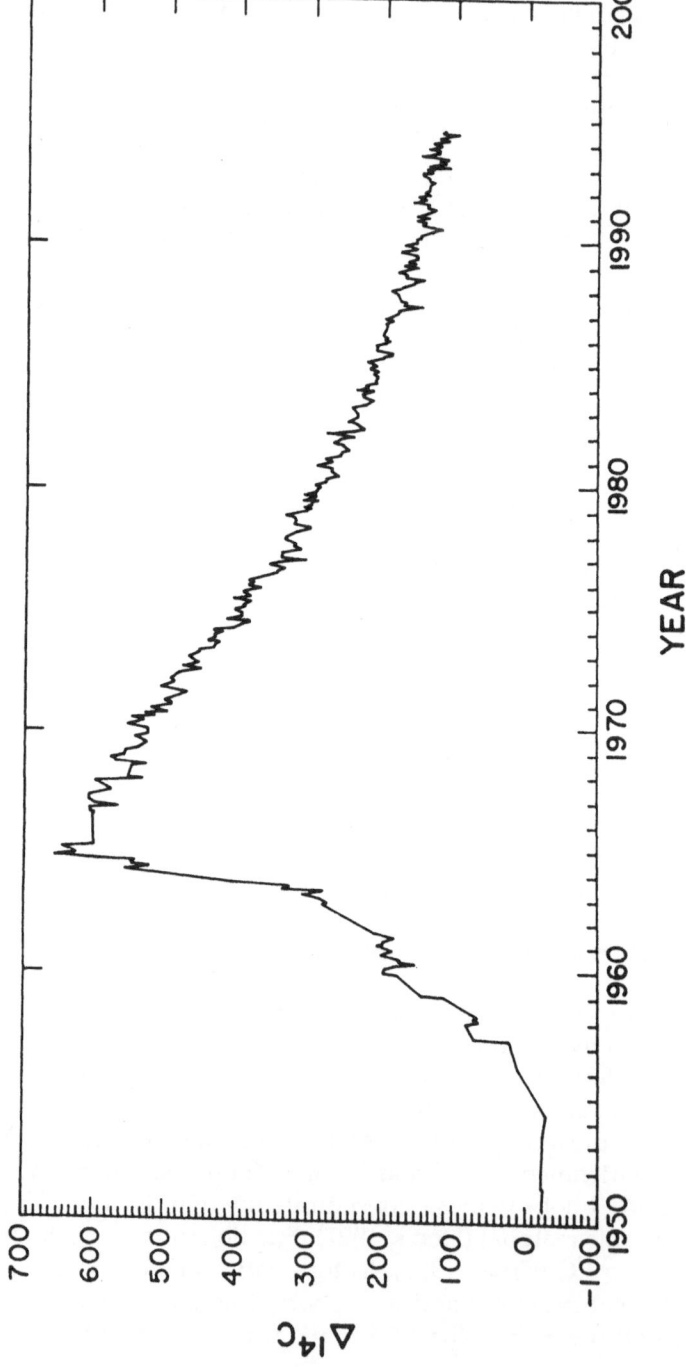

Fig. 1: Plot of D^{14}C $vs.$ year of measurement for atmospheric ^{14}C levels with respect to the ^{14}C Modern Standard, representative of the Southern Hemisphere.
Reproduced with the kind permission of Dr J.C. Vogel, Quaternary Research Dating Unit, EMATEK®CSIR, Pretoria, South Africa.

isotopic fractionation found in nature is shown in Figure 2. The isotopic fractionation factor is taken into account in the reporting of a conventional ^{14}C age (Stuiver and Polach, 1977), where the ^{14}C age results are normalised to the $\delta^{13}C$ value of -25 per mil, which is representative of wood (since wood from tree rings is the primary ^{14}C modern standard).

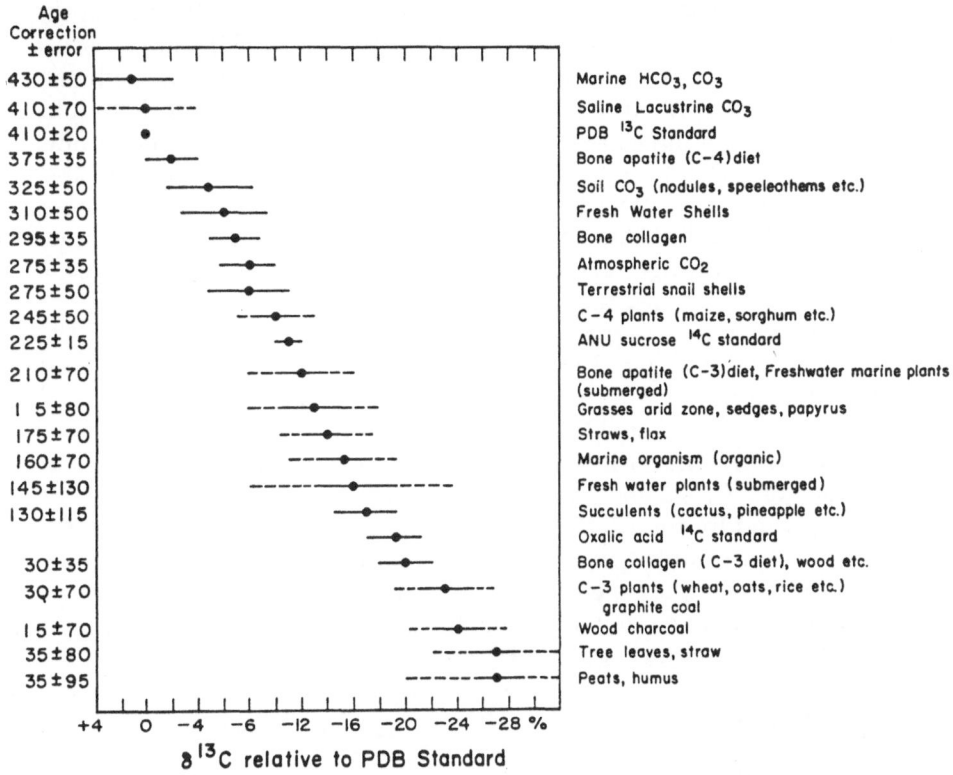

Fig. 2: $\delta^{13}C$ variations in nature and their significance for ^{14}C dating. These average values may be used in age calculation if measured values cannot be obtained (Stuiver and Polach, 1977; Gupta and Polach, 1985).

Isotopic fractionation caused by laboratory processing occurs more often than people would like to admit. The only safe way of ensuring that isotopic fractionation will not affect the sample measurement is to ensure virtually 100% yields during any chemical processing step. Most laboratories monitor sample processing closely in order to minimize the possibility of fractionation.

Measurement of ^{14}C ages

The measurement of ^{14}C activity or concentration tends to be rather difficult and time consuming because of the extremely low concentration of ^{14}C in the atmosphere and the relatively low energy of the β radiation emitted by the ^{14}C isotope. This means that measurements need to be carried out in conditions such that most natural radioactivity is shielded from the detectors. The two most effective measurement techniques for ^{14}C radioactivity have been gas proportional counting using CO_2, CH_4, or C_2H_2 as counting gas (Stuiver et al., 1979) and liquid scintillation spectrometry, using C_6H_6 as the counting liquid (Polach, 1992). Sample sizes can vary from approximately 50 mg carbon to 4 g carbon. With the smaller size samples, counting times can extend to one week (liquid scintillation), or months (gas proportional counting). Under normal counting conditions, using 2 - 5 g carbon, age limits can extend up to 65,000 years BP(Gupta and Polach, 1985; Kromer and Münnich, 1992). These limits are not usually attainable because of the extremely small amounts of incorporated younger carbon needed to produce much younger ages (Long and Kalin, 1992; Chappell et al., 1996).

Accelerator mass spectrometry was first used for measurement of ^{14}C in natural organic material in 1977 and since then has become the predominant measurement technique (Gove, 1992; Beukens, 1992). Negative carbon ions are accelerated to between 2 and 6 MeV,with loss of electrons consequently to form positive carbon ions with a charge state of +3 or +4. These ions of ^{12}C, ^{13}C and ^{14}C can be separated and detected. The most popular procedure for AMS laboratories is to obtain a $^{14}C/^{13}C$ ratio for both unknown samples and modern standards (Donahue et al., 1997). Carbon in the samples is converted to graphite, which is pressed into a target holder (sample size requirements varying from as little as 50 μg to 2mg carbon) (Klinedinst et al., 1994).

^{14}C Age Calculation

Since no technique used in the measurement of radioactive decay is 100% efficient, ^{14}C activity or concentration in a sample is measured as counts per minute (cpm). This value can be related to that obtained for the NBS (NIST) oxalic acid I modern standard. The actual modern standard value, normalised for isotopic fractionation is taken as

$$A_{ON} = 0.95 A\,OX\left[\frac{1 - 2(19 + \delta^{13}C)}{1000}\right]$$

The activity (concentration) of ^{14}C in a sample can then be related to that of the modern standard as

$$d^{14}C = \left(\frac{A_s}{A_{ON}} - 1 \right) 1000$$

expressed as per mil depletion or enrichment with respect to the modern standard. This value can then be normalised for isotopic fractionation.

$$D^{14}C = d^{14}C - 2(\delta^{13}C + 25)\left(1 + \frac{d^{14}C}{1000} \right)$$

The conventional ^{14}C age is then given by

$$t = -8033 \ln\left(1 + \frac{D^{14}C}{1000} \right)$$

where the value 8033 is the mean life of ^{14}C, based on the ^{14}C half-life value of 5568 yr. The conventional ^{14}C age is accompanied by a standard error, which can be calculated as a statistical sum of the standard errors associated with each calculation step. Hence, the ^{14}C age is reported as t ± one standard deviation. As an example, a conventional ^{14}C age reported as 1000 ± 100 yr BP implies that there is a 68.3% probability that the true age lies between 900 and 1100 yrs BP. A more realistic approach to the ^{14}C age for comparison or calibration purposes is to double the error, so that the above ^{14}C age becomes 1000 ± 200 yr BP implying that there is a 95% probability that the true age lies between 800 and 1200 yr BP. A much more detailed approach to the calculation of a conventional ^{14}C age has been presented by Stuiver and Polach (1977) and Gupta and Polach (1985). AMS laboratories relating the concentration of ^{13}C to ^{14}C have developed calculation procedures similar to those described by Donahue et al. (1990, 1997).

It is extremely useful to make an exception to the reporting of a conventional ^{14}C age when the ^{14}C activity (concentration) of the sample is greater than that of the modern standard. This occurs when the sample is less than forty years old since it contains 'bomb'-produced ^{14}C. Calculation of a conventional ^{14}C age results in a negative value (a future age?). If this value is reported as an enrichment with respect to the modern standard (positive $D^{14}C$), a comparison with plots of ^{14}C content in the atmosphere since 1950 (Fig.1) can pinpoint accurately the age since death of the sample.

Sample collection and submission

When collecting a sample for ^{14}C age determination, great care must be taken to avoid contamination by outside factors such as collection equipment, storage of sample etc. This becomes extremely important if the sample is to be submitted for accelerator mass spectrometry. It is also advisable that, for every sample submitted, the user documents his/her observations on the possibility of natural processes providing foreign carbon containing material that could have been incorporated into the sample. Examples of these processes are rootlet penetration, seasonal water table changes incorporation of soil organic matter, animal disturbances and so on. For this reason it is advisable that samples for dating be submitted with a very detailed report on the circumstances of collection and the type of environment in which the samples were found.

The complexity of the problem of post-depositional alteration of potential samples for ^{14}C age determinations becomes apparent when the laboratory scientist attempts to find specific patterns of alteration within each sample type. It needs to be taken into consideration that some of the most common materials submitted for dating are not necessarily homogeneous in their structural make-up and elemental content, thus making it difficult to specify a uniformly applicable pretreatment for any given type of material.

The merit of a particular type of pretreatment applied to a sample can only be assessed if information as to possible contaminants, contamination mechanisms and state of preservation of the material to be dated can be obtained. User interaction becomes very important. Liaison between the user and the dating facility can furnish valuable information such as the probable age of the sample, environment prior to collection and possible sources of sample contamination (Gupta and Polach, 1985).

Fossil Wood as ^{14}C Dating Material

Basic chemistry and environmental interactions

The wood cell wall is essentially made up of a meshwork of long, thin threads of cellulose embedded in a matrix of encrusting substances (the hemicelluloses and lignin) that are cross-linked polymers, disordered or amorphous in structure. Deposition of lignin and the hemicelluloses occurs at the last stages of cell wall development. Once lignification has been completed, the protoplasm of the cell is consumed, leaving it biologically dead. Extractable compounds (such as gums, resins, fats, waxes, tannins, sugars etc.) are deposited in cell cavities and cell walls. These compounds may have an effect on wood preservation by inhibiting some chemical reactions, or by inhibiting or attracting bacterial or fungal attack (Tsoumis, 1969).

Cellulose is essentially a linear polymer composed of glucose units linked together to form straight chains, with the number of glucose units in a chain ranging from 8000 to 10,000. They are insoluble in 17.5% NaOH solution and are sparingly soluble in strong acid solutions. The hemicelluloses are a group of carbohydrates consisting of a combination of various single carbohydrate units with different linkage types. These are soluble in NaOH solutions. Lignin can be considered as a generic name for a series of aromatic polymers whose composition varies according to tree species. These polymers are built up mainly from phenylpropane building blocks forming a three-dimensional network system. They are not hydrolysed by acids, are readily oxidised, and many are soluble in hot sodium hydroxide, sodium bisulphite and sodium chlorite solutions (Tsoumis, 1969, Head, 1979).

When wood is exposed to the outdoors above ground, a complex combination of factors such as solar radiation (ultraviolet, visible and infrared light), moisture (dew, rain, snow and humidity), temperature and oxygen, contribute to the weathering process. The photochemical degradation of wood due to the UV component of sunlight occurs quite rapidly on the exposed wood surface (0.05-2.2.5 mm), initially causing a yellowing or browning that proceeds to an eventual greying. These colour changes can be related to the decomposition of lignin and polyphenols, since they absorb light strongly at a wavelength band below 200 nm and have a strong peak at 280 nm with absorption through the visible spectrum. Absorbed energy causes the dissociation of bonds in the molecules of the wood constituents, thus producing free radicals as the primary photochemical products, which can lead to depolymerisation (Feist and Hon, (1984).

Further weathering processes can then be initiated by mechanisms such as the frequent exposure of the wood surface to rapid changes in moisture content. Water vapour is taken up directly by absorption under increased relative humidity: consequently the wood swells. Stresses are set up in the wood as it swells and shrinks due to moisture gradients between the surface and the interior. These induced stresses are greater the steeper the moisture gradient, and are usually largest near the surface of the wood (Stamm, 1964). As weathering continues, rainwater washes out degraded portions and further erosion takes place. Because of the different types of wood tissue on the surface, erosion and cracking differ in intensity and the wood surface becomes increasingly uneven. Hardwoods erode more slowly than do softwoods. Abrasion or mechanical action, such as wind, sand, and dirt, can be an important factor in the rate of surface degradation and removal of wood. Small particles such as sand can become lodged in surface cracks and, through swelling and shrinking weaken fibres in contact with the particles. Solid particles in combination with wind can have a sandblasting effect (Feist and Hon, 1984).

When wood is buried, it becomes part of a dynamic, open system and as such, becomes affected by many of the environmental mechanisms occurring within that system. The degree of interchange with the environment will often depend on the degree of weathering that has occurred before burial.

Methods of sample analysis and treatment for dating

Olson (1957) outlined the problems associated with possible humic acid contamination of samples for ^{14}C dating and Olson and Broecker (1958) proposed the separation of the holocellulose (carbohydrate) fraction as a useful pretreatment technique for fossil wood samples. Head (1979) concluded that the most inert and therefore the most reliable chemical component of wood for ^{14}C activity measurements is pure cellulose, provided unmodified cellulose can be shown to exist in a wood sample. The presence or absence of unmodified cellulose chains can be determined using X-ray diffraction techniques. These provide an indication of the degree of structural degradation in the wood which may not be obvious by visual inspection (Head, 1979, 1980, 1982). A portion of the wood sample is reduced in a Wiley mill and the size fraction ranging from 0.3 mm to 1.0 mm is separated by sieving. This wood sample is then pressed into a wafer, which can be used to produce X-ray diffraction patterns, examples of which are shown in Figure 3. If cellulose is found to be absent from the wood sample, then the reliability of any ^{14}C age obtained from the wood sample becomes questionable. One example where pure cellulose may not provide a sample with a ^{14}C age representative of the true age of the wood can occur in arid or semi-arid areas if the wood has come from multistem type species growing from a lignotuber structure (mallee). In this type of case, the lignotuber can store starch for relatively long periods of time and make it available for cellulose formation and stem growth (Wellington et al.,1979; Head and Lacey, 1988).

Another useful technique for testing the viability of wood samples for ^{14}C dating is to determine the %C, %H, %N and % ash contents of oven-dried fossil wood. From these values, atomic H/C and O/C ratios for the wood organic component can be determined. These relationships can also provide an indicator for the degree of degradation of the fossil wood sample, as well as indicate the possibility of incorporation of external organic compounds into the wood (Head, 1982).

The chemical pretreatment techniques adopted at the Australian National University (ANU) for fossil wood samples involve careful examination of the wood to recognise foreign materials that have become incorporated into the sample (eg. plant root material). Surfaces of the wood pieces are cleaned of soil and possible root remains are removed.

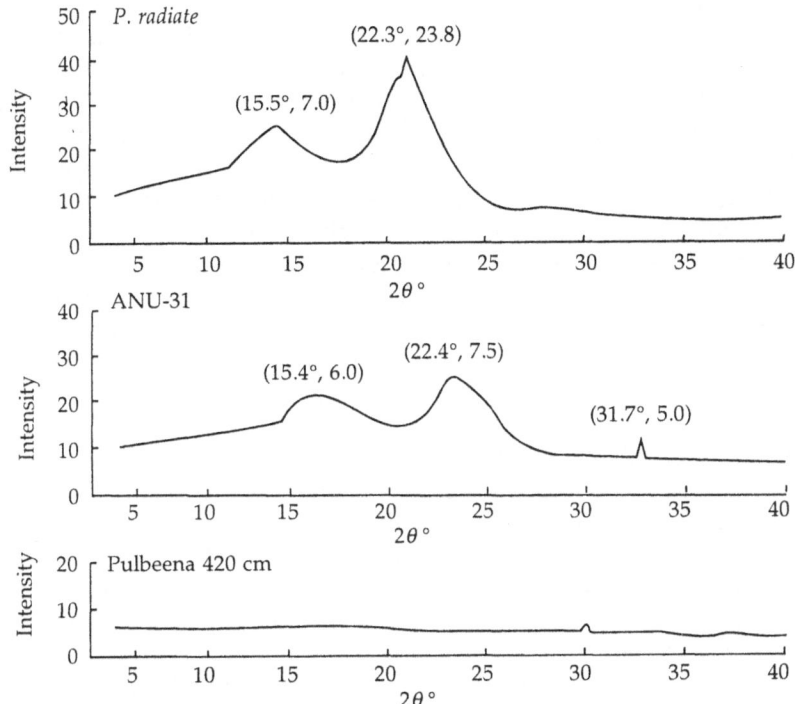

Fig. 3: X-ray diffraction patterns of undegraded partially degraded and fully degraded wood (Head 1979).

The wood is then broken up into matchstick size pieces, which are oven dried at about 100 °C. The wood fragments are transferred to a Wiley mill and reduced to fragments of size ranging between 1 mm and about 0.3 mm. Wood powder should not be used for pretreatment as uniform sample size is advantageous with regard to further pretreatment. The wood fragments are then placed in a Soxhlet extractor for solvent extraction to remove resins, fats, tannins, waxes, sugars etc., as it is possible for these types of substances to be translocated within the tree. (Long et al., 1979). Since the solvent extraction steps involve the use of organic solvents, which must be completely removed from the sample to avoid contamination, it is advisable to design the solvent extraction procedure so that a series of solvents is used ranging from slightly polar through to highly polar. Each successive solvent should also be miscible with the preceding one so that the more polar solvent washes the preceding less polar solvent from the sample. The final solvent needs to be distilled water, to ensure complete removal of organic solvents from the sample. The solvents most commonly used at the ANU are:—

    1. A 2:1 (v:v) mixture of chloroform with ethanol

2. Ethanol

3. Water

Extraction with each solvent mix is carried out until the solvent si-phoned from the wood sample is completely colourless. This may take a few hours or even a day. After each solvent extraction (not water), the wood is air dried. After water extraction, the sample can be carefully oven dried at about 60 °C. Sometimes, if there is a large amount of water-soluble material being extracted, it may be advisable to carry out a second extraction (Head, 1979; Gupta and Polach, 1985).

The extracted sample can then be placed in an Erlenmeyer flask with distilled water. A drop or so of HCl is added to bring the pH of the solution to about 3. The contents of the flask are then heated to ~75 °C, after which small amounts of potassium chlorite ($NaClO_2$) are slowly stirred in. This reaction needs to be carried out in a fume hood as chlorine gas is produced. Normally, about 7.5 g $NaClO_2$ is used per 25 g wood. The flask contents are then kept at 75 °C for about 3 hours (stir every 15 minutes or so) and the contents of the flask filtered thereafter. The wood should be basically white with a slight yellowish tinge: if it is not, the treatment needs to be repeated for a further 3 hours. The final product is known as holocellulose, containing basically cellulose plus hemicelluloses with a small amount of lignin, which is probably tightly bound to the hemicelluloses. In many cases this level of pretreatment is probably all that is necessary. With certain fossil wood samples, the holocellulose may be quite dark in colour, so it is worthwhile carrying out further leaching steps with NaOH solutions in a nitrogen atmosphere (Green, 1963; Head, 1979).

Charcoal

General considerations

Combustion of wood and plant material involves a complex series of physical transformations and chemical reactions that are further complicated by heterogeneity of the substrate. Wood and cellulose materials in general do not burn directly: under the influence of sufficiently strong heat sources they decompose to a mixture of volatiles, tarry complexes and highly reactive carbonaceous char. Gas-phase oxidation of the combustible volatiles and tarry products produces flaming combustion. Solid phase oxidation of the remaining char produces glowing or smouldering combustion, depending on the rate of oxidation (Shafizadeh, 1984).

The production of volatiles leaves a solid residue that is neither intact substrate nor pure carbon, but a different material at various stages of charring and carbonisation. The intermediate chars contain a high concentration of free radicals trapped in a rigid structure or stabilised by aromatic and olefinic structures, a large surface area and a high degree of reactivity (Shafizadeh, 1984). Hence charred material covers a wide spectrum of composition, from more or less pure carbon to the much more humic-like composition of partially burned and subsequently degraded plant remains. The 'charcoal' has a very high surface area and hence becomes an active site for incorporation of younger organic material transported by water moving down through the sedimentary profile in which the charcoal has been deposited. Measurement of H/C and O/C atom ratios, together with Fourier transform infrared (FTIR) spectrometry can supply valuable information with respect to the presence of possible organic contaminants in a charcoal sample (Head et al., 1996).

Methods of Pretreatment

The commonly used acid-alkali-acid treatment to remove possible organic contaminants (Olson and Broecker, 1958) can pose problems if the alkali treatment is carried out in the presence of air. There is strong evidence that CO_2 can react with OH groups in the 'charcoal' to produce carboxylate ions (Head et al., 1996).

Massive charcoal fragments (especially if heavily leached and brittle) are much more easily decontaminated. The only problem is that this type of charcoal may antedate the event associated with its collection, depending on the actual age of the wood before it was burned. Fine-grained charcoal poses another problem in that it is very easily washed down through a sandy profile, which means that it could be younger than the event it is supposed to be dating. The major difficulty in removing younger material from charcoal samples comes when the true age of the charcoal could be either close to or beyond the limits of the ^{14}C technique (Chappell et al., 1996). Gillespie et al. (1994) used a technique involving treatment with hydrofluoric acid (HF) to break down any clay left in the sample, then a cold oxidation with 10% sodium chlorate ($NaClO_3$) in 35% nitric acid (HNO_3). Bird et al. (1996) developed a similar technique involving extensive oxidation. These techniques look highly promising, though testing the various possible pretreatment techniques, and then analysing the remaining material, as well as looking at H/C and O/C ratios etc., really needs to be carried out.

Plant Macrofossils, Pollen and Peat

Basic problems

The structure of peaty materials varies from site to site, depending on climate, the plant species present and the mode of deposition. Since most peats are formed under acidic conditions, it may often be assumed that no pretreatment is necessary and, in many instances, this has been the case. The selection of pollen as a ^{14}C dating material depends on the assumption that the pollen has been deposited directly into a sediment matrix and has not been 'reworked', i.e., eroded from previously deposited sediments, then transported and redeposited in newly forming sediments. Vogel et al. (1989) showed that specific plant macrofossils that could be considered to have had an association with the event to be dated were ideal dating sources for lake sediments, and that total organics from lake sediments may consist of many different individual components, with quite different ^{14}C ages.

Peats formed in neutral or slightly alkaline conditions may pose problems because of enhanced mobility of humic material. An example of this is Pulbeena Swamp, north-western Tasmania. A 450 cm section was investigated and reported by Colhoun et al. (1982). A series of ages down to ca. 200 cm indicates a reasonable age/depth relationship. As discussed by Chappell et al. (1996), the ^{14}C ages for the sediments below 200 cm are indicative of chemical/physical mixing.

With the advent of accelerator mass spectrometry for ^{14}C age determinations, the dating of very specific components of organic sediments becomes possible. Andree et al. (1986) compared ^{14}C ages of gyttja versus remains of terrestrial plants (mainly fruits of birch trees) separated from a sediment core taken in Lake Lobsigen, a small lake on the western Swiss Plateau. The gyttja ages were consistently on average 800 years older than the corresponding ages for the terrestrial material, indicating that the gyttja was derived from aquatic plants which had taken their carbon from the dissolved CO_2 and bicarbonate in the lake water (contains bicarbonate originating from much older carbonaceous rock, rather than from the atmosphere). Similarly, a study by Marčenko et al. (1989) showed that living aquatic plants in lake systems in a karst environment can have an apparent ^{14}C age of at least 1500 years. These particular plants obtain their CO_2 from the bicarbonate in the lake rather than CO_2 from the atmosphere. A useful indicator in this case is the $\delta^{13}C$ value for each plant species.

Physical and chemical pretreatment

Physical 'contamination' of peat samples can be caused by vertical contamination and horizontal spread of younger root material. In order to guard against this possible contamination mechanism, a strategy of wet sieving and selecting the most suitable size fraction for dating is often employed. However, if the root material has degraded rapidly, a further separation into chemical fractions (acid/alkali/acid, with alkali treatment carried out in the absence of air) could be necessary. Williams (1989) carried out a series of pretreatment tests on freshwater peat collected from an exposure formed by tidal erosion of a bog located on the north side of Carrying Place Cove, South Lubec, Maine. The sampling was carried out 210 cm below the surface of the bog. The age of the material had previously been estimated to be 2000 BP. The only sources of contamination appeared to be post-depositional root penetration, bacterially generated humic acid and salt-water spray. The peat was manually broken into small pieces and slurried with deionised water in 4-l beakers. The peat was sieved, oven dried, ground to < 100 μm and homogenised. A series of leachings were carried out using various concentrations of HCl and NaOH. ^{14}C ages were obtained for both NaOH soluble and insoluble fractions of the peat, with the NaOH soluble ages being consistently significantly younger. The oldest ^{14}C age came from a sample of the peat that had undergone treatment with $NaClO_2$.

Brown et al. (1992) extracted pollen from peat samples using slight modifications in standard pollen separation techniques and isolated pollen in the 20-40 μm range for AMS ^{14}C determinations, while Long et al. (1992), developed a technique of hand picking pollen grains from standard pollen concentrate samples, also for AMS. The results from these two studies did not seem to deviate notably form stratigraphic chronologies obtained previously.

Phytoliths

Background

Opaline silica phytoliths consist of hydrous silica deposited within leaves, stems and roots of plants. The Si enters the plant as monosilicic acid dissolved in groundwater and is transported by the vascular system. Deposition in cells is as an amorphous silica gel. The sequence includes breakdown of the cell nucleus and organelles, thickening of the cell wall and formation of prism-like silica bodies within the cell. The cell eventually fills with Si, forming a solid body. Bits of the original cellular material are trapped in the body of the developing phytolith and persist as dark spots visible within the translucent microfossil. Phytoliths have been identified in most sediment types and their shape is controlled by the shape of the plant cells at the time of silica deposition. Hence, there is a potential for correlating phytolith shape with species of vegetation so that examination of phytolith assemblages may provide information similar to that obtained from pollen assemblages (Hart and Humphreys, 1997).

Carbon from the original cell material in phytoliths provides an alternative source for radiocarbon dating (Mulholland and Prior, 1992). Use of AMS techniques reduces the amount of carbon needed to the range where a kilogram or less of sediment will usually yield sufficient carbon for dating (~2 mg). Carbonaceous material enclosed in the phytolith also provides the advantage that the Si seals it from the soil environment. A cleaning procedure is used to ensure that organic material on the surface or infiltrated into cracks is removed before dating.

Pretreatment

Since phytoliths are mostly silt sized, the first step in their separation is to isolate the silt fraction. Warm (60°C) hydrogen peroxide is used to remove organic compounds that bind particles into aggregates. Disaggregation with sodium hexametaphosphate ensures clay particle dispersion before sieving. Sand grains larger than 88 μm are removed by sieving. Fine silt (< 20 μm) is removed by settling. A solution at specific gravity 2.3 is used to float phytoliths out of the sediment. The liquid is then diluted with water to a specific gravity of < 1.5 to settle the material. To increase phytolith recovery, the extraction step is repeated twice for a total of three times. The phytolith fraction is then subjected to extraction for an additional three times to remove any extraneous sediment particles. The resultant sample is composed almost exclusively of phytoliths. Cleaning of phytoliths is accomplished by a series of baths: hot (85°C) in chromic acid, warm (60°C) hydrogen peroxide and cold (22°C) 6N hydrochloric acid (Mulholland and Prior, 1992).

Organic Sediments and Soils

Problems

Radiocarbon dating of the organic component of soils, lake sediments etc., is complicated by contamination with recent organic matter. Most classic pretreatment techniques commonly used for removing recent organic material from samples of charcoal or wood have been shown by many researchers to be inadequate for eliminating contamination from the soil samples. Occasionally, charcoal or plant fragments may be found in the sediments and may provide a useful dating medium. However, material that was originally collected as charcoal has often been found to consist of a thin black organic layer covering a small rock fragment or clay aggregate. This phenomenon has been shown to be common in meadow soils or swampy sediments where aeration may have been restricted and degraded plant material reduced to a 'charcoal' like component (Kononova, 1966). Similarly, Skjemstad et al. (1994) examined layers of fine-grained charcoal fragments that had most probably moved down relatively loose-textured sandy soil profiles, and become concentrated at a specific level within the profile. These fragments examined under election microscopy, revealed a woody origin, yet ^{13}C CP MAS nuclear magnetic resonance spectrometry indicated basically a condensed aromatic structure.

The organic component of soils and other organic sediments is basically made up of organic substances of an individual nature (fats, waxes, resins, proteins, tannic substances and many others), and humic substances (Kononova, 1966). Humic substances are complex polymers formed from breakdown products of the chemical and biological degradation of plant and animal residues. They are dark coloured, acidic, predominantly aromatic compounds ranging in molecular weight from less than one thousand to tens of thousands (Schnitzer, 1976). They can be partitioned into three main fractions:

(i) Humic acid–soluble in dilute alkaline solution but can be precipitated by acidification of the alkaline extract.

(ii) Fulvic acid–soluble in alkaline solution but also soluble on acidification.

(iii) Humin–humic fraction that cannot be extracted from the soil or sediment by dilute acid or alkaline solutions.

The overriding problem for absolute dating of soils is that an exact numerical soil age may be unobtainable. Absolute ages can be linked only to ^{14}C dating of buried palaeosols, to charcoal or to wood fragments etc. All other ^{14}C dating of soils yields model age levels, expressed as Apparent Mean Residence Time of the dated organic carbon fraction (Scharpenseel and Becker-Heidmann, 1992). Geyh et al. (1971, 1983) indicated that age determinations of the organic component of soils originating in calcareous parent materials could be affected by an unknown amount of fossil CO_2 formed by the decomposition of soil carbonates, resulting in a ^{14}C age older than the true age. Alternatively, with non-calcareous soils, the ^{14}C age would be younger than the true age because of continuous incorporation of younger organic material coming from continuous surface decomposition of plant material. Scharpenseel and Schiffmann (1977) came to the same conclusion.

Wang et al. (1996) carried out an extensive ^{14}C dating study of total soil organic matter at a series of sites and derived a model for estimating soil ages by determining input and decay rates of organic matter. The model seemed to be reasonably successful provided the soil was not at steady state. Chichagova and Cherkinsky (1993) considered individual soil profiles to have their own dynamic properties, with the organic component separated into humic acid, fulvic acid and humin fractions dependent on the effects of the existing microclimate. Studies of the various chemical fractions provided data for the production of models of carbon exchange, which could be correlated with palaeogeographic data. Cherkinsky and Brovkin (1993) presented a model for humus accumulation in recent soils using the "atom bomb" ^{14}C atmospheric pulse as a tracer. Certainly, at this stage the complexity of the soil organic matter dynamic system seems to have foiled attempts to isolate a core organic component representative of the age of formation of soils or sediments.

Pretreatment techniques

Olson and Broecker (1958) considered that the most useful pretreatment technique to remove organic contamination from soils was to separate soil humic acids from the more inert chemical component of soil organic matter. The humic acid component was considered to be the most likely source of contamination because of its potential mobility within the soil. They outlined a technique whereby the soil sample was treated with boiling 2% NaOH solution: the black NaOH soluble material was then filtered off and acidified to a pH of <1. The humic component formed a precipitate at this pH and could be separated by filtration or centrifugation, rinsed and dried. The NaOH insoluble residue, or non-humic component of the sample, was treated with a boiling solution of 5% HCl, then filtered, rinsed and dried. This was assumed to be the most likely fraction to produce a reliable ^{14}C age.

A pretreatment strategy based on the above was used by Polach and Costin (1971) in an attempt to obtain ^{14}C ages for the formation of a palaeosol buried by periglacially induced slope mantle deposits 2.5 to 3.5 m thick in the Snowy Mountains area of southern New South Wales, at an altitude of 1000 to 2500 m. Remanent palaeosol material containing carbonised wood fragments was sampled from four sites. Visible contaminants such as plant rootlets were removed from each sample, then carbonised wood fragments which were used for dating, together with total soil organic matter, NaOH soluble and NaOH insoluble fractions. The carbonised wood fragments were considered to give the most reliable ^{14}C ages, with a mean age of 31,700 ± 1700 yr BP. A humic acid (NaOH soluble) fraction from one site gave the oldest ^{14}C age (32,050 ± 1650 yr BP), though the mean ^{14}C age for all humic acid fractions was much younger (24,300 ± 700 yr BP). The other two fractions gave still younger ^{14}C ages (~20,000 yr BP). The basic problem posed by this study was the fact that the fraction which most people would have expected to contain the majority of younger organic material gave the oldest ^{14}C age at one of the four sites.

Gilet-Blein et al. (1980) collected soil samples from a series of known-age soil profiles. Chemical treatment of these samples involved the following steps:
  (i) Elimination of most of the free organic matter (acid wash to remove carbonate, flotation, stirring, decantation, sieving).
 (ii) The samples were stirred for 15 hr in 0.1 M NaOH at room temperature. The insoluble fraction bound to the clay (humin) was acidified, rinsed and dried. The soluble fraction was acidified and the precipitate collected.
(iii) Progressively more concentrated acid solutions were used in 15-hr steps under reflux. The residue after the last hydrolysis was treated with 0.5 M NaOH producing an NaOH soluble fraction and leaving a final residue. The most resistant fraction (final residue) would generally have been assumed to be the oldest.

The ^{14}C dates from all sites were found to be too young. It was noted that for acid soils, the humin fraction produced older ^{14}C ages than the humic acid fraction. The reverse occurred with basic soils (carbonate soil). The conclusion reached was that most fossil soils could not be dated using the ^{14}C technique with organic matter obtained by the above fractionation techniques (Gilet-Blein et al., 1980).

Calderoni and Schnitzer (1984) extracted humic acids from six palaeosols in southern Italy by first shaking air-dried palaeosol with 1 M HCl at room temperature for 16 hr to remove carbonates. The supernatant solution was separated by centrifugation and the residue was neutralised with 1 M NaOH solution to pH 7. Dilute NaOH solution (0.1 M) was then added and the suspension shaken intermittently under N_2 at room temperature for 24 hr.

It has also been shown that extensive washing and a variety of other purification procedures fail to separate all carbohydrate and amino acid-containing substances from most soil humic acids. It is likely that these types of material are covalently linked to the core compounds in the humic acid macromolecules. Dry humic acids are difficult to re-wet and the hydrophobic nature of the dry material suggests that the less polar components orientate towards the outside of the structures during drying (Hayes, 1985).

Head et al. (1989) and Zhou et al. (1990) used a similar solvent extraction technique to that used in the pretreatment of wood (Head, 1979), after decalcification of sediment samples before using the normal alkali leaching technique for separation of humic acid and humin fractions from swamp sediment and palaeosol samples at Bei Zhuang Cun, China. The humid acid fractions of swamp sediment samples from the bottom of the profile gave ^{14}C ages that were not significantly different from ^{14}C ages of wood fragments found in the sediment sample, but gave significantly older ages than the corresponding humin fractions. For palaeosol samples in the upper part of the profile, the humin fractions gave older ages, but even these were younger than the ages of corresponding materials and historically dated pottery. Palaeosols from the Baxie loess/palaeosol sequence at Baxie and other sites in China (Zhou et al., 1992, 1994) gave similar results in that the humin fractions were older but were still significantly younger than corresponding thermoluminescence ages.

Table 1 illustrates more recent dating carried out at Bei Zhuang Cun which indicates the difficulty in interpreting ^{14}C results from soil chemical fractions (Head et al., 1997; Zhou et al., 1990, 1993).

Table 1: ^{14}C ages and δ^{13}C values of chemical fractions of the organic component of palaeosol samples collected from Bei Zhuang Cun, Shaanxi Province, China. In each case successive NaOH leaches (up to 3) were carried out to test the assumption that the NaOH soluble component yielded uniform material.

Sample Bei Zhuang Cun	Lab. Code	Fraction	δ^{13}C(%)	^{14}C Ages (yr BP)
Depth 300 cm				
86-C-28	ANU-6201	NaOH insol		3270 ± 200[*]
	ANU-6201	NaOH sol (l)		2540 ± 250[*]
	ANU-6201	NaOH sol (1)		1680 ± 50[**]
	OZB-315U	NaOH sol (3)	−22.3 ± 0.2	2200 ± 200[***]
Depth 360 cm				
86-C-27	ANU-6202	NaOH insol		3420 ± 360[*]
	ANU-6202	NaOH sol (1)		3010 ± 360[*]
	ANU-6202	NaOH sol (1)		3040 ± 100[**]
	OZB-283U	NaOH sol (1)	−19.6 ± 0.2	3400 ± 180[***]
	OZB-316U	NaOH sol (2)	−20.4 ± 0.2	2580 ± 80[***]
	OZB-211U	NaOH sol (3)	−19.1 ± 0.2	3970 ± 280[***]
Depth 950 cm				
86-C-25	XLLQ-106	NaOH insol		14,650 ± 190[*]
	ANU-6393	NaOH insol		13,460 ± 780[*]
	ANU-6393	NaOH sol		14,000 ± 170[*]
	OZB-213U	NaOH sol (1)	−23.7 ± 0.2	16,080 ± 1910[***]
	OZB-284U	NaOH sol (3)	−24.9 ± 0.2	16,970 ± 1240[***]

[*]Age determinations carried out using liquid scintillation spectrometry at the Xian Laboratory of Loess and Quaternary Geology, Chinese Academy of Sciences (XLLQ) and the Quaternary Dating Research Centre, ANU (ANU) (Head et al., 1989; Zhou et al., 1990).

[**]Age determinations carried out using accelerator mass spectrometry at the Department of Nuclear Physics, Research School of Physical Sciences and Engineering, ANU (Zhou et al., 1993).

[***]Age determinations carried out using accelerator mass spectrometry at ANSTO, Lucas Heights, NSW (Head et al., 1997).

From Table 1, Palaeosol sample 86-C-28, collected at 300 cm gave a series of significantly different results, with the NaOH insoluble humin giving the oldest age, which is still younger than the expected age for the sample. The first NaOH soluble fraction gave the oldest ^{14}C result for the humic acid fractions. The two NaOH (1) samples are not duplicates, but have undergone separate chemical treatments. Hence, the results show non-uniformity within the humic acid fractions. Palaeosol sample 86-C-27, collected at 360 cm has its most reliable age at 3970 ± 280 yrs BP [NaOH sol (3)]. This also indicates non-uniformity of the humic acid fraction but fits the sequence slightly better that the other results, though it is probably still slightly young. Sample 86-C-25 is a sandy mud and indicates a similar pattern to 86-C-27 in that the third NaOH leach provided the better ^{14}C age. This result seems to fit the age/depth profile for the site. These results indicate that the chemical separations carried out on palaeosol material from the Bei Zhuang Cun sequence did not achieve their purpose and did not produce uniform chemical fractions.

A further evaluation of separation techniques for humic and non-humic components was recently carried out following a technique suggested by Tan (1996). The method used was as follows:

   (i) The sample is washed with hot dilute HCl to remove carbonates, then rinsed and air dried.

   (ii) 0.1 M NaOH solution is added and the slurry is shaken for 24 hr in a nitrogen atmosphere.

  (iii) The dark-coloured supernatant solution is collected by centrifugation and the soil rinsed with distilled water. The residue is acidified, rinsed and dried as the NaOH insoluble (humin) fraction.

  (iv) The black liquid is then acidified to pH 2 and the resultant precipitate (humic acid) is separated and rinsed by centrifugation.

   (v) The humic acid precipitate is redissolved with 0.1 M NaOH under nitrogen and then centrifuged to separate any undissolved fraction, which is discarded. The dissolved humic acid is then reacidified to pH 2 and centrifuged. The supernatant liquid is discarded.

  (vi) The humic acid precipitate is then shaken with a HCl/HF mixture to make sure no colloidal clays have managed to stay with the sample. The humic acid is washed with distilled water and centrifuged, then dried at a low temperature (30°C).

 (vii) The solvent extraction procedure mentioned previously (Head 1979) is carried out at this stage, and the humic acid sample dried at a low temperature.

(viii) The humic acid is again redissolved using 0.1 M NaOH solution under nitrogen, then centrifuged. Any solid material is discarded.

  (ix) The humic acid solution is purified further by passing it through an H-saturated cation exchange column (Dowex 50-X8), then freeze-dried.

An interesting result of the ion exchange step is that the pure acid form of the humic acid is produced, which is soluble at acid pH, as against the sodium salt, which is insoluble at acid pH. Preliminary analyses indicate that this humic fraction may represent a much purer form of humic acid and may be a potentially useful dating fraction, as against the acid and alkali insoluble humin fraction which is still tightly bound within the soil clay fraction. A complete evaluation of the above separation technique has not yet been achieved.

Carbonates

Formation and reliability

Detailed mechanisms of the various types of carbonate formation in soils, lake sediments, loess deposits etc., are not well understood. For this reason ^{14}C determinations on these materials have mostly been interpreted with reasonable success only when stratigraphic relationships could be set up with sequences dated using more reliable materials. An appreciation of the environment in which deposition took place is needed before any interpretation can be made.

Calcretes have been defined to include all accumulations of carbonate within the regolith at any stage of development. The formation of calcretes usually involves cementation and/or replacement of a pre-existing regolith or host material by carbonate precipitated from the soil water or groundwater, primarily within the vadose zone (Goudie, 1972, Netterberg, 1978). Calcretes are usually composed of two distinct phases:—

(i) Original host material usually consisting of quartz, feldspars and clay minerals, plus rock fragments such as limestone and calcrete.

(ii) An authigenic cementing and replacing medium, predominantly composed of calcium carbonate in the form of microcrystalline calcite.

There may be both several ages of calcification and several calcretes in different stages of development in the same profile. The host material may also be composed of fractions of different ^{14}C ages. Thus the age of a nodular calcrete is the age of the centres of the largest nodules; the age of a honeycomb calcrete is the age of the matrix joining the nodules together, the age of hardpan is the age of matrix finally closing the large voids in the honeycomb structure. For calcareous and calcified gravels, the age of pebble coatings would define the age of the calcareous gravel. The age of the matrix farthest away from the coating around the pebbles represents the age of the calcified gravel. In order to obtain a useful ^{14}C age of a calcrete one must selectively sample it, since most calcretes exhibit facies changes (Netterberg, 1978).

Chen and Polach (1982) indicated four possible sources of authigenic soil carbonate:—

(i) from soil CO_2;
(ii) from dissolved carbonate in groundwater;
(iii) from calcareous dust derived from eroded soil exposure;
(iv) from limestone or shell detritus in parent sediments.

In this study they also carried out a literature survey, finding 82 pairs of ^{14}C dates where carbonate ages could be cross-checked against coexisting organic carbon. They found that:

(i) approximately 1/2 of carbonate ^{14}C ages were older and 1/2 were younger than the organic carbon ages;
(ii) 90% of the pairs differed from each other by 1000 to 3000 yr, 10% differed by 3000 to 20,000 yr;
(iii) most of the dates from drill cores obtained from lake deposits or marine sediments differed by < 1000 yr.

It could be expected that many calcretes would give ^{14}C ages significantly younger than the parent sediment since they form around and within the sediment matrix, usually during relatively humid-warm periods.

Bowler and Polach (1971) devised a study for which calcretes were collected from six sites in a transect across Victoria, corresponding to a steep climatic gradient from an area of longitudinal dunes in the semi-arid mallee environment in the north-west, across the riverine plain in northern central Victoria, to the temperate higher rainfall region of the Keilor Terrace in the Maribyrnong Valley, near Melbourne. Profiles were selected for which ^{14}C or stratigraphic data were already available so that a measure of independent control on the age of soil development could be obtained. Only soft and porous concretions were used. They concluded that:

(i) The ^{14}C ages of soil carbonates are younger than those of sediments on which they have formed.
(ii) In south-eastern Australia, low activity effects due to limestone dilution are rapidly counteracted by uptake of younger [14]C.
(iii) Uptake of modern carbon sometimes continues long after initial carbonate segregation. While the carbonate remains porous and lies within the zone of frequent wetting, continued exchange may be expected.
(iv) Constancy of $\delta^{13}C$ values, despite variations in the carbonate source, vegetation and climate indicate that fractionation is not controlled by these factors alone.
(v) The results highlight the need for cautious and critical evaluation.

Drysdale and Head (1994) dated stream tufas formed in spring water from a karst area in north-western Queensland, evaluating the reservoir effect by precipitating carbonate from the stream onto an inert matrix. This material gave an apparent age of approximately 2000 years. Similar studies were carried out by Pazdur et al. (1995), Srdoč et al. (1986) and Krajcar Bronič et al. (1992). Evaluation of reservoir effects needs to be carried out at every deposition site if older material is to be dated. If possible, cross-checking against other materials in the same deposition layer, or the use of other dating techniques such as U/Th dating as a cross-check is highly desirable. Speleothems pose similar problems for dating (Pazdur et al., 1995) and need to be examined for resolution effects.

Shells

Reservoir effect problems

^{14}C ages from marine molluscs are very easy to obtain but these ages are usually difficult to interpret because of the effect of mixing of deep water which is not in equilibrium with the atmosphere. Typical reservoir effects produce an apparent age around 400-600 years, depending on the locality (Heier-Nielsen et al., 1995). Shells from island lagoons or estuaries can either have greater apparent ages or can be living in equilibrium with the atmosphere, depending on the source and flow rate of the water (Head, 1991; Spennemann and Head, 1996). The origin of these shells can be determined from δ^{13}C values. Pretreatment usually involves the removal of shell surfaces either by the action of dilute HCl, or by use of a small drill. Marine reservoir effects are best determined by dating museum collections of shells collected live before 1950, as bomb effect corrections have not been satisfactory.

Freshwater and terrestrial molluse and snail shells are likely to have incorporated older limestone derived bicarbonate into their shell structure. Stable carbon isotope measurements (δ^{13}C) often provide valuable information as to the origin of carbonate forming these shells. Most of these species produce shells consisting of aragonite. Aragonite and calcite can be visually distinguished and separated using a small drill or grinding wheel. The aragonite can be verified using the X-ray diffraction technique. If post-depositional recrystallisation has occurred, calcite is the favoured form of the reworked calcium carbonate.

Evin et al. (1980) indicated that an estimation of the validity of ^{14}C ages obtained from snail shells depends on:
(i) A precise knowledge of the stable carbon isotopic composition of shells formed under controlled conditions.
(ii) ^{14}C determinations of shell from modern gastropods sampled from various natural environments.

(iii) Comparison of ^{14}C results from snail shells with other datable materials from the same horizons.

Treatment of the terrestrial shell material involved:

(i) careful hand scraping of the shell surfaces;

(ii) leaching with hydrogen peroxide;

(iii) washing by ultrasonic agitation;

(iv) selection of the most nacreous shells;

(v) brief leaching with dilute acid to remove the outer portion of the shells.

Macumber and Head (1991) dated melanopsis shells from sedimentary sequences within the Wadi al-Hammeh, Jordan. One deposit was associated with archaeological evidence of a living site, and plant fragments from the same horizon could be dated. Both shells and plant remains gave the same ^{14}C age - approximately 11,000 yr. BP. Yet living shells of the same type gave an apparent age of ~5000 yrs B.P. The reason for this enormous discrepancy is that the shells live on the edge of surface springs and their carbonate reflects the apparent age of the water and the time of living. Hence the apparent age of spring water in the area 11,000 yr ago was virtually zero, but the apparent age of the spring water at present is above 5000 years. Terrestrial snail shells can be correlated with living shells of the same species provided the micro-environment is similar. It is possible to relate living snail shells with the atmospheric ^{14}C values at the time of collection (Goodfriend, 1987; Goodfriend and Hood, 1983).

As can be seen, the dating of terrestrial shells needs much work in terms of site observation and interpretation before the reliability of results can be assessed.

Bone

Problems and pretreatment

Contemporary bone consists of about 18% collagen with about 2.5% of other proteins and fats. The rest of the structure is basically composed of crystals of calcium phosphate with the structure of hydroxy-apatite, containing a small amount of carbonate that has been incorporated into the crystal structure. These crystals form a relatively loose matrix with collagen fibres acting as reinforcing agents. Collagen is built up from 18 amino acids, the most prevalent being proline, hydroxyproline, glycine and alanine.

Younger carbon can be incorporated easily into both inorganic and organic components of fossil bone, though most pretreatment techniques have concentrated on the isolation of bone collagen, since this material is the easiest to purify. Longin (1971) extracted pure collagen from bone as gelatin by successive hydrolyses using different concentrations of HCl. The presence of dolomite in fossil bone as a secondary mineral often poses problems because of the difficulty in dissolving dolomite in dilute HCl. Collagen from fossil bone degrades quite rapidly in arid regions, often making it very difficult to obtain enough material for dating. It is then necessary to use accelerator mass spectrometric methods.

Most laboratories have never seriously considered dating the bone apatite fraction because of the distinct possibility of continuous exchange between the carbonate in the bone and that in the surrounding environment, especially if the bone has been submerged or subjected to alternate wetting and drying conditions. Haynes (1968) suggested a technique for treating crushed bone with 50% acetic acid to remove secondary carbonate, than treating with dilute HCl and collecting the evolved CO_2 from the more acid-resistant carbonate held in the crystal lattice for dating.

Hassan et al., (1977) used modern and fossil bone apatite samples in an investigation using X-ray diffraction patterns and infrared spectrometry to trace the changes in mineral structure which usually occur during fossilisation. IR absorption spectra from bone are usually a composite of both protein and mineral absorption bands. Spectra of bone apatite are characterized by vibration bands produced by H_2O, OH, PO_4 and CO_3 groups. Generally CO_3 exists in bone apatite at three different positions:

(i) absorbtion onto the crystal surface,

(ii) at OH⁻ sites,

(iii) at PO_4^{3-} sites.

About 1/3 to 2/3 of the total carbonate occurs at the surface. The remainder is distributed between PO_4^{3-} and OH⁻ sites with the greater majority in PO_4^{3-} sites. The spectra and frequencies for IR absorbance for bone indicate that during fossilisation bone apatite experiences some changes. The relative carbonate content of fossil bone apatite decreases sharply with triammonium citrate and acetic acid treatments. Bone apatite crystals are extremely minute with the smallest dimension less than 50 Å and the largest dimension much less than 1000 Å, paralleling the C-axis. The calculated unit cell dimensions, even though not very accurate, show a distinction between contemporary and fossil bone apatite. The slight decrease in the A-axis with fossilisation is consistent with the observed increase in flourine and CO_3 content.

Haas and Banewicz (1980) carried out investigations on fresh bone and fossil bone apatite, and a carbonate similar to the secondary carbonate expected to occur in fossil bone. The method developed consists of pretreating a sufficiently large bone sample with acetic acid and hydrazine, heating it slowly in oxygen to 600°C, then raising the temperature rapidly to 800°C. Actual sample CO_2 collection takes place during a third temperature rise to 950°C. The studies indicate that a small sample of CO_2 gas can be collected above 800°C, which stems mainly from original bone apatite. Saliége et al. (1995) successfully dated the apatite fraction of bone from archaeological sites in the southern Sahara and developed an analytical technique to justify their use of this fraction. They attributed the success of this technique to the extreme aridity of the area and hence the waterproof nature of the graves.

Water flowing through sediment helps remove both organic and inorganic components of bone, leaving a porous, high surface area structure which can adsorb organic matter produced in the soil. Soil chemical processes may cause precipitation of new minerals (calcite). Bone specimens for ^{14}C dating exist in a continuum of quality, ranging from well-preserved, uncontaminated with more recent carbon, to completely devoid of original amino acids, and consequently undatable (Long et al., 1989).

To be declared datable, a bone specimen must pass a sequence of tests:
(i) visual appearance under optical microscope;
(ii) proportion of original protein remaining;
(iii) collagen-like appearance of the amino acid chromatogram;
(iv) level of exotic amino acids appearing in the chromatogram.

Severe degradation of bone protein and addition of exogenous protein or amino acids can alter the relative proportions of amino acids in the chromatogram. A distorted pattern of amino acids indicates that a date on the total amino acids may not be trustworthy. It is risky to select a single amino acid for dating a bone when its chromatogram is suspect and degree of preservation poor (Long et al., 1989).

Stafford et al. (1987) carried out the following chemical pretreatment on fossil bones. First the bones are washed in tap water to remove sediments, then broken into 1 to 3 cm fragments that are ultrasonically cleaned in tap water, followed by distilled water. Physically cleaned bone is ground to < 63 μm or left intact if grinding losses need to be avoided. Inorganic carbon is extracted from the OH-apatite phase by hydrolysing bone powder with 95% phosphoric acid. Bone powder is either untreated or extracted for 24 h with 1 M acetic acid under vacuum. Organic carbon phases are concentrated by decalcifying bone powder in 0.6 M HCl at 4°C. Acid insoluble collagenous residue is separated by centrifugation from the acid soluble phase. The acid soluble fraction is filtered through 0.45 μm teflon millipore membranes and rotary evaporated. The acid soluble phase can be further purified by passing through XAD resin, used to remove fulvic acids. The acid insoluble collagen is lyophylised, then hydrolysed and converted to gelatin. Protein is hydrolysed by heating ca. 10 mg protein per 1 ml distilled 6 M HCl for 24 hr at 110°C. Teflon-sealed tubes purged with nitrogen are used for the hydrolysis and hydrolysate solution is filtered before being passed through XAD resin. Gelatin is extracted from the weak acid insoluble residue by heating 10 mg protein and 10 ml water at pH 3, at 90°C for 3 to 4 hr. Hydrolysis tubes are purged with nitrogen prior to sealing. The gelatin solution is centrifuged and filtered before lyophylisation. The freeze-dried gelatin is hydrolysed and purified with XAD resin. Fulvic acids are removed from the gelatin and collagen hydrolysates by passing the 6 M HCl through a column of XAD resin. 10 to 50 ml hydrolysate is passed through 1 cm by 40 cm glass column of 20 to 50 mesh XAD-2 resin. Pretreatment before use. 500 gm XAD-2 resin is washed exhaustively with acetone, methanol and water and the resin is extracted 3 times alternately with 3 M HCl and 3 M NaOH. The resin is finally washed with 1 M HCl. A bed of resin 20 to 30 cm high is poured, capped with glass wool and equilibrated with 3 bed volumes of 6 M HCl. The protein hydrolysate is passed through the resin at 100 μl/min, or at a flow rate slow enough to adsorb fulvic acids in the upper third of the resin bed. XAD purified hydrolysates are filtered and rotary evaporated. Fulvic acids separated from the hydrolysed protein can be eluted from the resin by washing with distilled water until the eluate pH is between 1 and 2. A 1 M ammonium hydroxide solution is used to desorb the fulvic acids, immediately acidified with HCl before drying and combustion. Ratios of $^{14}C/^{13}C$ were measured by TAMS at the University of Arizona. Even with this technique, most of the results still showed younger contamination. Hedges and Van Klinken (1992) did an evaluation of similar techniques for separation of collagen and reported limited success.

Discussion

It can be seen that radiocarbon dating is not straightforward. However, with careful observation and sample treatment, disasters can be minimised and useful samples can be selected to provide maximum information. This paper has just scratched the surface of the literature but hopefully has provided the reader with an insight into the complexities of radiocarbon dating and some background material for further successful research.

References

Andree, M. Oeschger, H. Siegenthaler, U. Riesen, T., Moell, M., Ammann, B. and Tobolski, K. (1986). ^{14}C dating of plant macrofossils in lake sediment. In: Proc. 12th Internat. Radiocarbon Conf. M. Stuiver and R. Kra (eds.) *Radiocarbon* 28 (2A): 411-416.
Arnold, J.R. and Libby, W.F. (1949). Age determinations by radiocarbon content: Checks with samples of known age. *Science* 110: 678-680.

Becker, B. (1992). The history of dendrochronology and radiocarbon calibration. In: *Radiocarbon After Four Decades. An Interdisciplinary Perspective*, R.E. Taylor, A. Long and R.S. Kra, (eds), pp. 34-49. Springer-Verlag, New York.

Beukens, R.P. (1992). Radiocarbon accelerator mass spectrometry: Background, precision and accuracy. In: *Radiocarbon after Four Decades. An Interdisciplinary Perspective* R.E. Taylor, A. Long and R.S. Kra (eds), pp. 214-229. Springer-Verlag, New York.

Bird, M.I., Chivas, A.R. and Head, J. (1996). A latitudinal gradient in carbon turnover times in forest soils. *Nature*, 381: 143-146.

Bowler, J.M. and Polach, H.A. (1971). Radiocarbon analysis of soil carbonates: An evaluation from palaeosols in Southeastern Australia. *Palaeopedology—Origin, Nature and Dating of Palaeosols* D. Yaalon, (ed.), pp. 97-108. Jerusalem, Israel.

Brown, T.A., Farwell, G.W., Grootes, P.M. and Schmidt, F.H. (1992). Radiocarbon AMS dating of pollen extracted from peat samples. In: Proc. 14th Internat. Radiocarbon Conf. Tucson, Arizona, 1991. A. Long and R.S. Kra (eds.) *Radiocarbon*, 34 (3): 550-556.

Calderoni, G. and Schnitzer, M. (1984). Effects of age on the chemical structure of palaeosol humic acids and fulvic acids. *Geochim. Cosmochim. Acta* 48: 2045-2051.

Chappell, J., Head, J. and Magee, J. (1996). Beyond the radiocarbon limit in Australian archaeology and quaternary research. *Antiquity* 70: 543-552.

Chen, Y. J. and Polach, H.A. (1982). Validity of ^{14}C ages of carbonates in sediments In: Proc. 12th Internat. Radiocarbon Conf. M. Stuiver and R. Kra, (eds.), *Radiocarbon*. 28, (2A): 464-472.

Cherkinsky, A. E. and Brovkin, V.A. (1993). Dynamics of radiocarbon in soils. *Radiocarbon* 35(3): 363-368.

Chichagova, O.A. and Cherkinsky, A.E. (1993). Problems in radiocarbon dating of soils. *Radiocarbon* 35 (3): 351-362.

Colhoun, E.A., Van de Geer, G. and Mook, W.G. (1982). Stratigraphy, pollen analysis, and palaeoclimatic interpretation of Pulbeena Swamp, northwestern Tasmania. *Quat. Res.* 18: 108-126.

Craig, H. (1959). Carbon 13 in plants and the relationship between carbon 13 and carbon 14 variations in nature. *J. Geology* 62: 115-143.

Donahue, D.J., Beck, J. W., Biddulph, D., Burr, G.S., Courtney, C., Damon, P.E., Hatheway, A.L., Hewitt, L., Jull, A.J.T., Lange, T., Lifton, N., Maddock, R., McHargue, L.R., O'Malley, J.M. and Toolin, L.J. (1997). Status of the NSF-Arizona AMS Laboratory. *Nuclear Instruments and Methods in Physics Research* B 123: 51-56.

Donahue, D.J., Linick, T.W. and Jull, A.J.T. (1990). Isotope-ratio and background corrections for accelerator mass spectrometry radiocarbon measurements. *Radiocarbon* 32, (2): 135-142.

Dorn, T.F., Fairhall, A.W., Schell, W.R. and Takashima, Y. (1962). Radiocarbon dating at the University of Washington, I. *Radiocarbon* 4: 1-12.

Drysdale, R. and Head. M.J. (1994). Geomorphology, stratigraphy and ^{14}C chronology of ancient tufas at Louie Creek, Northwest Queensland, Australia. *Geographie Physique et Quaternaire* 48 (3): 285-296.

Duplessy, J.C., Arnold, M., Bard, E., Labeyrie, L., Duprat, J. and Moyes, J. (1992). Glacial-to-Interglacial changes in ocean circulation In: *Radiocarbon After Four Decades. An Interdisciplinary Perspective* R.E. Taylor, A. Long and R.S. Kra (eds.), pp. 62-74. Springer-Verlag, New York.

Evin, J., Marechal, J., Pachiaudi, C. and Piussegur, J.J. (1980). Conditions involved in dating terrestrial shells. *Radiocarbon*, 22 (2): 545-555.

Feist, W.C. and Hon. D.N.S. (1984). The chemistry of weathering and protection In: *The Chemistry of Solid Wood* R. Rowell; (ed.), pp. 401-454. Am. Chem. Soc. Washington, D.C.

Geyh, M.A., Benzler, J.H. and Roeschmann, G. (1971). Problems of dating Pleistocene and Holocene soils by radiometric methods. In: *Paleopedology-Origin, Nature and Dating of Palaeosols* D. Yaalon, (ed.) pp. 63-76, Jerusalem, Israel.

Geyh, M.A., Roeschmann, G., Wijmstra, T.A. and Middeldrop. S.A. (1983). The unreliability of ^{14}C dates obtained from buried sandy podzols. *Radiocarbon*, 25 (2): 409-416.

Gilet-Blein, N., Marien, G. and Evin. J. (1980). Unreliability of ^{14}C dates from organic matter of soils. In: *Proc. 10th Int. Radiocarbon Conf.* M. Stuiver and R.S. Kra, (eds.). *Radiocarbon*, 22 (3): 919-929.

Gillespie, R., Prosser, I.P., Dlugokencky, E., Sparks, R.J., Wallace, G. and Chappell, J.M.A. (1992). AMS dating of alluvial sediments on the South Tablelands of New South Wales, Australia. *Radiocarbon* 34 (1): 29-36.

Goodfriend. G.A. (1987). Radiocarbon age anomalies in shell carbonate of land snails from semi-arid areas. *Radiocarbon* 29 (2): 159-168.

Goodfriend, G.A. and Hood, D.G. (1983). Carbon isotope analysis of land snail shells: implications for carbon sources and radiocarbon dating. *Radiocarbon* 25 (3): 810-830.

Goudie, A. (1972). The chemistry of world calcrete deposits. *J Geol.* 80: 449-463.

Gove, H.E., (1992). The history of AMS, its advantages over decay counting: applications and prospects. In: *Radiocarbon After Four Decades. An Interdisciplinary Perspective* R.E. Taylor, A. Long and R.S. Kra (eds.), pp. 214-229. Springer-Verlag, New York.

Green, J.W. (1963). Wood cellulose. In: *Methods in Carbohydrate Chemistry* R.L. Whistler, (ed.), Vol. III, pp. 9-21, Academic Press, New York.

Gupta, S.K. and Polach, H.A. (1985). *Radiocarbon Dating Practices at ANU*. Monograph, ANU, Canberra.

Haas, H. and Banewicz, J. (1980). Radiocarbon dating of bone apatite using thermal release of carbon dioxide. *Radiocarbon*, 22 (2): 537-544.

Hart, D.M. and Humphreys, G.S. (1997). Plant opal phytoliths: an Australian perspective. *Quaternary Australasia* 15 (1): 17-25.

Hassan, A.A., Termine, J.D. and Haynes. C.V., Jr. (1977). Mineralogical studies on bone apatite and their implications for radiocarbon dating. *Radiocarbon* 19(3): 364-374.

Hayes, M.H.B. (1985). Extraction of humic substances from soil. In: *Soil, Sediment and Water. Geochemistry, Isolation and Characterisation* G.R. Aiken, D.M. McKnight, R.L. Wershaw and P. McCarthy (eds.), pp. 329-362. John Wiley and Sons, New York.

Haynes, C.V. (1968). Radiocarbon: Analysis of inorganic carbon of fossil bone and enamel *Science* 161: 687-688.

Head, M.J. (1979). Structure and properties of fresh and degraded wood: Their effects on radiocarbon dating measurements. Thesis for Master of Science, ANU, 103 pp. Unpub.

Head, M.J. (1980). Structural characteristics of fossil wood. *Inst. Cons. Cult. Mat. Bull.* 6: 17-23.

Head, M.J., (1982). The degree of degradation of fossil material from archaeological sites: can the influence of past environments be defined? In: *Archaeometry: An Australian Perspective* W.R. Ambrose and P. Duerden, (eds.) pp. 220-227. Dept. Prehistory. Res. School Pacific Studies. Australian Nat. Univ.

Head, J. (1991). The radiocarbon dating of fresh water and marine shells. In: *Proc. Quaternary Dating Workshop, Canberra, 1990* R. Gillespie (ed.), pp. 16-18.

Head, M.J. and Laccy, C.J. (1988). Radiocarbon age determinations from lignotubers. *Aust. J. Botany* 36: 93-100.

Head, M.J., Zhou, W.J. and Zhou, M.F. (1989). Evaluation of ^{14}C ages of organic fractions of palaeosols from loess-paleosol sequences near Xian, China. In: Proc. 13th Internat. Radiocarbon Conf. A. Long, R.S. Kra and D. Srdoc (eds.) *Radiocarbon* 31(3): 680-696.

Head, M.J., Jacobsen, G. and Tuniz, C. (1996). Assessment of the AAA pretreatment technique for charcoal and other organic materials used for ^{14}C AMS studies. *Radiocarbon*, 36, (1): 46.

Head, M.J. Zhou, W.J., An. Z.S. and Tuniz, C. (1997). New approaches to the ^{14}C dating of organic components of paleosols from the Loess Plateau in China using accelerator mass spectrometry. In: *The Changing Face of East Asia during the Tertiary and Quaternary*. H. Jablonski (ed.), pp. 38-51. Centre of Asian Studies, University of Hong Kong.

Hedges, R.E.M. and Van Klinken, G.J. (1992). A review of current approaches in the pre-treatment of bone for radiocarbon dating by AMS. In: Proc. 14th Internat. Radiocarbon Conf. Tucson, Arizona. 1991 A. Long and R.S. Kra (eds.). *Radiocarbon* 34 (3): 279-291.

Heier-Nielsen, S., Conradsen, K., Heinemeier, J., Knudsen, K.L., Nielsen, H.L., Rud, N. and Sveinbjörnsdõttir, Ã.E. (1995). Radiocarbon dating of shells and foraminifera from the Skagen cave, Denmark: Evidence of reworking. In: *Proc. 15ᵗʰ Internat. Radiocarbon Conf.* G.J. Cook, D.D. Harkness, B.J. Miller and E.M.S. Scott (eds). *Radiocarbon* 37(2): 119-130.

Hoefs, J. (1997). *Stable Isotope Geochemistry*. Springer-Verlag, Berlin-Heidelberg, 4th ed., pp. 38-42.

Houtermans, J., Suess, H.E. and Munk, W. (1967). Effect of industrial fuel combustion on the carbon-14 level of atmospheric CO_2. In: *Radioactive Dating and Methods of Low-level Counting*. Internat. Atomic Energy Agency, Vienna, pp. 57-68.

Karlén, I., Olsson, I.U., Kallberg, P. and Kilieci, S. (1966). Absolute determination of the activity of two ^{14}C dating standards. *Arkiv Geofysik* 6: 465-471.

Klinedinst, D.B., McNichol, A.P., Currie, L.A., Schneider, R.J., Klouda, G.A., von Redden, K.F., Verkouteren, R.M. and Jones, G.A. (1994). Comparative study of Fe-C bead and graphite target performance with the National Ocean Science AMS (NOSAMS) facility recombinator ion source. *Nuclear Instruments and Methods in Physics Research* B92: 166-171.

Kononova, M.M. (1966). *Soil Organic Matter*. Pergamon Press, Oxford, pp. 200-201.

Krajcar Bronic, I., Horvatincic, N., Srdoc, D. and Obelic, B. (1992). Experimental determination of the ^{14}C initial activity of calcareous deposits. *Proc. 14th Internat. Radiocarbon Conf.*, Tucson, Arizona, 1991. A. Long and R.S. Kra (eds.) In: *Radiocarbon* 34 (3): 593-601.

Kromer, B. and Münnich, K.O. (1992). CO_2 gas proportional counting in radiocarbon dating—review and perspective. In: *Radiocarbon After Four Decades. An Interdisciplinary Perspective* R.E. Taylor, A. Long and R.S. Kra (eds.), pp. 184-197. Springer-Verlag, New York.

Lal, D. (1992). Cosmogenic in situ radiocarbon on the Earth. In: *Radiocarbon After Four Decades. An Interdisciplinary Perspective* R.E. Taylor, A. Long and R.S. Kra (eds.), pp. 146-162. Springer-Verlag, New York.

Levin, I., Münnich, K.O. and Weiss, W. (1980). The effect of anthropogenic CO_2 and ^{14}C sources on the distribution of ^{14}C in the atmosphere. In: Proc. 10th Internat. Radiocarbon Conf. M. Stuiver and R.S. Kra (eds.) *Radiocarbon* 22 (2): 379-391.

Levin. I., Bösinger, R., Bonani, G., Francey, R.J., Kromer, B., Münnich, K.O., Suter, M., Trivett, N. B.A., and Wölfli, W. (1992). Radiocarbon in atmospheric carbon dioxide and methane: Global distribution and trends. In: *Radiocarbon After Four Decades. An Interdisciplinary Perspective* R.E. Taylor, A. Long and R.S. Kra (eds.), pp. 503-518. Springer-Verlag, New York.

Libby, W.F., Anderson E.C. and Arnold, J.R. (1949). Age determination by radiocarbon content: world-wide assay. *Science* 109: 227-228.

Long, A., Arnold, L.D., Damon, P.E., Ferguson, C.W., Lerman, J.C., and Wilson, A.T. (1979). Radial translocation of carbon in bristlecone pine. In: *Radiocarbon Dating* R. Berger, and H.E. Suess (eds.) pp. 532-537. Univ. California Press, Berkeley.

Long, A., Wilson, A.T., Ernst, R.D. and Gore, B.H. (1989). AMS radiocarbon dating of hones at Arizona. In: *Proc. 13th Int. Radiocarbon Conf.* A. Long, R.S. Kra and D. Srdoc (eds.) *Radiocarbon* 31 (3): 231-238.

Long, A. and Kalin, R.M. (1992). High sensitivity radiocarbon dating in the 50,000 to 70,000 BP range without isotopic enrichment. In: *Proc. 14th Int. Radiocarbon Conf.*, Tueson, Arizona, 1991. A. Long and R.S. Kra (eds.), *Radiocarbon* 34 (3): 351-359.

Long, A., Davis, O.K. and De Lanois, J. (1992). Separation and ^{14}C dating of pure pollen from lake sediments: Nanofossil AMS dating. In: Proc. 14th Internat. Radiocarbon Conf., Tucson, Arizona, 1991. A. Long and R.S. Kra (eds.). *Radiocarbon*, 34: 557-560.

Longin, R. (1971). New method of collagen extraction for radiocarbon dating. *Nature* 230: 241-242.

Macumber, P.G. and Head, M.J. (1991). Implications of the Wadi al-Hammeh sequences for the terminal drying of Lake Lisan, Jordan. *Palaeogeog. Palaeoclim., Palaeoecol.* 84: 163-173.

Marčenko, E., Srdoc. D., Golubic, S., Pezdič, J. and Head, M.J. (1989). Carbon uptake in aquatic plants deduced from their natural ^{13}C and ^{14}C content. In: Proc. 13th Internat. Radiocarbon Conf. A. Long, R.S. Kra and D. Srdoc (eds.). *Radiocarbon* 31(3): 785-794.

Mulholland, S.C. and Prior, C. (1992). Processing of phytoliths for radiocarbon dating by AMS. *Phytolitharien*, 7 (2): 17-19.

Netterberg, F. (1978). Dating and correlation of calcretes and other pedocretes. *Trans. Geol. Soc. S. Afr.* 81: 379-391.

Olson, E.A. (1957). Problem of humic acid contamination in radiocarbon dating. *Geol. Soc. Amer. Bull.* 69: 1625.

Olson, E.A. and Broecker, W.S. (1958). Sample contamination and reliability of radiocarbon dates. *Trans. NY Acad. Sci., Ser.* 11, 20 (7): 593-604.

Pandow, M., MacKay, C. and Wolfgang. R. (1960). The reaction of atomic carbon with oxygen: significance for the natural radiocarbon cycle. *J. Inorg. Nuclear Chem.* 14: 153-158.

Pazdur, A., Fontugne, M., Goslar, T. and Pazdur, M.F. (1995). Late Glacial and Holocene water level changes of the Gosciaz Lake, Central Poland, derived from carbon isotope studies of laminated sediment. *Quat. Sci. Rev.* 14: 125-135.

Pazdur, A., Pazdur, M.F., Pawlyta, J., Gorny, A. and Olszewski, M. (1995). Palaeoclimatic implications of radiocarbon dating of speleothems from the Kracow-Wielun Upland, southern Poland. In: Proc. 15th Internat. Radiocarbon Conf., G.T. Cook, D.D. Harkness, B.F. Miller and E.M. Scott (eds.) *Radiocarbon*, 37 (2): 103-110.

Polach, H.A. (1992). Four decades of progress in ^{14}C dating by liquid scintillation counting and spectrometry In: *Radiocarbon After Four Decades. An Interdisciplinary Perspective* R.E. Taylor, A. Long and R.S. Kra (eds.), pp. 198-213. Springer-Verlag, New York.

Polach, H.A. and Costin, A.B. (1971). Validity of soil organic matter radiocarbon dating: Buried soils in Snowy Mountains, southeastern Australia as Example, In: *Palaeopedology—Origin, Nature and Dating of Palaeosols* D. Yaalon (ed.), pp. 89-96. Jerusalem, Israel.

Saliege, J-F., Person, A. and Paris, F. (1995). Preservation of ^{13}C/^{12}C original ratio and ^{14}C dating of the mineral fraction of human bones from Saharan Tombs, Niger. *J. Arch. Sci.* 23: 301-312.

Scharpenseel, H.W. and Schiffmann, H. (1977). Soil radiocarbon analysis and soil dating. *Geophys. Surv.* pp. 143-156.

Scharpenseel, H.W. and Becker-Heidmann, P. (1992). Twenty-five years of radiocarbon dating soils: Paradigm of erring and learning. In: *Proc. 14th Int. Radiocarbon Conf.*, Tucson, Arizona, 1991. A. Long and R.S. Kra (eds.), *Radiocarbon*, 34 (3): 541-549.

Schnitzer, M. (1976). The chemistry of humic substances. In: *Environmental Geochemistry* J.O. Nriagu (ed.), pp. 89-108. Ann Arbor Science, Ann Arbor, Michigan.

Shafizadeh, F. (1984). The chemistry of pyrolysis and combustion. In: *The Chemistry of Solid Wood* R. Rowell, (ed.) pp. 489-529. Am. Chem. Soc., Washington D.C.

Skjemstad, J.O., Oades, J.M., Taylor, J.A., Clarke, P. and McClure, S.G. (1994). Aromatic carbon in soils. *Abstracts, Organic Matter in Soils, Sediments and Waters Conference, University of Adelaide.*

Sonett, C.P. (1992). The present status of understanding of the long-period spectrum of radiocarbon. In: *Radiocarbon After Four Decades. An Interdisciplinary Perspective* R.E. Taylor, A. Long, R.S. Kra (eds.), pp. 50-61. Springer-Verlag, New York.

Spennemann, D.H.R. and Head, M.J. (1996). Reservoir modification of radiocarbon signatures in coastal and near-shore waters of Eastern Australia: The state of play. *Quat. Australasia* 14 (1): 32-39.

Srdoc, D., Krajcar Bronic, L., Horvatíncic, N. and Obelic, B. (1986). Increase of ^{14}C activity of dissolved inorganic carbon along a river course. In: *Proc. 12th Int. Radiocarbon Conf.* M.Stuiver and R. Kra (eds.), *Radiocarbon* 28, (2A): 495-502.

Stafford, Jr., T.W., Jull, A.J.T., Brendel, K., Duhamel, R.C. and Donahue, D.J. (1987). Study of bone radiocarbon dating accuracy at the University of Arizona NSF Accelerator Facility for Radioisotope Analysis. *Radiocarbon* 29 (1): 24-44.

Stamm, A.J. (1964). *Wood and Cellulose Science.* The Ronald Press Co., New York, pp. 142-165.

Stuiver, M. and Polach, H.A. (1977). Discussion: Reporting of ^{14}C Data. *Radiocarbon,* 19(3): 355-363.

Stuiver, M., Robinson, S.W. and Yang, I.C. (1979). ^{14}C dating to 60,000 years BP with proportional counters. In: *Radiocarbon Dating: Proc. 9th Internat. Conf., Los Angeles and La Jolla, 1976* R. Berger and H.E. Suess (eds.), pp. 202-215. Univ. California Press, Berkeley-Los Angeles.

Stuiver, M. and Kra, R. (1986). Proc. 12th Internat. Radiocarbon Conf. (Calibration Issue). *Radiocarbon,* 28 (2B): 805-1022.

Stuiver, M. and Pearson, G.W. (1992). Calibration of the radiocarbon time scale, 2500-5000 BC. In: *Radiocarbon After Four Decades. An Interdisciplinary Perspective* R.E. Taylor, A. Long and R.S. Kra (eds.), pp. 19-33. Springer-Verlag, New York.

Stuiver, M., Long, A. and Kra, R.S. (1993). Calibration 1993. *Radiocarbon* 35 (1): 239 pp.

Suess, H.E. (1955). Radiocarbon concentration in modern wood. *Science,* 122 (3166): 415-417.

Tan, K.H. (1996). *Soil Sampling, Preparation, and Analysis.* Marcel Dekker, New York, pp. 232-241.

Tsoumis, G. (1969). Chemical composition and ultrastructure of wood. In: *Wood as Raw Material.* Pergamon Press, London, pp. 60-63.

Vogel. J.S., Briskin, M., Nelson, D.E. and Southon, J.R. (1989). Ultra-small carbon samples and the dating of sediments. In: *Proc. 13th Int. Radiocarbon Conf.* A. Long, R.S. Kra and D. Srdoc (eds.). *Radiocarbon* 31 (3): 601-609.

Vries, H. de. (1958). Variation in concentration of radiocarbon with time and location on earth *Koninkl. Ned. Akad. Wetenschap. Proc.* B6: 94-102.

Wang. Y., Amundson, R. and Trumbore, S. (1996). Radiocarbon dating of soil organic matter. *Quat. Res.* 45: 282-288.

Wellington, A.B., Polach, H.A. and Noble, I.R. (1979). Radiocarbon dating of lignotubers from mallee forms of eucalypts. *Search* 10: 282-283.

Williams, J.B. (1989). Examination of freshwater peat methodology. In: *Proc. 13th Int. Radiocarbon Conf.* A. Long, R.S. Kra and D. Srdoc. (eds.), *Radiocarbon* 31 (3): 269-275.

Zhou, W. J., Zhou, M.F. and Head, J. (1990). ^{14}C chronology of Bej Zhuang Cum sedimentation sequence since 30,000 years BP. *Chinese Sci. Bull.* 35: 567-572.

Zhou W. J., An, Z.S. and Head, M.J. (1994). Stratigraphic division of Holocene loess in China. *Radiocarbon* 36 (1): 37-46.

Zhou, W.J., An, Z.S., Lin, B.H., Xiao, J., Zhang, J.Z., Xie, J., Zhou, M.F., Porter, S.C., Head, M.J. and Donahue, D.J. (1992). Chronology of the Baxie Loess Profile and the history of monsoon climates in China between 17,000 and 6000 years BP. In: *Proc. 14th Int. Radiocarbon Conf.*, Tueson, Arizona, 1991. A. Long and R.S. Kra (eds.). *Radiocarbon* 34 (3): 818-825.

Zhou, W.J., Landsberg. J.P., Fifield, L.K., Davje, R.F., and Head, M.J. (1993). The use of ^{14}C accelerator mass spectrometry and ^{14}C liquid scintillation spectrometry to interpret apparent differences in humic fractions of organic sediments from Shaanxi Province, China. *Collected Oceanic Works* 16: 37-42.

The Editors

Ashok Kumar Singhvi is a Professor in the Earth Science Division at the Physical Research Laboratory, Ahmedabad, India. A physicist turned geoscientist, he specializes now in geochronology and paleoclimatology in arid environments. He has been a Ford Foundation Fellow, Alexander von Humboldt Fellow and German Research Foundation Visiting Professor and has worked at Oxford, St. Louis, Heidelberg, and Freiberg, and is currently a Visiting Professor and Leverhulme Fellow at Sheffield University in England. He is a Fellow of the Indian Geophysical Union and Fellow of the Geological Society of India and is on the Editorial Advisory Board of Quaternary Science Reviews and Ancient TL. He was awarded the Krishnan Gold Medal of the Indian Geophysical Union for pioneering the concept of absolute dating of desert sands using luminescence techniques. He is currently coordinating a major Indian programme on the Thar Desert and has contributed extensively to the chronology of loess and desert sands. He has authored and edited a book on the Thar Desert, edited five international conference proceedings and has contributed over 90 research publications.

Edward Derbyshire is Research Professor in Quaternary Science in the Department of Geography at Royal Holloway, University of London, England, and Honorary Professor of the Academy of Sciences of Gansu Province, China. He is a holder of the Antarctic Service Medal of the United States, and the back Award of the Royal Geographical Society. He has experience of universities in the United Kingdom, Australia, New Zealand, Canada, The Netherlands and China. He has conducted field research in all six continents, in pursuit of his interests in the geomorphology and sedimentology of glacial and wind-blown deposits. He was President of Section E (Geography) of the British Association for the Advancement of Science in 1990, Secretary-General of the International Union for Quaternary Research (INQUA) 1991-95, and is currently Chairman of the Scientific Board of the International Geological Correlation Programme (IUGS-UNESCO). He is Editor of Quaternary Perspectives and is on the editorial board of Glacial Geology and Geomorphology. He has authored or edited more than 200 scientific articles, including 6 books and several conference volumes.